D1261312

About the editors

Robert J. Naiman, presently an Associate Scientist at Woods Hole Oceanographic Institution, has previously been a Research Associate at Oregon State University and a Postdoctoral Fellow/Visiting Scientist at the Pacific Biological Station, Nanaimo, British Columbia. He holds M.A. and Ph.D. degrees in zoology from UCLA and Arizona State University, respectively. Dr. Naiman, author of *The Natural History of Native Fishes in the Death Valley System*, was the 1974-1976 recipient of a National Research Council of Canada Postdoctoral Fellowship.

David L. Soltz is presently Associate Professor of Biology at California State University, Los Angeles where he teaches courses in ecology, environmental biology, and ichthyology. He received a B.A. degree from the University of California, Berkeley and a Ph.D. from UCLA. Dr. Soltz authored *The Natural History of Native Fishes in the Death Valley System.* His research interests are the evolutionary ecology and life history strategies of fishes in arid regions.

FISHES IN NORTH AMERICAN DESERTS

FISHES IN NORTH AMERICAN DESERTS

Edited by

ROBERT J. NAIMAN
Woods Hole Oceanographic Institution
Woods Hole, Massachusetts

and

DAVID L. SOLTZ
California State University
Los Angeles, California

A Wiley-Interscience Publication
JOHN WILEY & SONS, New York • Chichester • Brisbane • Toronto

Library of Congress Cataloging in Publication Data

Main entry under title:

Fishes in North American deserts.

"A Wiley-Interscience publication."
Includes index.
1. Fishes—Habitat. 2. Desert fauna—United States.
3. Fishes—United States—Habitat. 4. Cyprinodon.
I. Naiman, Robert J. II. Soltz, David L.
QL627.F5 597.0973 81-202
ISBN 0-471-08523-5 AACR2

Printed in the United States of America

10 9 8 7 6 5 4 3 2 1

Contributors

ROBERT J. BEHNKE, Department of Fishery and Wildlife Biology, Colorado State University, Fort Collins, Colorado 80523

GERALD A. COLE, Department of Zoology, Arizona State University, Tempe, Arizona 85281

GEORGE D. CONSTANTZ, Division of Limnology and Ecology, Academy of Natural Sciences, Philadelphia, Pennsylvania 19103

C. ROBERT FELDMETH, Joint Science Department, Claremont College, Claremont, California 91711

SHELBY D. GERKING, Department of Zoology, Arizona State University, Tempe, Arizona 85281

STANLEY D. HILLYARD, Department of Biological Sciences and Desert Biology Research Center, University of Nevada, Las Vegas, Nevada 89154

MICHAEL F. HIRSHFIELD, Academy of Natural Sciences, Benedict Estuarine Research Laboratory, Benedict, Maryland 20612

ASTRID KODRIC-BROWN, Department of Ecology and Evolutionary Biology, University of Arizona, Tucson, Arizona 85721

ROBERT R. MILLER, Museum of Zoology, University of Michigan, Ann Arbor, Michigan 48109

ROBERT J. NAIMAN, Woods Hole Oceanographic Institution, Woods Hole, Massachusetts 02543

EDWIN P. PISTER, California Department of Fish and Game, Bishop, California 93514

ALLAN A. SCHOENHERR, Division of Life Sciences, Fullerton College, Fullerton, California 92634

GERALD R. SMITH, Museum of Paleontology, University of Michigan, Ann Arbor, Michigan 48109

MICHAEL L. SMITH, Museum of Zoology, University of Michigan, Ann Arbor, Michigan 48109

DAVID L. SOLTZ, Department of Biology, California State University, Los Angeles, California 90032

JAMES D. WILLIAMS, U.S. Fish and Wildlife Service, Office of Endangered Species, Washington, D.C. 20240

Preface

In this last decade many aspects of the biology of fishes in desert regions have received considerable study. Scientists from a variety of disciplines find these aquatic ecosystems interesting for several reasons. For evolutionary ecologists desert fishes often have small populations isolated for "known" lengths of time in a diverse array of frequently harsh environments. These systems are ideal for research on genetic and behavioral divergence and speciation. Physiologists are attracted by the seemingly harsh environmental extremes, particularly of temperature, salinity, and dissolved gases, to which these fishes have adjusted. Characteristically these aquatic ecosystems have depauperate fish faunas, simple food webs, and large numbers of a few species at each trophic level, attributes that make them ideal ecosystems for study by community and ecosystem ecologists. Yet surprisingly few studies have been attempted. The characteristics of these aquatic ecosystems, combined with their limited number and restricted distributions, create a situation in which the fauna and flora are highly susceptible to disturbance by man. In particular, the ever increasing demand for water by an expanding human population in arid regions and the changing land-use patterns have drastically altered many habitats in the last 25 years. In response to these changes great strides have been made in managing the desert fauna and flora and in the conservation of aquatic ecosystems.

Because of the scientific and managerial advances made in the last decade, we considered it necessary to summarize the value of this work, attempt a synthesis of some of the more extensively studied disciplines, and evaluate the state of knowledge about desert fishes. The 15 chapters presented in this volume cover a large proportion of the recent research on the biology of fishes in the deserts of North America. It was necessary, and perhaps somewhat unfortunate, to restrict the geographical scope of the symposium to accomplish the stated goals. Partly because of this the coverage of biological themes is not uniform, but that reflects the current state of knowledge rather than the biases of the editors. The large number of chapters that deal with pupfishes (*Cyprinodon*) is a consequence of their preponderance in desert fish faunas, their diversity of ecological and physiological adaptations, and their resultant intensive study. The limited information on community ecology and ecosystem dynamics reflects a lack of extensive research in these areas. The information presented is exciting, however, and the directions for future productive and useful research are clear.

The symposium was sponsored by the American Society of Ichthyologists and Herpetologists as part of the 60th annual meeting, June 15 to 20, 1980, at Texas Christian University, Fort Worth, Texas. We are grateful to the Endangered Species Office of the U.S. Fish and Wildlife Service for providing partial funding for the symposium. We thank Dr. Gary Ferguson, Chairman of the Local Committee fo the A.S.I.H. for his excellent institutional and logistic support that led to a well organized and smoothly conducted meeting. We also thank all the participants for a stimulating symposium, for preparing manuscripts, and for adhering to editorial deadlines.

Special thanks are due to the many anonymous reviewers of individual chapters for their careful critiques. The quality of a book like this is dependent, in large part, on the efforts and expertise of these reviewers. Finally, we sincerely thank Dianne Steele and Elaine Ellis of the Woods Hole Oceanographic Institution for their valuable assistance in typing the manuscripts, and assisting with editorial problems.

ROBERT J. NAIMAN
DAVID L. SOLTZ

Woods Hole, Massachusetts
March 1981

Contents

FISHES IN NORTH AMERICAN DESERTS

1 Fishes in Deserts: Symposium Rationale

DAVID L. SOLTZ
California State University, Los Angeles

ROBERT J. NAIMAN
Woods Hole Oceanographic Institution
Woods Hole, Massachusetts

ABSTRACT

In the late 1960s, upon entering the University of California as graduate students, we were keenly aware of the extensive studies of North American deserts by Carl Hubbs and Robert Miller. However, neither of us had any intention of studying desert fishes. Afterall, for students wanting to learn about fisheries ecology the desert seems like an odd place to start. That attitude could not have been more wrong and ended with only one trip to Death Valley, California. This visit determined a large part of our lives for the next decade. The research questions and conservation issues we encountered in a few days were enough to occupy a lifetime. We were shown endemic fishes isolated for thousands of years in small pools, often separated by many kilometers of sand in one of the driest deserts in the world; fish surviving in the outflow of thermal springs while others experienced annual temperature ranges of 40°C and daily changes of 15°C; fish in hypersaline water but able to survive direct transfer to freshwater; a diversity of breeding systems within the same species; fish living in the entrance to a waterfilled cave; the disastrous effects of introduced fishes on native populations; ecosystems so intrinsically interesting but different and begging to be studied; cattle and wild burros destroying fragile habitats essential for survival of the fish and other native wildlife; riparian vegetation being removed to tap limited aquifers for irrigation of soils that could support crops for only a few years; and pumps installed directly in springs or next to them, thus threatening unique habitats and fishes with extinction. Shortly after came the realization this situation was not confined to the Death Valley area but was occurring actively throughout all deserts of North America (Deacon 1979; Minckley and Deacon 1968; Pister 1979).

Fortunately, others recognized the enormous research and aesthetic potential represented by desert fishes and their aquatic habitats, and actively cared about their conservation. The next few years were a flurry of research and conservation activity by numerous people throughout the United States and Mexico. These proceedings summarize part of that enormous effort to understand and protect the aquatic habitats of North American deserts.

The most appropriate definition of a desert is perhaps that used by Goodall (1976): areas in which "biological potentialities are severely limited by lack of water." So defined, the desert aquatic ecosystems considered in these symposium proceedings lie within the Basin and Range Province of southwestern North America and include the Great Basin, Mojave, Sonoran, and Chihauhuan deserts. All are of recent origin; formed in the 10,000 to 12,000 years since the last pluvial period (Axelrod 1979; Van Devender and Spaulding 1979; see also M. C. Smith, this volume). These areas constitute less than 10 percent of the total land mass of North America, yet contain numerous aquatic organisms. Within these deserts is a variety of fishes physiologically and ecologically adapted to environmental diversity. The purpose of this symposium is to synthesize earlier studies and to evaluate our state of knowledge about the fishes and their habitats.

Deserts characteristically contain limited amounts of surface water; therefore most aquatic ecosystems are restricted to small areas. Basically, waters are of two types: (1) rivers and streams (often intermittent) originating in adjacent mesic regions and flowing into arid areas (Figure 1a), and (2) springs in which groundwater, also originating in distant areas, is driven to the surface along geological faults (Figure 1b; see also Naiman, and Cole, this volume). At one time marshes associated with streams and springs provided important habitats for aquatic organisms. Lakes are not common in deserts; the few that occur are found only in the Great Basin. The aquatic habitats range from harsh, variable environments to spring heads that can be remarkably stable in temperature, salinity, and flow.

Extant desert waters are isolated remnants of extensive interconnected waterways of preceding pluvial periods (Hubbs and Miller 1948; Hubbs et al. 1974; several authors in this volume). Length of habitat isolation can be placed at various interpluvials with some confidence (see Miller, this volume) but much of the isolation has been recent, occurring during the present period of increasing aridity (10,000 to 12,000 years BP).

In comparison to mesic regions fish faunas in deserts tend to be depauperate, with many springs and some small streams containing only a single species. Endemism within isolated habitats is high; for example, more than 20 species of pupfish (*Cyprinodon*) occur allopatrically in the deserts of North America and many species, particularly endemics, have a restricted distribution, often occurring in a single location.

Figure 1. Typical habitats of fishes in southwestern deserts: (a) Canello Hills Cienega, Arizona, habitat of five native fishes (photo from The Nature Conservancy files); (b) Big Springs, Ash Meadows, Nevada, habitat of three native fishes, one is now extinct (photo by D. L. Soltz).

Desert fishes, particularly those species that inhabit small streams and associated marshes, often have broad physiological tolerances. They are found at temperatures to 44°C, the warmest tolerated by any fish, and many species can adjust to water temperatures near 0°C and dissolved oxygen concentrations of <1.0 ppm (Soltz and Naiman 1978). Some species, such as the Cottonball Marsh pupfish (*Cyprinodon milleri*), tolerate salinities up to 3X seawater and high concentrations of unusual ions such as boron. At the same time other species isolated in temperate spring habitats retain the broad physiological tolerances of the ancestral stock, which presumably lived in the dessicating, contracting habitats common since the last pluvial period (Brown and Feldmeth 1971; Hirshfield et al. 1980).

It is clear that fishes inhabiting North American deserts are only a small subsample of the fauna of adjacent mesic areas (see Constantz, this volume; M. L. Smith, this volume). Among the four dominant families in our deserts, minnows (Cyprinidae) and suckers (Catostomidae) predominate in rivers and streams. A few large forms, such as the Colorado River squawfish (*Ptychocheilus lucius*), which reaches nearly 2 m standard length (Figure 2a), persist in the one remaining large river, the Colorado. Numerous intermediate-sized species, such as the Tahoe sucker (*Catostomus tahoensis*) (Figure 2b) of the Lahontan System and pluvial forms of cutthroat trout (*Salmo clarki*) (see Behnke Chapter 4, Figure 4, this volume), occur in desert rivers and larger streams. Small desert streams support only small fishes (<6 cm SL) such as speckled dace (*Rhinichthys osculus*) and longfin dace (*Agosia chrysogaster*). Springs and marshes are also predominantly inhabited by small fishes of the families Cyprinodontidae (Figure 2c) and Poeciliidae (Figure 2d).

Human demand for water in the American Southwest, especially from the population centers of Los Angeles, Las Vegas, and Phoenix, has affected most aquatic habitats, particularly the large rivers. This is the most significant long term threat to desert fishes and other organisms that rely on the water for survival (see Pister, this volume). Disturbance of fragile aquatic ecosystems, combined with the restricted distributions

Figure 2. Representatives of the four predominant families of fishes in North American deserts: (a) Cyprinidae, Colorado River squawfish (*Ptychocheilus lucius*); (b) Catostomidae, Tahoe sucker (*Catostomus tahoensis*). Photos by (a) D. L. Soltz and (b) E. P. Pister.

Figure 2. (cont) (c) Cyprinodontidae, Owens pupfish (*Cyprinodon radiosus*); (d) Poeciliidae, Gila topminnow (*Poeciliopsis occidentalis*). Photos by (c) A. Heller, and (d) A. A. Schoenherr.

of many endemic forms, has placed many species in a precarious situation. Approximately 70 percent (30 species and subspecies) of fishes on the official United States Federal List of Endangered and Threatened Species live only in deserts (see Williams, this volume).

Fishes in North American deserts have been studied extensively in the last decade. Researchers have investigated the systematics, evolutionary biology, population ecology, behavior, and physiology of many forms. This symposium is the first attempt to synthesize this work and evaluate our state of knowledge. Contributors were selected for their expertise in the areas of desert fish biology, ecosystem dynamics, limnology, and conservation and management in an attempt to obtain representative, state-of-the-art reviews. Any bias in the coverage of topics in these proceedings is not intentional; rather it reflects the popularity of certain fish species with numerous interesting adaptations that are easily studied both in the field and the laboratory. Subjects receiving minimum coverage in this volume are those that have received little comprehensive study (e.g., community ecology).

Fishes in deserts provide excellent systems for investigating general questions of physiology, behavior, and ecology. Significant advances at the general and specific levels are reviewed in these proceedings and directions for future work have been laid out by most contributors. Our increased knowledge of the biology of fishes in North American deserts can and must be used for conservation and by resource managers to protect the many endangered and threatened forms. In so doing these fragile aquatic ecosystems will be preserved for future study, aesthetic enjoyment and to maintain the essential biotic diversity of the ecosphere.

REFERENCES

Axelrod, D. I. 1979. Age and origin of Sonoran Desert vegetation. Occasional Papers of the California Academy of Sciences 132:1-74.

Brown, J. H. and C. R. Feldmeth. 1971. Evolution in constant and fluctuating environments: thermal tolerances of desert pupfish (*Cyprinodon*). Evolution 25:390-398.

Deacon, J. E. 1979. Endangered and threatened fishes of the west. Great Basin Naturalists Memoirs 3: 41-64.

Goodall, D. W. 1976. Introduction. *In*: D.W. Goodall (Ed.), Evolution of Desert Biota. University of Texas Press, Austin, pp. 3-5.

Hirshfield, M. F., C. R. Feldmeth and D. L. Soltz. 1980. Genetic differences in physiological tolerances of Amargosa pupfish (*Cyprinodon nevadensis*) populations. Science 207:999-1001.

Hubbs, C. L. and R. R. Miller. 1948. The zoological evidence: correlation between fish distribution and hydrographic history in the desert basins of western United States. In: The Great Basin, with emphasis on glacial and postglacial times. Bulletin of the University of Utah 38(20), Biological Series 10:17-166.

Hubbs, C. L., R. R. Miller and L. C. Hubbs. 1974. Hydrogaphic history and relict fishes of the north-central Great Basin. Memoirs of the California Academy of Sciences 7:1-259.

Minckley, W. L. and J. E. Deacon. 1968. Southwestern fishes and the enigma of "endangered species." Science 159:1424-1432.

Pister, E. P. 1979. Endangered species: Costs and benefits. Great Basin Natualist Memoirs 3:151-158.

Soltz, D. L. and R. J. Naiman. 1978. The Natural History of Native Fishes in the Death Valley System. Natural History Museum of Los Angeles County, Science Series 30:1-76.

Van Devender, T. R. and W. G. Spaulding. 1979. Development of vegetation and climate in the southwestern United States. Science 204:701-710.

2 Late Cenozoic Fishes in the Warm Deserts of North America: A Reinterpretation of Desert Adaptations

MICHAEL LEONARD SMITH
University of Michigan, Ann Arbor

ABSTRACT

Deserts in North America appear to have considerable antiquity as locally restricted habitats. Fossil plants and animals from the area of the present warm deserts show, however, that expansion of xeric habitats to form regional desert environments is geologically new.

The isolated drainage basins and depauperate fish faunas that are characteristic of southwestern North America today are not characteristic of general conditions during the late Cenozoic. The fish fauna that persists in the desert region has endured desert climates of relatively brief duration which have been interrupted by longer episodes characterized by pluvial climates. The fishes of North American deserts are best interpreted as generalists that have accumulated in the desert region rather than as specialized autochthonous organisms. Coastal marine habitats may have served as a source of groups of fishes that are able to tolerate the stress of periodic desert climates.

INTRODUCTION

Although deserts could be defined on the basis of physical parameters, it is perhaps most useful to regard them as areas in which "biological potentialities are severely limited by lack of water" (Goodall 1976). So defined, deserts can be described in terms of their biotas and boundaries can be drawn for desert regions that are reasonably discrete (e.g., Rzedowski 1973; Shreve 1942). In studies of deserts as biotic units the most valuable contribution of paleontological data is the ability to address questions that involve a time component. Did ancient desert regions correspond to present ones? Were earlier desert climates like modern ones? Where and when did desert environments originate?

The high level of endemism and unique adaptations of many modern desert organisms have been interpreted as evidence of the great antiquity and continuity through time of the warm deserts (Blair 1976; Blair et al. 1976; Rzedowski 1962). The fossil record reveals, however, that the area of the present deserts has supported a succession of vegetation (Axelrod 1958, 1979) and that plants and animals usually regarded as characteristic of deserts have been associated with nondesert taxa at times in the past (King and Van Devender 1977; Van Devender and Mead 1978; Van Devender et al. 1977).

Questions about the age of desert habitats raise others about the history of presumed desert adaptations. The fishes of aquatic habitats in deserts are subject to environmental conditions that often represent extreme records for freshwater fishes. Responses to severe environmental regimes may involve unusual physiological tolerances and life-history strategies which are documented elsewhere in this volume. Are the abilities of desert fishes to tolerate environmental extremes the result of evolution in, and adaptation to, desert environments, or have deserts accumulated fish taxa with appropriate physiological and reproductive abilities that actually evolved elsewhere? The answers to these questions are of fundamental significance to evolutionary and systematic studies of desert fishes.

The objective of this chapter is to summarize the known paleontological evidence in regard to the origin, affinities, and ecological history of fishes in the warm deserts of North America. Concepts relevant to current evolutionary and ecological studies are discussed in light of the time perspective provided by the fossil record. The area of concern includes the Mohave, Sonoran, Chihuahuan, and Hidalgan deserts and intervening areas that have or have had an arid climate (see Miller, Figure 1, this volume).

PALEOENVIRONMENTS OF THE WARM DESERTS

The development of freshwater habitats is controlled principally by structural geologic events that determine the amount of relief in an area and the morphology of the land surface. It is controlled subordinately by climate that affects the regimes of lakes and streams. Both factors have played prominently in the history of fishes in the North American warm deserts. Tectonic activity was intense during much of Cenozoic time and hydrographic patterns can often be related to specific tectonic or volcanic events (e.g., Albritton 1958; Tamayo and West 1964). The area has been subject to a succession of changing climates which have included pluvial episodes and periods of severe aridity. The geologic and climatic history of the warm deserts is summarized as background for the interpretation of the fossil record.

Regional Geologic Setting

The tectonic and igneous activity that has dominated the development of the surface configuration of western North America can be related to crustal plate interactions most clearly documented in adjacent sections of the Pacific basin floor (Allison 1964; Atwater

1970; Malfait and Dinkelman 1972; Menard 1955). The American and Pacific plates appear to be moving past one another parallel to the San Andreas fault, and much of the late Cenozoic tectonic activity of western North America may be related to boundary deformation between the two plates (Atwater 1970).

The development of present surface features has its beginning in the early Cenozoic. It appears that lithospheric units in the area became separated in the Eocene or early Oligocene and subsequently were further subdivided (Atwater 1970; Malfait and Dinkelman 1972). The relative motion of plates, reoriented during this time, resulted in a complex system of fracture zones and faults in the eastern Pacific basin, some of which extend into south-central Mexico (Gastil and Jensky 1973; Menard 1960).

Cenozoic geology in the area of the warm deserts can be understood as a complex of events related to the development of the middle North American Cordillera between the Transverse Ranges of southern California and the Neovolcanic Axis of central Mexico. Geologic reconstructions of this area before Miocene rifting (Gastil and Jensky 1973; Karig and Jensky 1972) indicate that California south of the Transverse Ranges, including all of Baja California, was approximately 300 km to the southeast of its present position, lying in an area of relatively low relief against the present coast of the Mexican mainland. Major changes in the area began with the eruption of the ignimbrite sheets that compose the Sierra Madre Occidental. Radiometric dates suggest that most of the ignimbrites were produced 22 to 30 million years ago (Karig and Jensky 1972), although the southernmost sequence, near Gudalajara, has been dated at 8.5 to 9.5 (Watkins et al. 1971). Continued volcanic activity built up a thick volcanic pile, the Sierra Madre Occidental, primarily during the Oligocene and Miocene (de Cserna 1960; Eardley 1962; P. B. King 1942; R. E. King 1939) although local uplift and volcanism continued into the Pleistocene in some areas (Albritton 1958; R. E. King 1939).

To the south a second episode of volcanism which began in the Miocene produced the lava flows and associated volcaniclastics of the Mexican Mesa Central (de Cserna 1975). Although volcanism in this area reached its greatest intensity during the Pleistocene, it continues today in volcanic structures developed in historical times (Wilcox 1954; Williams 1945, 1950).

A fault lying beneath the Quaternary volcanic rocks of the trans-Mexican volcanic belt has been inferred by Gastil and Jensky (1973) from offsets in the structural lineament of the Sierra Madre Occidental, petrographic and potassium-argon belts, and western Mexican gravity gradient. Recent geophysical studies in areas bordering the Gulf of California (Gastil and Jensky 1973; Karig and Jensky 1972) indicate that the trans-Mexican fault system may be a continuation of the San Andreas system (see Maldonado-Koerdell 1964, Figure 11). Gastil and Jensky (1973) inferred that approximately 175 km of movement occurred along the fault in late Cretaceous or early Tertiary, followed by an additional 260 km of similar movement in Miocene or Pliocene time. Although strike-slip displacement is the primary component of movement on the fault, a vertical component is indicated by the fractured and folded lacustrine sediments of the Chapala Formation which apparently were uplifted during the Pleistocene (M. L. Smith, 1980).

The California peninsula, which first separated from the mainland in the Miocene (Karig and Jensky 1972), has been rafted to its present position as a result of Pliocene strike-slip displacement and seafloor spreading. Magnetic anomaly patterns indicate that the present episode of strike-slip motion in the San Andreas fault system originated less than 30 million years ago (Atwater 1970). At the mouth of the Gulf of California most of the spreading occurred within the last four million years (Larson et al. 1968).

Development of the surface configuration of southwestern North America therefore involves events of the last 30 million years. Major relief in the region first began to develop in the Oligocene but is younger (Miocene or Pliocene) in most areas. Topography is generally controlled by boundary interactions between the Pacific and American plates. Although the boundary is usually drawn as a single break between two rigid plates, western North America could be considered as a broad transform fault zone. Pure strike-slip motion occurs along faults that parallel the San Andreas; faults at other angles show opening, which results in basin-and-range topography, and closing, which results in transverse ridges (Atwater 1970).

The intense tectonic and volcanic activity of the area is associated with instability and isolation of drainage units. Large Pliocene lakes with long lacustrine histories (M. L. Smith 1980; Uyeno and Miller 1965) were superseded by episodic lakes in isolated basins (see Miller, this volume). An analog of the Pleistocene lakes of western United States still exists in central Mexico, where a series of endorheic basins coincides with the tectonically active Neovolcanic Axis (M. L. Smith and Miller 1980).

Climatic History

The present climate of the warm deserts is influenced by factors operating at several different levels. Between the latitudes of 23 and 30° (a range that includes most of the Sonoran and Chihuahuan deserts) weather is controlled by permanent high-pressure cells; air arriving in the area is subject to adiabatic heating and consequently is low in moisture. In much of the remaining regional desert, particularly inland, aridity is controlled or accentuated by the rain-shadow effect of high mountains. This factor is important in the Chihuahuan Desert which is bordered by mountain ranges that block moisture from the east and west and in the Mohave Desert, which is bounded by great relief on its windward side to the west. In coastal areas of northern Baja California aridity results from the condensation effect of cold off-shore currents that reduce the moisture content of air moving inland.

Under the control of these factors deserts have expanded, contracted, and moved over the present desert region. During the late Cenozoic floras and faunas in southwestern North America were repeatedly displaced altitudinally and latitudinally in response to changes in climate (Axelrod 1979; King and Van Devender 1977;

Metcalf 1978; W. E. Miller 1977; Van Devender 1976, 1977; Van Devender and Spaulding 1979; Van Devender and Worthington 1978; Wells 1966, 1978). Climatic fluctuations have also been inferred from geomorphological data, such as abandoned beaches and wavecut terraces which record past pluvial episodes in interior basins that are now arid (Blackwelder and Ellsworth 1936; Buwalda 1914; Reeves 1969). The evidence indicates a general trend toward increasing aridity in a region that, although interrupted during pluvial episodes, has resulted in a dry climate presently more severe than at any earlier time (Axelrod 1979).

Late Cenozoic climatic events are also reflected in the present hydrography of the Mexican Central Plateau. Most of the streams of the Chihuahuan Desert have been tributaries to the Rio Grande system, as indicated by their fish faunas (Meek 1902, 1904). With the development of a desert climate in the last 11,500 years or less (Van Devender 1977; Van Devender and Spaulding 1979; Wells 1978) and presumably between late Cenozoic pluvial periods the volume of flow in these streams has been reduced to the extent that many of them no longer achieve exterior drainage. The occurrence in Central Mexico of relict fossil fishes related to species of the Rio Grande also indicates former hydrographic connections across the present desert (M. L. Smith 1980). Although piracy of headwaters may account for a reduction in flow in some Chihuahuan Desert streams (e.g., the Rio Nazas; Albritton 1958), regional climatic fluctuations and rain-shadow effects consequent on uplift of the Sierra Madre Occidental are certainly also involved.

The extent to which climatic fluctuations extended southward is uncertain. Clements (1963) inferred pluvial stages of Lake Chapala (20°15'N latitude) on the basis of purported wave-cut terraces. These structures have not been relocated, but if they do exist they could be explained plausibly by inferred drainage changes (M. L. Smith 1980) without invoking pluvial episodes. Extension of glaciers on the high peaks around the Valley of Mexico has been inferred from end moraines (White 1954, 1956, 1962), but there is no evidence to show that these glacial advances were associated with significant climatic changes below about 3000 m in elevation. On the basis of mid-to-late Pleistocene diatom stratigraphy in the Valley of Mexico, Bradbury (1971) concluded that "climatic changes known to have occurred in more northerly latitudes do not seem to have been reflected in this region of Mexico." The persistence of many diatom species throughout the sediments examined by Bradbury and their common distribution in the basin today suggest that truly pluvial conditions did not reach the Valley of Mexico (between 19 and 20°N latitude).

The most detailed picture of late Cenozoic climatic change in the area of the warm deserts is provided by the paleobotanical record. Recent reviews of fossil floras (Axelrod 1979; Van Devender 1977; Van Devender and Spaulding 1979) lead to three conclusions that are critically important to the interpretation of the ecological history of desert organisms:

1. The expansion of xeric habitats to form regional desert environments is geologically new, having occurred only during Pleistocene interglacials and especially after the Wisconsinan glaciation.
2. Climates have been more rigorous since the beginning of the last glaciation than at any earlier time; the Wisconsinan was the most extreme glacial period and the last few thousand years have been the most extreme interglacial.
3. Segregated biotic communities, which are typical of the present, are less typical of most of the Cenozoic and may be a response to pronounced seasonality.

The evidence is summarized below.

The warm deserts have been considered ancient because they harbor unusual organisms of obvious antiquity (relicts of the late Cretaceous) which display unique adaptations to xeric habitats. Fossil floral assemblages, however, suggest to Axelrod (1979 and earlier works cited therein) that ancient desert plants originated in edaphically or seasonally dry sites or local dry basins; they were preadapted in form and function to regional deserts that emerged much later. The novelty and cyclic nature of desert climates in North America are corroborated by evidence from the middens of packrats (*Neotoma* spp.). Plant macrofossils, pollen, and vertebrate remains from middens and cave fills appear to exclude the possibility of a regional treeless, desert-scrub formation during the last glaciation; instead, they indicate widespread woodland communities (Van Devender 1976, 1977; Van Devender and Spaulding 1979; Wells 1966, 1978). A change from woodland to desert or grassland occurred about 8000 years ago in the Chihuahuan, Sonoran, and Mohave deserts (Van Devender 1977).

The development of progressively severer climates in North America is documented by sequences of plant and animal communities (Axelrod 1979; Hibbard 1960; Hibbard et al. 1965; Taylor 1965; Van Devender 1977; Van Devender and Spaulding 1979); for example, the area of the present Sonoran Desert has supported a progression of vegetation (savanna, dry tropic forest, short-tree forest, woodland-chaparral, and thorn forest) which has led Axelrod (1979) to conclude that the regional desert environment "now has its greatest extent under a dry climate whose severity has not previously been equalled." Whereas the current postpluvial period may provide the most rigorous desert climate, the last pluvial episode may represent the most extreme glacial climate. Pollen cores from the beds of pluvial lakes in Arizona bear Wisconsin-age pine and spruce pollen thought to reflect invasion of the now arid San Augustin Plains by boreal conifers (Clisby and Sears 1956). The significant feature of this record is not the fact that a spruce invasion occurred, but that it appears to have happened only during the Wisconsinan and not during earlier pluvial episodes (Martin and Mehringer 1965). The view that the last glaciation was a time of maximal climatic stress is also supported by Wisconsinan extinctions of many animals that had survived earlier climatic changes (Hibbard 1960; Taylor 1965).

The midden data further indicate that regional vegetation during the Wisconsinan glaciation was a mixture of woodland and desert species in areas that are presently deserts (King and Van Devender 1977; Van Devender and Mead 1978; Van Devender et al. 1977; Wells

1978). Late Wisconsinan remains of the desert tortoise, Gopherus agassizi, are associated with juniper and pinyon macrofissils, although the species does not live near woodlands today (Van Devender and Wiseman 1977). A late Pleistocene fauna from the lower Grand Canyon of Arizona represents a mixture of present desert reptiles and woodland forms that live in a flora that was a mixture of species now segregated by a vertical separation of several hundred meters (Van Devender et al. 1977). In examining Neotoma deposits from Texas, Wells (1978) similarly found that characteristic "Chihuahuan Desert" species were represented in close contemporaneous association with the remains of woodland species before the emergence of the Holocene desert climate. A mixed community of fishes occurred on the Mesa Central of Mexico, where the remains of a bass are associated with those of trouts (see page 25).

To explain mixed communities of desert and non-desert species Van Devender et al. (1977) postulated a climate with cooler summer temperatures that allowed woodland species to descend into the desert and winter temperatures moderate enough not to eliminate the desert species. Pleistocene assemblages of vertebrates from the Great Plains also suggest that extreme seasonal differences in temperature are a Holocene phenomenon, not representative of general conditions of the Quaternary (Hibbard 1960; Hibbard et al. 1965). Even during the peak of glaciation, when the continental ice sheet reached heights of 3300 m, winters may have been milder than today in southern North America because arctic air would have undergone compression and adiabatic warming as it descended from the ice cap (Bryson and Hare 1974).

FISHES OF THE WARM DESERTS

Present Fauna

The fishes of the warm deserts are heterogeneous and reflect the diversity of aquatic habitats in southwestern North America. The occurrence of many primary freshwater fishes in the area of the warm deserts today is a result of their persistence in high-elevation streams above the desert zone or in large rivers that may not differ markedly in ecology or fauna from their counterparts in nondesert areas. The fauna of the desert region therefore includes such characteristically "nondesert" fishes as gars, sturgeons, trout, redhorse suckers, sunfishes, and darters (cf. Meek 1902, 1904; Miller 1972, 1978).

The warm deserts also contain aquatic habitats that exhibit physical and chemical characteristics that may represent extremes for inland bodies of water. They are variously characterized by high temperatures, intense radiation, high mineralization, limited habitat volume, unusual chemistry, and, in some cases, great fluctuations in these and other parameters (Cole 1968, this volume; Deacon and Minckley 1974). In these respects aquatic habitats in deserts are most closely approximated by marginal marine habitats (e.g., Simpson and Gunter 1956). Inland bodies of water whose regimes are closely controlled by a desert climate (mostly small streams, marshes, and isolated springs) are often dominated by cyprinodontoid fishes, the distribution of which is centered in coastal marine and estuarine habitats. The fact that ecological similarities are paralleled by phylogenetic affinities suggests that many adaptations of fishes in deserts may be the heritage of an estuarine origin.

Late Cenozoic Fauna

Fossil fish remains ranging in age from Miocene to Holocene are known from 29 localities widely distributed over the region of the warm deserts (Figure 1). Although the total assemblage includes representatives of 12 families, most sites have yielded limited material, often representing only a single species. Significant concentrations of fossils occur in two regions, southern California and central Mexico. The legend to Figure 1 provides a list of original references to fossil fishes of the region.

Mohave Area. The remains of a stickleback, *Gasterosteus haynesi*, and at least six species of cyprinodonts (in

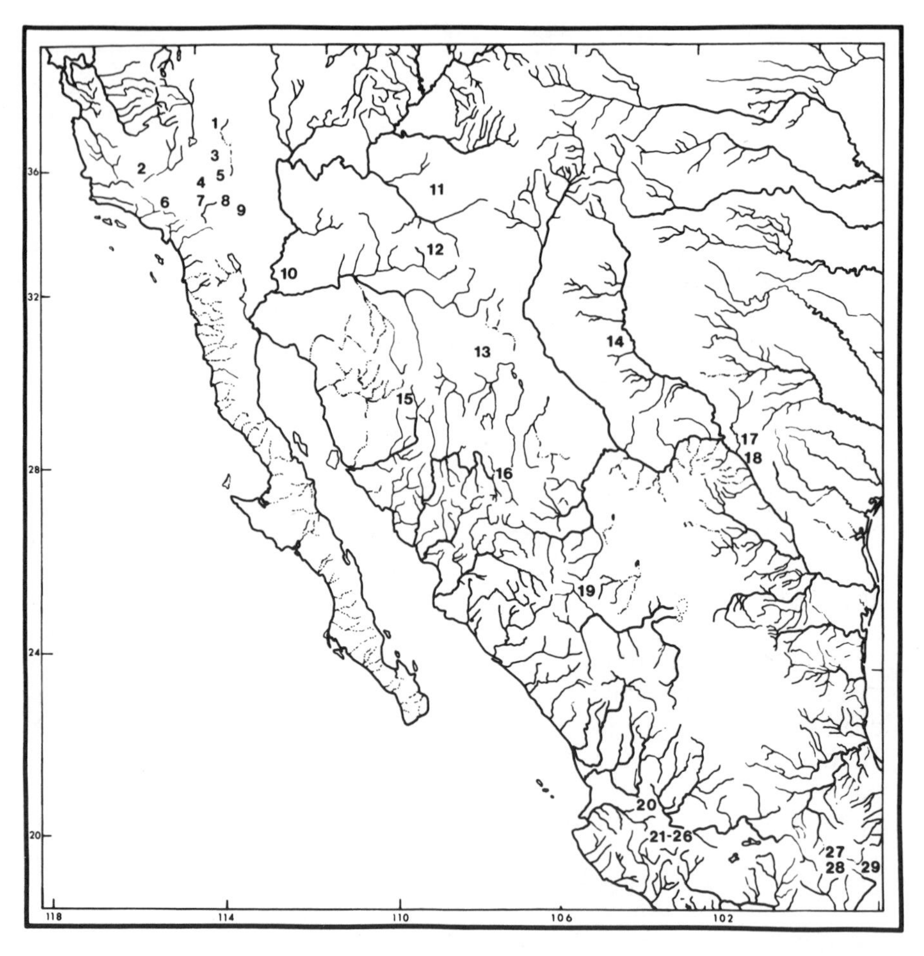

FIGURE 1. Late Cenozoic fish localities in the desert region of southwestern North America. Fossil sites are indicated by numbers that correspond to the legend below:

1. Furnace Creek, late Tertiary. Cyprinodontidae. Miller 1945.
2. Lake Tulare, Pleistocene(?). Cyprinidae. Jordan 1927.
3. Titus Canyon, late Tertiary. Cyprinodontidae. Miller 1945.
4. White Hills, Pleistocene (Kansan). Miller 1965.
5. Searles Lake, Pleistocene (Wisconsinan). Cyprinidae. Flint and Gale 1958.

6. Ridge Formation, Pliocene. Cyprinodontidae, Gasterosteidae. Bell 1973; David 1945; Jordan 1924; Uyeno and Miller 1962.
7. Black Mountain, late Tertiary. Cyprinodontidae. Miller 1945.
8. Barstow, Miocene. Cyprinodontidae. Pierce 1959; Uyeno and Miller 1962.
9. Lake Manix, Pleistocene (Wisconsinan). Cyprinidae. Blackwelder and Ellsworth 1936; Buwalda 1914.
10. Bouse Formation, Pliocene. Atherinidae. Todd 1976.
11. Bidahochi Formation, Pliocene. Cyprinidae. Uyeno and Miller 1965.
12. Snowflake, Pleistocene. Cyprinidae, Catostomidae. Uyeno and Miller 1963.
13. Howell's Ridge Cave, Holocene. Cyprinidae. Van Devender and Wiseman 1977; Van Devender and Worthington 1978.
14. Gatuna Formation, Plio-Pleistocene. Clupeidae. Miller, unpublished data.
15. Rancho La Brisca, late Pleistocene. Catostomidae, Cyprinidae, Poeciliidae. Unpublished data.
16. Yepomera, Pliocene. Cyprinidae, unidentified cyprinodontoid. Unpublished data.
17. Baker's Cave, Holocene. Catostomidae, Ictaluridae, Centrarchidae. Unpublished data.
18. Damp and Centipede Caves, Holocene. Lepisosteidae, Catostomidae, Ictaluridae, Centrarchidae. Lundelius 1963.
19. Rio Zape, Holocene. Cyprinidae, Catostomidae, Ictaluridae. Brooks et al. 1962.
20. Santa Rosa, Miocene. Goodeidae. Alvarez and Arriola 1972; Smith 1980.
21. Lake Atotonilco, late Pleistocene. Ictaluridae. Smith 1980.
22. Lake San Marcos, late Pleistocene. Salmonidae, Cyprinidae, Ictaluridae. Miller 1972; Smith 1980.
23. Lake Zacoalco, late Pleistocene. Cyprinidae, Ictaluridae. Alvarez 1966, 1974; Smith 1980.
24. Jocotepec Quarry, Pleistocene. Catostomidae, Cyprinidae, Ictaluridae, Goodeidae, Centrarchidae, Atherinidae. Smith et al. 1975; Smith 1980.

25. Ajijic, late Pleistocene. Salmonidae, Centrarchidae. Cavender and Miller, unpublished data.
26. Chapala Formation, Pliocene. Cyprinidae, Goodeidae, Atherinidae. Alvarez 1966, 1974; Smith 1980.
27. Tlapacoya, late Pleistocene. Cyprinidae, Goodeidae, Atherinidae. Alvarez and Moncayo 1976.
28. Lake Texcoco, late Pleistocene. Atherinidae. Bradbury 1971.
29. Valsequillo Formation, late Pleistocene. Cyprinidae. Miller, unpublished data.

the genera _Fundulus_, _Cyprinodon_, and _Empetrichthys_) have been recovered from Mio-Pliocene deposits in the Mohave Desert and nearby areas (David 1945; Uyeno and Miller 1962). These fishes, all extinct, are related to living species distributed in coastal habitats and low-gradient streams. Their widespread distribution in southern California and the occurrence of an additional Pliocene stickleback and cyprinodont as far inland as Hazen, Nevada (Bell 1974), suggest a terrain with gentle relief drained by streams flowing to the Pacific. A broad coastal plain in the area of the present Great Basin was inferred by Axelrod (1958; 1962) on the basis of Mio-Pliocene floras and reconstructed stream profiles.

Primary freshwater fishes do not appear in the fossil record of the Mohave Desert until the late Pleistocene. Remains of chubs (genus _Gila_) have been recovered from Wisconsinan lacustrine sediments of pluvial Lakes Searles and Manix (Blackwelder and Ellsworth 1936; Buwalda 1914; Flint and Gale 1958).

Lower Colorado River Drainage. The area of the present lower Colorado River has been subject to extensive marine transgressions (Miller, this volume). The Pliocene Bouse Formation was deposited by a former embayment of the Gulf of California, indicated by a marine-to-brackish-water sequence of organisms that included foraminiferans, mollusks, ostracodes, charaphytes, and barnacles (Metzger 1968; P. B. Smith 1970). The only vertebrate known from the formation is _Colpichthys regis_, an atherinid now restricted to the northern Gulf of California (Todd 1976).

A large freshwater lake or system of lakes covered much of northeastern Arizona and extended into New Mexico in early-to-middle Pliocene time (Lance 1954; Repenning and Irwin 1954). The extensive lacustrine habitat supported a diverse community of cyprinid fishes (Uyeno and Miller 1965) which included an extinct genus (*Evomus*) and moderate-sized chubs probably aligned with the genus *Gila*. The top carnivore in the community was a large minnow, *Ptychocheilus prelucius*, whose elongate jaws and pharyngeal bones were adapted for piscivory. Except for *Evomus*, whose affinities are unclear, the Bidahochi fishes are now represented by living relatives in the large-stream habitat of the lower Colorado River.

The Mesa Central. The most diverse fossil-fish assemblage known from southwestern North America is a complex lake fauna from the western Mesa Central, Jalisco, Mexico. Remains of 22 species in seven families have been collected from seven sites that range in age from Miocene to late Pleistocene (M. L. Smith 1980). The remains of two extinct trout (genus *Rhabdofario*) have been discovered in late Pleistocene lacustrine sediments. Catostomid remains from primarily fluvial Pleistocene sediments are a representative of the *Moxostoma robustum* species group and an extinct species of *Moxostoma*. Cyprinids occur at all Plio-Pleistocene and younger localities; two genera, *Algansea* and *Yuriria*, are represented in the fauna. The most abundantly represented species in sediments younger than the Plio-Pleistocene boundary is an extinct catfish of the *Ictalurus punctatus* group. The Miocene goodeid, *Tapatia occidentalis*, is the oldest freshwater fossil fish known in the present desert region. Goodeids are most diverse in Plio-Pleistocene to late Pleistocene sediments; at least six of the currently recognized genera are represented. The atherinid genus *Chirostoma* has left remains of about six species. A large centrarchid, *Micropterus relictus*, appeared in late Pliocene.

Geological and paleontological data indicate that the western Mesa Central has had a long lacustrine history. An extensive system of lakes still persists in the area but at a much reduced volume, judged from the distribution of lake sediments. Although most of the extinct Pleistocene fishes are represented in the area today by possible descendent forms, the salmonids and centrarchid are not. These fishes were large top carnivores associated with lacustrine sediments. Their extinction during the late Pleistocene is synchronous with reduction in the volume of lacustrine habitats in the western Mesa Central.

Age of the Desert Fish Fauna

The major groups of freshwater fishes do not appear in the fossil record of North America until Eocene time (Estes 1970); their absence in available freshwater deposits may indicate that they did not enter the continent, or at least its lower latitudes, until early-to-middle Tertiary. Many peripheral freshwater fishes, including salmonids and cyprinodonts, appeared in North America by the Miocene (Estes 1970).

The oldest remains known from the area of the warm deserts are those of fishes with marine affinities. Cyprinodonts were widespread under conditions of low-to-moderate relief that persisted over much of the area until middle-to-late Tertiary. Goodeids have had a long history in central Mexico; they were present in the area of the present Mesa Central in the Miocene and their unique reproductive specializations had appeared by that time. Fundulines may have entered the area of the present deserts in the early Cenozoic; they have been widespread since the Miocene. The only known fossil pupfish, *Cyprinodon breviradius*, comes from deposits of uncertain age but is presumed to be late Tertiary.

Primary freshwater fishes are not known in southwestern North America from deposits older than middle Pliocene. The earliest remains are those of a chub (close to *Gila*) from Pliocene deposits in Chihuahua, Mexico (unpublished data), and the small but diverse cyprinid fauna from Pliocene Lake Bidahochi (Uyeno and Miller 1965).

DISCUSSION

The aquatic habitats of deserts provide some of the severest environments occupied by fishes. Because these habitats are special in some of their characteristics, the ability of certain fishes to tolerate them have been regarded as specializations that result from evolution in a desert environment. However, the fossil record reveals that modern desert environments are not typical of the ecological history of the desert-fish fauna.

Reconstructions of past floras from plant macrofossils and pollen deposits indicate that woodland communities occupied much of the present Chihuahuan, Sonoran, and Mohave deserts under "pluvial climates" of the late Cenozoic (Axelrod 1979; Van Devender and Spaulding 1979; Wells 1978). Mild winters and cool summers produced plant and animal associations that were unusual compared with those of today. Warm-desert species persisted at low elevations in association with woodland species or in local dry sites. Large lakes occupied intermontane basins throughout the present desert region (Miller, this volume), probably as a consequence of lower evaporation rates associated with cooler summers (G. R. Smith, this volume).

Fishes established wide ancestral distributions during periods of greater hydrographic continuity, indicated by relict populations of darters, minnows, suckers, and trouts in streams now isolated. The reduction of lacustrine habitats during interpluvials and after the Wisconsinan glaciation is associated with reduction in the fish fauna. On the basis of limited fossil data extinction appears to have been higher among larger fishes. Large piscivorous salmonids and bass disappeared from the Mesa Central, although smaller species have persisted or are represented in the area today by close living relatives that include possible descendent forms (M. L. Smith 1980). Large cyprinids became extinct in Arizona (Lake Bidahochi; Uyeno and Miller 1962).

The flora of the North American warm deserts contains taxa with unique adaptations and isolated taxonomic position that may be relicts of the late

Cretaceous (Axelrod 1979). The fish fauna of the same region includes no such elements that can be documented as ancient autochthons. Two factors account for this contrast:

1. During periods of climatic amelioration, desert refugia are available to terrestrial desert-adapted species but not to the aquatic. Edaphically dry sites and local dry basins influenced by rain-shadow effects provide xeric habitats that have been continuous, though sometimes much reduced, during the general climatic fluctuations of the Quaternary. No such analogs of island habitats are available to the inhabitants of desert springs and marshes during pluvial episodes. Rather the isolated aquatic habitats on the floors of desert basins have been repeatedly inundated by pluvial lakes.

2. Although the ranges of terrestrial organisms adjust readily to climatic fluctuations, distributional shifts by freshwater organisms are limited by the relatively slow process of hydrographic change. Terrestrial, desert-adapted organisms have tracked the expansion and contraction of desert environments. Aquatic plants and animals whose distributions are more effectively limited by barriers have survived climatic changes in place or suffered extinction.

The fossil record shows that desert climates are not typical of the ecological history of the fishes that occur in deserts today. The durations of Quaternary pluvial climates have been on the order of 100,000 years, whereas interpluvial (desert) climates have lasted only about 10,000 to 20,000 years, according to oxygen-isotope and pollen-stratigraphic data (Van Devender and Spaulding 1979 and references cited therein). During the greater part of the Pleistocene the fishes of southwestern North America may have occupied habitats similar to present headwater streams in the Sierra Madre of Mexico; for example, in the upper Conchos, Nazas, and Aguanaval basins pupfishes occur in openstream habitat in association with as many as 19 additional native species (unpublished data).

The ecological history of fishes in North American deserts suggests that these fishes should be considered generalists rather than specialized autochthonous organisms. Although some adjustments to changing climates can be demonstrated, the broad ecological tolerances of many fishes that persist in the warm deserts may be the heritage of evolution in estuarine habitats.

ACKNOWLEDGMENTS

I thank R. R. Miller, G. R. Smith, and T. R. Van Devender for comments and discussions. My interest in some of the problems considered here results from insightful questions that G. R. Smith asked me at my doctoral examination. My field experience has been supported by NSF DEB 77-17315 (to R.R.M.) and grants from the New York Zoological Society and the Scott Turner Awards in Earth Science. Permission to collect fossils in Mexico was arranged by Ismael Ferrusquia-Villafranca Instituto Nacional de Geologia.

REFERENCES

Albritton, C. C. Jr. 1958. Quaternary stratigraphy of the Guadiana Valley, Durango, Mexico. Bulletin of the Geological Society of America 69:1197-1215.

Allison, E. C. 1964. Geology of areas bordering Gulf of California. In: T. H. Van Andel and G. G. Shor, Jr. (Eds.), Marine Geology of the Gulf of California. American Association of Petroleum Geologists Memoirs 3, pp. 3-29.

Alvarez, J. 1966. Contribucion al conocimiento de los bagres fosiles de Chapala y Zacoalco, Jalisco, Mexico. Paleoecologia 1:1-26.

Alvarez, J. 1974. Contribucion al conocimiento de los peces fosiles de Chapala y Zacoalco (aterinidos y cyprinidos). Anales del Instituto Nacional de Antropologia e Historia 4:191-209.

Alvarez, J. and J. Arriola. 1972. Primer goodeido fosil procedente del Plioceno jalisciense (Pisces, Teleostomi). Boletin de la Sociedad de Ciencias Naturales de Jalisco 6:6-15.

Alvarez, J. and M. E. Moncayo. 1976. Contribucion a la paleoictiologia de la Cuenca de Mexico. Anales del Instituto Nacional de Antropologia e Historia 6:191-242.

Atwater, T. 1970. Implications of plate tectonics for the Cenozoic tectonic evolution of western North America. Bulletin of the Geological Society of America 81:3513-3536.

Axelrod, D. I. 1958. Evolution of the Madro-Tertiary geoflora. Botanical Review 24:433-509.

Axelrod, D. I. 1962. Post-Pliocene uplift of the Sierra Nevada, California. Bulletin of the Geological Society of America 73:183-198.

Axelrod, D. I. 1979. Age and origin of Sonoran Desert vegetation. Occasional papers of the California Academy of Sciences 132:1-74.

Bell, M. A. 1973. Pleistocene three-spine sticklebacks, *Gasterosteus aculeatus*, (Pisces) from southern California. Journal of Paleontology 47: 479-483.

Bell, M. A. 1974. Reduction and loss of the pelvic girdle in *Gasterosteus* (Pisces): a case of parallel evolution. Contributions in Science, Natural History Museum of Los Angeles Cunty 257:1-36.

Blackwelder, E. and E. W. Ellsworth. 1936. Pleistocene lakes of the Afton Basin, California. American Journal of Science 31:453-463.

Blair, W. F. 1976. Adaptation of anurans to equivalent desert shrub of North and South America. *In:* W. Goodall (Ed.), Evolution of Desert Biota. University of Texas Press, Austin, pp. 197-222.

Blair, W. F., A. C. Hulse and M. A. Mares. 1976. Origin and affinities of the vertebrates of the North American Sonoran Desert and the Monte Desert of northwestern Argentina. Journal of Biogeography 3:1-18.

Bradbury, J. P. 1971. Paleolimnology of Lake Texcoco, Mexico. Evidence from diatoms. Limnology and Oceanography 16:180-200.

Brooks, R. H., L. Kaplan, H. S. Cutler and T. W. Whitaker. 1962. Plant material from a cave on the Rio Zape, Durango, Mexico. American Antiquity 27: 356-369.

Bryson, R. A. and F. K. Hare. 1974. Climates of North America. In: R. A. Bryson and F. K. Hare (Eds.), World Survey of Climtology, Vol. II. Elsevier, New York, pp. 1-47.

Buwalda J. P. 1914. Pleistocene beds at Manix in the eastern Mohave Desert region. University of California Publications, Department of Geology Bulletin 7:443-464.

Clements, T. 1963. Pleistocene history of Lake Chapala, Jalisco, Mexico. In: Essays in Marine Geology in Honor of K. O. Emery. University of Southern California Press, Los Angeles, pp.35-49.

Clisby, K. H. and P. B. Sears. 1956. San Augustin Plains-Pleistocene climatic changes. Science 124: 537-539.

Cole, G. A. 1968. Desert limnology. In: Desert Biology, Vol. I. Academic, New York, pp. 423-486.

David, L. R. 1945. A Neogene Stickleback from the Ridge Formation of California. Journal of Paleontology 19:315-318.

Deacon, J. E. and W. L. Minckley. 1974. Desert fishes. In: Desert Biology, Vol. II. Academic, New York, pp. 385-488.

de Cserna, Z. 1960. Orogenesis in time and space in Mexico. Geologische Rundschau 50:595-605.

de Cserna, Z. 1975. Mexico. In: R. H. Fairbridge (Ed.), The Encyclopedia of World Regional Geology, Part I. Dowden, Hutchinson and Ross, Stroudsburg, PA, pp. 348-360.

Eardley, J. A. 1962. Structural Geology of North America, Second Edition. Harper and Row, New York.

Estes, R. 1970. Origin of the recent North American lower vertebrate fauna: an inquiry into the fossil record. Forma et Functio 3:139-163.

Flint, R. F. and W. A. Gale. 1958. Stratigraphy and radiocarbon dates at Searles Lake, California. American Journal of Science 256:689-714.

Gastil, R. G. and W. Jensky. 1973. Evidence for strike-slip displacement beneath the trans-Mexican volcanic belt. Stanford University Publications in Geological Science 13:171-190.

Goodall, D. W. 1976. Introduction. In: D. W. Goodall (Ed.), Evolution of Desert Biota. University of Texas Press, Austin, pp. 3-5.

Hibbard, C. W. 1960. An interpretation of Pliocene and Pleistocene climates in North America. Michigan Academy of Sciences Report 62:5-30.

Hibbard, C. W., D. E. Ray, D. E. Savage, D. W. Taylor and J. E. Guilday. 1965. Quaternary mammals of North America. In: H. E. Wright, Jr. and D. G. Frey (Eds.), The Quaternary of the United States. Princeton University Press, Princeton, NJ, pp. 509-525.

Jordan, D. S. 1924. Description of Miocene fishes from southern California. Bulletin of the Southern California Academy of Science 23:42-50.

Jordan, D. S. 1927. The fossil fishes of the Miocene of Southern California. Stanford University Publications University Series, Biological Science 5:89-99.

Karig, D. S. and W. Jensky. 1972. The proto-Gulf of California. Earth and Planetary Science Letters 17:169-208.

King, J. E. and T. R. Van Devender. 1977. Pollen analysis of fossil packrat middens from the Sonoran Desert. Quaternary Research 8:191-204.

King, P. B. 1942. Tectonics of northern Mexico. Proceedings of the Eighth American Science Congress 4:395-398.

King, R. E. 1939. Geological reconnaissance in northern Sierra Madre Occidental of Mexico. Bulletin of the Geological Society of America 50: 1625-1722.

Larson, R. L., H. W. Menard and S. M. Smith. 1968. Gulf of California: a result of ocean-floor spreading and transform faulting. Science 161: 781-784.

Lance, J. F. 1954. Age of the Bidahochi Formation, Arizona (abstract). Bulletin of the Geological Society of America 65:1276.

Lundelius, E. L. 1963. Non-human skeletal material. In: J. F. Epstein (Ed.), Centipede and Damp Caves: Excavations in Val Verde County Texas, 1958. Texas Archaeological Society Bulletin Numer 33, pp. 127-129.

Maldonado-Koerdell, M. 1964. Geohistory and paleogeography of Middle America. In: R. Wauchope and R. C. West (Eds.), Handbook of Middle American Indians, Vol. I. University of Texas Press, Austin, pp. 3-32.

Malfait, B. T. and M. G. Dinkelman. 1972. Circum-Caribbean tectonic and igneous activity and the evolution of the Caribbean plate. Bulletin of the Geological Society of America 83:251-272.

Martin, P. S. and P. J. Mehringer. 1965. Pleistocene pollen analysis and biogeography of the Southwest. In: H. E. Wright, Jr. and D. G. Frey (Eds.), The Quaternary of the United States. Princeton University Press, Princeton, NJ, pp. 433-451.

Meek, S. E. 1902. A contribution to the ichthyology of Mexico. Field Columbian Museum Publication 65 (Zoology) 3:63-128.

Meek, S. E. 1904. The fresh-water fishes of Mexico north of the Isthmus of Tehuantepec. Field Columbian Museum Publication 93 (Zoology) 5:1-252.

Menard, H. W. 1955. Deformation of the northeastern Pacific basin and the west coast of North America. Bulletin of the Geological Society of America 66: 1149-1198.

Menard, H. W. 1960. The East Pacific rise. Science 132:1737-1746.

Metcalf, A. L. 1978. Some Quaternary molluscan faunas from the northern Chihuahuan Desert and their paleoecological implications. In: R. H. Wauer and D. H. Riskind (Eds.), Transactions of the Symposium on the Biological Resources of the Chihuahuan Desert Region, United States and Mexico. U.S. National Park Service Transactions and Proceedings Series Number 3, pp. 53-66.

Metzger, D. G. 1968. The Bouse Formation (Pliocene) of the Parker-Blythe-Cibola area, Arizona and California. United States Geological Survey, Professional Papers 600D:126-136.

Miller, R. R. 1945. Four new species of fossil cyprinodont fishes from eastern California. Journal of the Washington Academy of Sciences 35: 315-321.

Miller, R. R. 1959. Origin and affinities of the freshwater fish fauna of western North America. In: C. L. Hubbs (Ed.), Zoogeography. American Association for the Advancement of Science Publication 51, pp. 187-222.

Miller, R. R. 1965. Quaternary freshwater fishes of North America. In: H. E. Wright, Jr. and D. G. Frey (Eds.), The Quaternary of the United States. Princeton University Press, Princeton, NJ, pp. 569-581.

Miller, R. R. 1972. Classification of the native trouts of Arizona with the description of a new species, *Salmo apache*. Copeia 1972:401-422.

Miller, R. R. 1973. Two new fishes, *Gila bicolor snyderi* and *Catostomus fumeiventris*, from the Owens River basin, California. Occasional Papers of the Museum of Zoology, University of Michigan 667:1-19.

Miller, R. R. 1978. Composition and derivation of the native fish fauna of the Chihuahuan Desert Region. In: R. H. Wauer and D. H. Riskind (Eds.), Transactions of the Symposium on the Biological Resources of the Chihuahuan Desert Region, United States and Mexico. U. S. National Park Service Transactions and Proceedings Series Number 3, pp. 365-381.

Miller, W. E. 1977. Pleistocene terrestrial vertebrates from southern Baja California. Geological Society of America, Abstracts 9:468.

Pierce, D. P. 1959. Silicified eggs of vertebrates from Calico Mountains nodules. Bulletin of the Southern California Academy of Sciences 58:79-83.

Reeves, C. C., Jr. 1969. Pluvial Lake Palomas, northwestern Chihuahua, Mexico. In: D. A. Cordoba, S. A. Wengard and J. Shomaker (Eds.), Guidebook of the Border Region. New Mexico Geological Society Field Conference Guidebook 20, pp. 143-154.

Renfro, J. L. and L. G. Hill. 1971. Osmotic acclimation in the Red River pupfish, *Cyprinodon rubrofluviatilis*. *Comparative Biochemistry and Physiology* 40A:711-714.

Repenning, C. A. and J. H. Irwin. 1954. Bidahochi Formation of Arizona and New Mexico. Bulletin of the American Association of Petroleum Geologists 38:1821-1826.

Rzedowski, J. 1962. Contribuciones a la fitogeographia floristica e historica de Mexico. I. Algunas consideraciones acerca del elemento endemico en la flora mexicana. Boletin de la Sociedad Botanica de Mexico 27:52-65.

Rzedowski, J. 1973. Geographical relationships of the flora of the Mexican dry regions. *In*: A. Graham (Ed.), *Vegetation and Vegetational History* of Northern Latin America. Elsevier, New York, pp. 61-72.

Shreve, F. 1942. The desert vegetation of North America. Botanical Review 8:195-246.

Simpson, D. G. and G. Gunter. 1956. Notes on habitats, systematic characters and life history of Texas salt water Cyprinodontes. Tulane Studies in Zoology 4:115-134.

Smith, G. R. 1975. Fishes of the Pliocene Glenns Ferry Formation, Southwest Idaho. University of Michigan Papers on Paleontology 14:1-68.

Smith, M. L. 1980. The evolutionary and ecological history of the fish fauna of the Rio Lerma basin, Mexico. Ph.D. Dissertation, University of Michigan.

Smith, M. L., T. M. Cavender and R. R. Miller. 1975. Climatic and biogeographic significance of a fish fauna from the late Pliocene-early Pleistocene of the Lake Chapala basin (Jalisco, Mexico). University of Michigan Papers on Paleontology 12:29-38.

Smith, M. L. and R. R. Miller. 1980. *Allotoca maculata*, a new species of goodeid fish from western Mexico, with comments on *Allotoca dugesi*. Copeia 1980:408-417.

Smith, P. B. 1970. New evidence for a Pliocene marine embayment along the lower Colorado River area, California and Arizona. Bulletin of the Geological Society of America 81:1411-1420.

Tamayo, J. L. and R. C. West. 1964. The hydrography of Middle America. In: R. Wauchope and R. C. West (Eds.), Handbook of Middle American Indians, Vol. 1. Univ. Texas Press, Austin, pp. 84-121.

Taylor, D. W. 1965. The study of Pleistocene non-marine mollusks in North America. In: H. E. Wright, Jr. and D. G. Frey (Eds.), The Quaternary of the United States. Princeton University Press, Princeton, NJ, pp. 597-611.

Todd, T. N. 1976. Pliocene occurrence of the recent atherinid fish *Colpichthys regis* in Arizona. Journal of Paleontology 50:462-466.

Uyeno, T. and R. R. Miller. 1962. Relationships of *Empetrichthys erdisi*, a Pliocene cyprinodontid fish from California, with remarks on the Fundulinae and Cyprinodontinae. Copeia 1962:520-532.

Uyeno, T. and R. R. Miller. 1963. Summary of late Cenozoic freshwater fish records for North America. Occasional Papers of the Museum of Zoology, University of Michigan 631:1-34.

Uyeno, T. and R. R. Miller. 1965. Middle Pliocene cyprinid fishes from the Bidahochi Formation, Arizona. Copeia 1965:28-41.

Van Devender, T. R. 1976. The biota of the hot deserts of North America during the last glaciation: the packrat midden record. American Quaternary Association, Fourth Biennial Meeting, Abstracts, pp. 62-67.

Van Devender, T. R. 1977. Holocene woodlands in the southwestern deserts. Science 198:189-192.

Van Devender, T. R. and J. I. Mead. 1978. Early Holocene and late Pleistocene amphibians and reptiles in Sonoran Desert packrat middens. Copeia 1978: 464-475.

Van Devender, T. R., A. M. Phillips, III and J. I. Mead. 1977. Late Pleistocene reptiles and small mammals from the lower Grand Canyon of Arizona. Southwestern Naturalist 22:49-66.

Van Devender, T. R. and W. G. Spaulding. 1979. Development of vegetation and climate in the southwestern United States. Science 204:701-710.

Van Devender, T. R. and F. M. Wiseman. 1977. A preliminary chronology of bioenvironmental changes during the Paleoindian Period in the Monsoonal Southwest. In: E. Johnson (Ed.), Paleoindian Lifeways. West Texas Museum Association, The Museum Journal Number 17, pp. 13-27.

Van Devender, T. R. and R. D. Worthington. 1978. The herpetofauna of Howell's Ridge Cave and the paleoecology of the northwestern Chihuahuan Desert. In: R. H. Wauer and D. H. Riskind (Eds.), Transactions of the Symposium on the Biological Resources of the Chihuahuan Desert Region, United States and Mexico. U.S. National Park Service Transactions and Proceedings Series Number 3, pp. 85-106.

Watkins, N. D., B. M. Gunn, A. K. Baksi, D. York, and J. Ade-Hall. 1971. Paleomagnetism, geochemistry and potassium-argon ages of the Rio Grande de Santiago volcanics, Central Mexico. Bulletin of the Geological Society of America 82:1955-1968.

Wells, P. V. 1966. Late Pleistocene vegetation and degree of pluvial climatic change in the Chihuahuan Desert. Science 153:970-975.

Wells, P. V. 1978. Post-pluvial origin of the present Chihuahuan Desert less than 11,500 years ago. In: R. H. Wauer and D. H. Riskind (Eds.), Transactions of the Symposium on the Biological Resources of the Chihuahuan Desert Region. U.S. National Park Service Transactions and Proceedings Series Number 3, pp. 67-83.

White, S. E. 1954. The firn-field on the volcano Popocatepetl. Journal of Glaciology 16:389-392.

White, S. E. 1956. Probable substages of glaciation on Ixtaccihuahtl, Mexico. Journal of Geology 64: 289-295.

White, S. E. 1962. Late Pleistocene glacial sequence for the west side of Ixtaccihuahtl, Mexico. Bulletin of the Geological Society of America 73: 935-958.

Wilcox, R. E. 1954. Geology of Paricutin volcano, Mexico. United States Geological Survey Bulletin 965C:281-349.

Williams, H. 1945. Geologic setting of Paricutin volcano. Transactions of the American Geophysical Union 26:255-256.

Williams, H. 1950. Volcanoes of the Paricutin region, Mexico. United States Geological Survey Bulletin 969B:165-279.

3 Coevolution of Deserts and Pupfishes (Genus *Cyprinodon*) in the American Southwest

ROBERT RUSH MILLER
University of Michigan, Ann Arbor

ABSTRACT

The pupfishes of the genus *Cyprinodon* constitute about 30 species of small, oviparous fishes that are typical inhabitants of the deserts of southwestern United States and northern Mexico. Their arid-land distribution extends from the Owens Valley-Death Valley region of California and Nevada, eastward and southward to New Mexico, western Texas, and northern Mexico. About 10 species live outside arid regions.

The known fossil record is poor, with only one species, probably of Miocene age, described from near Death Valley. Inferences drawn from distributional patterns, geological history, and species relationships suggest that the genus is not very old and that much of its differentiation probably took place before

Wisconsinan glaciation during the pluvial-interpluvial fluctuations of the Pleistocene, some even as late as post-Wisconsinan time, after the final recession of pluvial waters.

In Plio-Pleistocene times the area of the current regional deserts of the Basin and Range Province was dotted with many lakes and streams. As these bodies of water shrank synchronously, in correlation with increasing climatic deterioration, extinction occurred, but remnant pupfish populations persisted and were able to maintain their often precarious gene pools over millenia and to differentiate into numerous distinctive populations that are now often the sole vertebrate inhabitants of isolated desert waters. Evidently, pupfishes were preadapted to survive in habitats that showed extreme chemical and physical characteristics (see M. L. Smith, this volume) and by further evolution to tolerate salinities three to four times that of sea water and temperatures to 45°C. Thus they have been able to adjust physiologically where most other fishes failed to survive and to avoid competition from other teleosts.

The paleohydrology of the Mohave, Sonoran, and Chihuahuan deserts is discussed and mapped, these deserts (and the Great Basin Desert) are compared, and routes of dispersal, chiefly of Plio-Pleistocene age, are hypothesized to explain the present broad distribution of desert pupfishes. The biogeography and differentiation of *Cyprinodon* in the Owens-Death Valley region, and Sonoran and Chihuahuan deserts are treated and the current threats to pupfish survival are noted.

INTRODUCTION

There is no common agreement on what constitutes a desert. Nor is it possible to define a desert fish or other aquatic vertebrate as typical of arid habitats. Nevertheless, when we talk about the fishes that inhabit the North American deserts there is little room for disagreement about the general areas under discussion. The boundaries of the arid and semiarid regions of North America, in which the annual mean rainfall is generally under 250 to 300 mm, lie principally within the Basin and Range Physiographic Province (Fenneman 1931), and those mapped are generally satisfactory to most students of desert biotas (Figure 1). Among these deserts I consider only the Mohave (which some include in the Great Basin Desert), Sonoran, and Chihuahuan because these three are the only deserts known to be inhabited by pupfishes. (The fish fauna of the Great Basin, a high, cool desert, has been treated by Hubbs et al. 1974; Hubbs and Miller 1948; and G. R. Smith 1978 and this volume.)

Coevolution, as used here, refers to the historical relationship between evolving desert landscapes and the evolutionary history of a monophyletic group of fishes, the genus _Cyprinodon_.

The most important adaptation needed for desert life is resistance to or avoidance of desiccation. It is usually rather difficult for fish to live where there is no water, although in relatively humid, seasonally dry regions (Beadle 1943; Turner 1964) some fishes have been successful (e.g., cyprinodontids, such as the South American _Cynolebias_ and African _Nothobranchius_, and the South American and African lungfishes _Lepidosiren_ and _Protopterus_). Desiccation of aquatic habitats with the onset of postpluvial aridity no doubt exterminated many fishes and other aquatic organisms that were unable to aestivate or move to other waters. Another important feature of most desert environments is their unpredictability. Aquatic organisms in these arid regions have to face floods as well as sparse water supplies and must often endure extreme temperature and chemical fluctuations. Hence, in addition to physiological adaptations, behavioral responses are important to their survival.

The genus _Cyprinodon_ constitutes about 30 known species (Figure 2), of which 20 occur in deserts or semideserts. Although widespread in the brackish and salt waters of eastern North America, the greatest known concentration of species in a single biotic province occurs in the Chihuahuan Desert region of southern New Mexico, western Texas, and northern Mexico, where 15 kinds are now known (Miller 1976, 1978; Miller and Echelle 1975; Smith and Miller 1980). A secondary

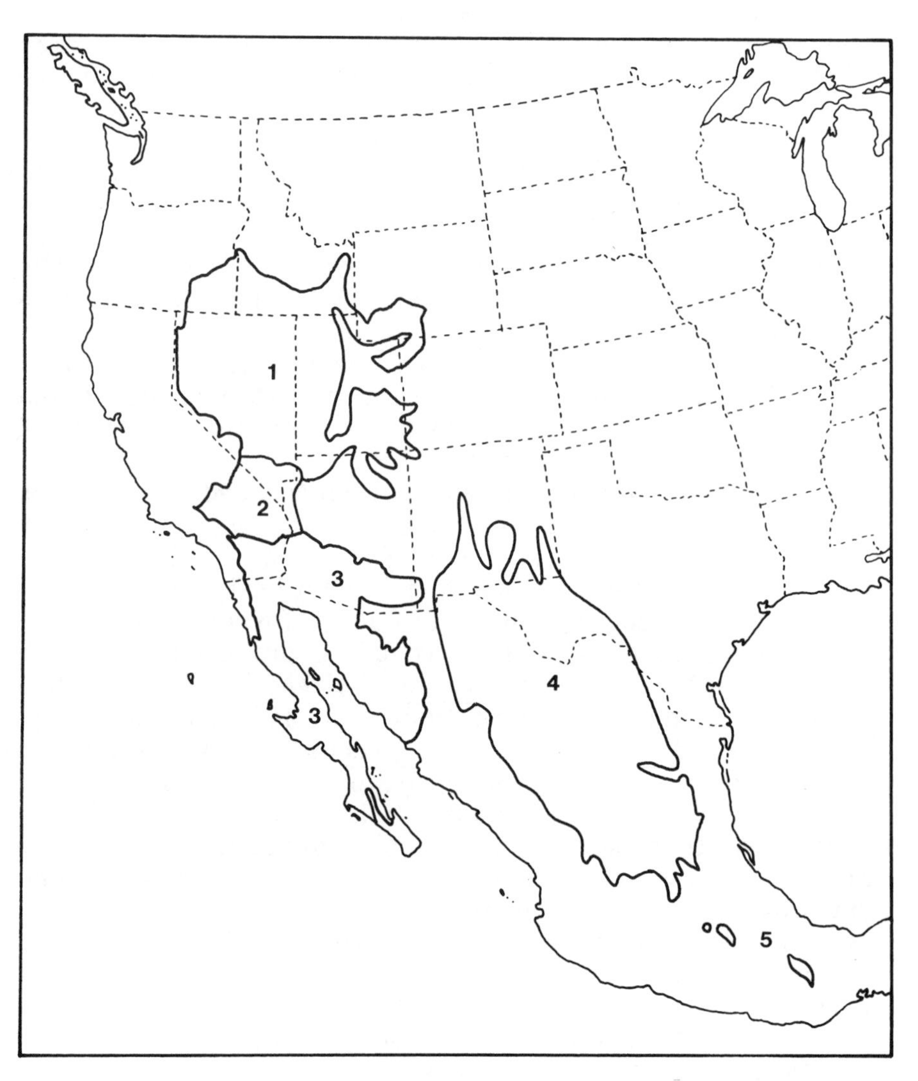

Figure 1. North American deserts of the Basin and Range Physiographic Province and vicinity (from Rzedowski 1973; Sonoran Desert from Shreve 1942, as modified by Lowe 1955; Chihuahuan Desert from Miller 1978, modified in 1980 for this Chapter, see text). 1: Great Basin, 2: Mohave, 3: Sonoran, 4: Chihuahuan, 5: Hidalgan (not discussed). Number 2 is often considered part of 1, and 3 includes the localized Colorado Desert of southeastern California.

center of speciation is in the Owens River-Death Valley region of southeastern California and southwestern Nevada (LaBounty and Deacon 1972; Miller 1948, 1950; Soltz and Naiman 1978). Only one fossil species, *Cyprinodon breviradius*, is known (Miller 1945), occurring near Death Valley in beds whose age is not known precisely but probably is late Miocene (James McAllister, personal communication 1973). Pupfishes rarely attain total lengths that exceed 7 cm; the variation is from 2 cm in the Devils Hole pupfish (*C. diabolis*) to 10 cm in the Maya pupfish (*C. maya*) of Laguna Chichancanab, Yucatan (Humphries and Miller 1981).

Cyprinodon can live in habitats that present extremely severe osmotic and thermal problems, and their ability to adapt to such stress environments reduces or eliminates competition from other fishes and from aquatic or semiaquatic vertebrates in general. Pupfishes thus appear to be what may be called "stress tolerant" species. *Cyprinodon* is not necessarily a poor competitor (e.g., see *C. bifasciatus* in Miller 1968; Deacon and Minckley 1974:452); its high resistance to stress habitats enables pupfishes to survive where other fishes have become extinct (Deacon and Minckley 1974:442). Although a large number of essentially freshwater animals have an upper limit of salinity tolerance between 3 and 10 ppt (Bayly 1972), many species of *Cyprinodon* readily tolerate 90 ppt and one has even survived at 142 ppt[1] (Simpson and Gunter 1956:124). At Berkeley in 1936 I found that *Cyprinodon macularius* could adjust rapidly to direct transfer back and forth from fresh to full sea water. Temperature tolerance in *Cyprinodon* ranges from below 0° to 45°C (Echelle et al. 1972; Smith and Chernoff 1981) and pupfishes can live in oxygen concentrations as low as 0.13 mg O_2/liter (Lowe et al. 1967), the lowest for any fishes restricted to branchial respiration. The ability

---

[1]It is of more than passing interest that the closely related Old World genus *Aphanius* (once referred to *Cyprinodon*), which inhabits salt marshes in the Red Sea-Mediterranean region, has a salinity range of almost fresh water to about 145 ppt (Lotan 1971).

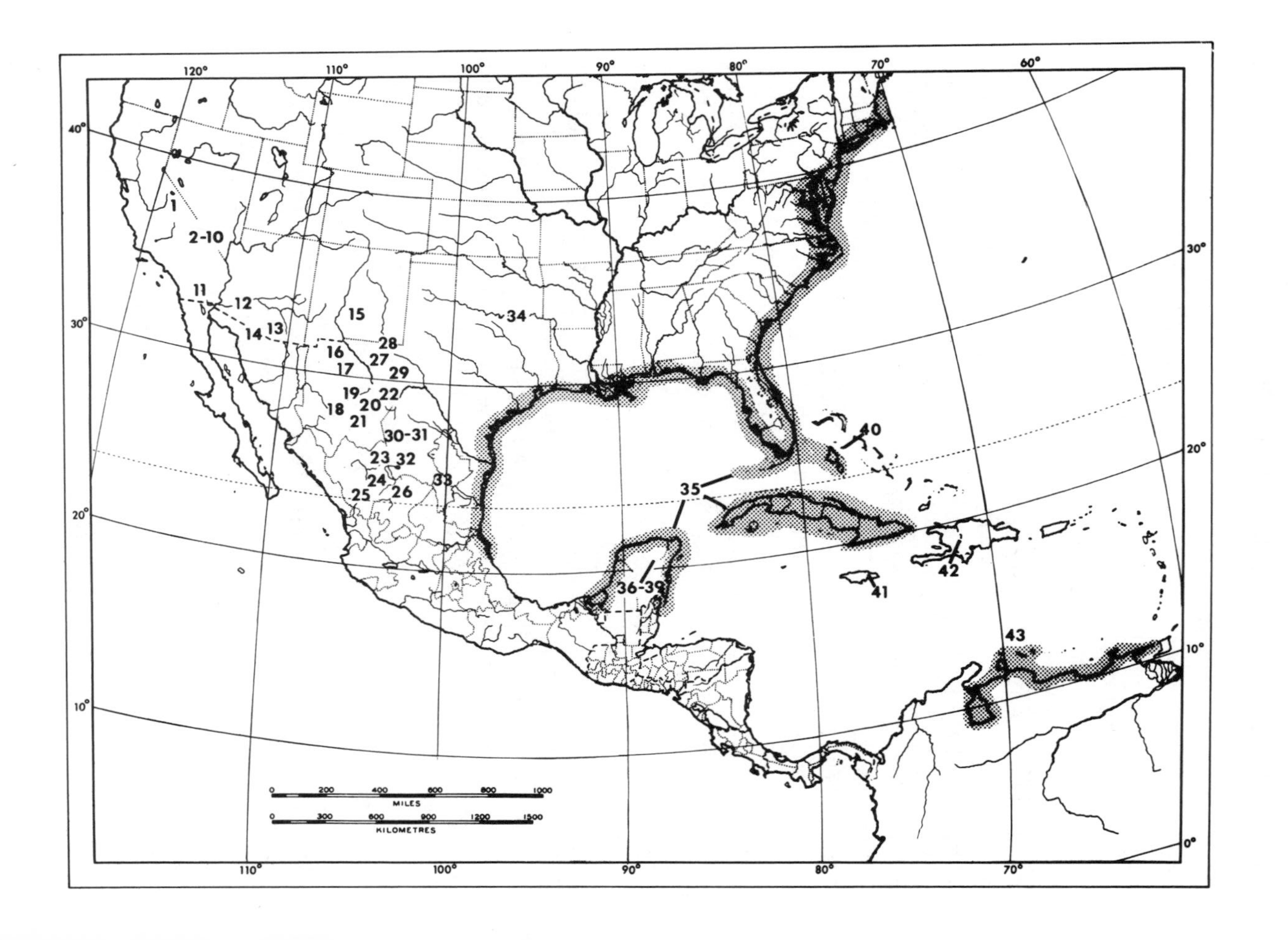

1
2-10
11
12
13
14
15
16
17
18
19
20
21
22
23
24
25
26
27
28
29
30-31
32
33
34
35
36-39
40
41
42
43
MILES
KILOMETRES

Figure 2. General distribution of pupfishes of the genus *Cyprinodon*. The status of named West Indian forms and the disjunct mainland population of *C. variegatus* around the Yucatan Peninsula is uncertain. The fish described as *Cyprinodon amazona* by Eigenmann (1894: 627-628) is a characin, probably *Nannostomus marginatus* (Weitzman 1966:32). The status of *Cyprinodon martae* Steindachner, 1875, from Colombia remains an enigma, but it is not a species of *Cyprinodon*. 1. *C. radiosus*; 2-10. *C. diabolis*, *C. salinus*, *C. nevadensis*, and subspecies; 11-12, 14. *C. macularius* and subspecies; 13. *C.* n. sp. (extinct); 15. *C. tularosa*; 16. *C.* n. sp.; 17. *C. fontinalis*; 18. *C.* n. sp.; 19. *C.* n. sp.; 20. *C.* n. sp. (San Diego); 21. *C. macrolepis*; 22. *C. eximius*; 23-24, 26. *C. nazas*; 25. *C. meeki*; 27. *C. elegans*; 28. *C. pecosensis*; 29. *C. bovinus*; 30-31. *C. atrorus* and *C. bifasciatus*; 32. *C. latifasciatus* (extinct); 33. *C. alvarezi*; 34. *C. rubrofluviatilis*; 35. *C. variegatus*; 36-39. *C. beltrani*, *C. labiosus*, *C. maya*, *C. simus*; 40. *C. laciniatus*; 41. *C. jamaicensis*; 42. *C. bondi*; 43. *C. dearborni*.

to osmoregulate and to tolerate wide thermal variations is considerably reduced at sexual maturity; thus young or juvenile pupfishes can tolerate greater salinities and higher temperatures than adults.

It is therefore not surprising to find that species of Cyprinodon frequently live alone or occur with only one or two other fishes. During Quaternary pluvial periods many now relict pupfish populations must have been part of more complex fish communities in more mesic habitats (M. L. Smith, this volume). Nearly all the described species are allopatric.

The following are a few noteworthy exceptions to the foregoing statements: Cyprinodon bifasciatus of the Cuatro Cienegas bolson in Coahuila, Mexico, is highly sensitive to thermal changes and is sympatric with eight families and 12 species of fishes, including such predatory genera as Astyanax, Micropterus, and Cichlasoma; in Laguna Chichancanab, an isolated lake basin in the interior of the Yucatan Peninsula, there is a species flock of four pupfishes whose habitats overlap broadly and thus are syntopic as well as sympatric (Humphries and Miller 1981); also, the Palomas pupfish, Cyprinodon n. sp. (Figure 2, No. 16), is inferred to have been sympatric, in different parts of its range, with C. fontinalis and C. n. sp. (Figure 2, No. 18).

ORIGIN AND NATURE OF NORTH AMERICAN DESERTS

The Basin and Range Physiographic Province includes all the deserts considered here. It extends through approximately 18° of longitude and latitude, from southern Idaho and eastern Oregon to the Mexican states of Coahuila, Nuevo Leon, and Durango (de Cserna 1975; Fenneman 1931; Thornbury 1965). Following Raisz (1964), de Cserna uses the term Mesa Central in Mexico for the higher, southern extension of this province, where the Sierra Madre Oriental veers abruptly westward near the southern boundary of Coahuila, south of a line connecting Monterrey and Torreon. Pupfishes barely enter the Mesa Central at El Potosi, Nuevo Leon (see Paleohydrology).

The distinctive topographic features of this province are isolated, nearly parallel mountain ranges, commonly aligned north-south, with intervening valleys or basins composed of sediments derived from the mountains. Many of the intermontane basins are endorheic bolsons and have playas at their lowest parts. A fault-block origin of the basin-and-range landform is widely accepted, at least in the north. Most of the present topography of the province was developed during the last 20 million years (my) or so (Hunt 1969; G. R. Smith 1978). In the Mexican extension of the province (i.e., the Chihuahuan Desert), the parallel ranges are not so orderly; they mostly lack obvious fault scarps and drainage to the sea is proportionately greater here than in the more northerly parts of the province where endorheic basins abound. Progressively increasing aridity during Tertiary time has led to the present dry climates of the lowland parts of these deserts (Axelrod 1979; Mehringer 1977; Wells 1978). Even the vast central desert of Australia developed only about 20 my ago (Raven and Axelrod 1975).

In a masterful and cogent analysis using plate tectonics and a detailed review of paleobotany since the late Cretaceous, Axelrod (1979) has shown that the regional North American deserts are not ancient, as claimed, but are the product of the interglacials and especially of postglacial time (approximately the last 10,000 years). These deserts developed gradually over a span of some 70 my, accumulating floral taxa preadapted to progressively increasing and spreading arid conditions. Paleobotanical evidence demonstrates, for example, that in late Cretaceous to Tertiary times the present Sonoran Desert contained plants that are ancestral to, and in some cases identical with, those now inhabiting this desert and its bordering regions. At that time these now arid areas had a more equable climate. It is the antiquity of these unique floral elements (at the generic and familial levels) that led investigators to invoke great antiquity for modern deserts. Abundant evidence counters this idea (Axelrod 1979).

An attempt has been made (Table 1) to provide a general comparison of some of the important physical

TABLE 1. Some Biological and Physical Contrasts in North American Deserts[a]

Characteristic	Great Basin	Mohave	Death Valley Region[b]	Sonoran	Chihuahuan
Major valley-floor vegetation	Sagebrush (*Artemisia*)	Creosotebush (*Larrea*)	Shadscale, desert holly, or none	Creosotebush	Creosotebush
General nature of vegetation	Shrubs, *Artemisia* and *Atriplex*	Shrubs, *Larrea* and *Ambrosia*	Shrubs or none	Arboreal, numerous small trees; diversified	Shrubs and semishrubs, few arboreal; succulents abundant
Creosotebush					
Spacing	None	Widely spaced[c]	Like Mohave when present	Similar to Mohave	Closely spaced
Size	None	Smaller[c]	Like Mohave	Similar to Mohave	Larger[d]
Ocotillo (*Fouquieria*)	None	None	None	On slopes	On valley floors and slopes
Plant diversity	Least	Intermediate	Less than Mohave; typically barren on floors	Most	Considerable
Rainy season	Winter-spring	Winter (except E part)	Winter	Winter-summer	Spring-summer

Months	November–March (snow regular), April to May; driest in June	November–March (snow frequent at higher elevations)	November–January (~ 76 mm) (snow scarce)	February–March July–August (frosts rare)	May–October (65 to 85 present in June–September) (frosts common to south)
Duration	More uniform throughout year than other deserts	3 to 4 months; none for years in some places	2 to 3 months	4 months; none for years in some places	6 months
Average elevation	1500 m	1000 m	400 m	650 m	1500 m
Fault scarps	Conspicuous	Conspicuous	Conspicuous	Uncommon	Virtually absent
Climate	Cold (temperate)	Warm (temperate)	Very hot (temperate)	Hot (subtropical–tropical)	Warm (tropical–subtropical)

[a]Axelrod (1979) and Shreve (1942) were the chief references used.

[b]Including Searles–Panamint–Amargosa Valley (from Ash Meadows region south).

[c]In many large areas *Larrea* grows only to 38 to 45 cm tall and covers only 3 present of the surface (Shreve 1942).

[d]*Larrea* grows to 1 to 1.5 m high in many areas in the southern part of the Chihuahuan Desert.

and biological contrasts between North American deserts by relying on the literature and personal observations. These deserts lack sharp boundaries and those chosen are of necessity arbitrary to some extent. "The flora of the desert is very distinct as to its species but strongly related to adjacent regions as to its genera" (Shreve 1942:197). Important dominants are _Larrea_ (creosotebush), _Artemisia_ (sagebrush), _Fouquieria_ (ocotillo), _Prosopis_ (mesquite), and _Acacia_ (acacia). Sharp contrasts exist between the Great Basin (cold, temperate) and the Sonoran (hot, subtropical) deserts and significant differences are evident between the Sonoran and Chihuahuan deserts. These differences have played an important role in the evolution and survival of arid-land fishes.

In a comparison of the deserts inhabited by _Cyprinodon_ the dominant shrubs of the Mohave are _Larrea divaricata_ and _Ambrosia_ (_Franseria_) _dumosa_; only a small part of this desert lies below 300 m and about three-fourths, between 640 and 1250 m. Rainfall is variable, generally less than 127 mm; some stations have gone without precipitation for more than two years. Rains occur in winter and early spring over most of this desert, but along its eastern margin there is a wet season in summer.

The Sonoran Desert is characterized by arboreal and diversified vegetation; elevations are chiefly below 640 m. The mean annual rainfall of 51 to 102 mm gradually rises to 305 to 356 mm at increasing distances from the Gulf of California on the mainland of Mexico. Precipitation is biseasonal throughout the part that lies east and north of the Gulf of California (where pupfishes live); rain occurs in late winter and midsummer. Temperatures are higher than in the Mohave or Chihuahuan deserts (except locally in Death Valley) and there are few undrained basins (Salton Sea is a notable exception).

The Chihuahuan Desert has less diversified vegetation than the Sonoran but averages much higher in elevation (nearly half is more than 1250 m and some of the highest parts are 1760 m); limestone is far more abundant than in any of the other deserts. Annual rainfall averages about 76 mm in some undrained bolsons of northern Coahuila to as high as 305 to 406 mm at more elevated sites near the western and southern boundaries. Summer precipitation (June to September) equals 65 to 85 percent of the total and light rain falls from October to December (January through April is very dry). Summer daytime temperatures are 5.5 to 11°C lower than those of the Sonoran Desert (except at low elevations along the Rio Grande). "There is a single growing season that is both longer and more certain than the summer season of plant activity in the other North American deserts" (Shreve 1942:236). In a list prepared by Shreve of the 12 most common and charactristic plants of the Sonoran and Chihuahuan deserts only three are shared: *Larrea*, *Prosopis*, and *Fouquieria*.

The most detailed analysis of late Cenozoic climatic change in the American deserts is reflected in the paleobotanic record (Axelrod 1979; Van Devender and Spaulding 1979). In the late Pliocene all these desert regions experienced a marked increase in precipitaton and a decrease in temperature, thus providing a moister climate and a greater development of aquatic ecosystems, probably on a gradient that decreased in a generally north-south direction over the Basin and Range Province. Also during the moist pluvial ages much of the desert environment disappeared. Probably only during the interglacials did deserts comparable to the current ones come into existence and then attained their maximum extent in postglacial time (during the last 8000 to 10,000 years).

PALEOHYDROLOGY

In the following discussion the term paleohydrology is restricted to that part of the hydrologic cycle of the past that involved the occurrence and movement of water over the earth's surface. Reference to behavior of water in the atmosphere is mostly incidental. As in other attempts to decipher historical events, I apply the principle that existing relations between climate and hydrology can be extrapolated into the past (Schumm 1965).

Abundant physicochemical evidence and biotic information support the existence of many Pleistocene lakes and waterways in what are now the North American deserts (Hubbs and Miller 1948; Hubbs et al. 1974; Feth 1961, 1964; Mifflin and Wheat 1979; G. R. Smith 1978; Snyder et al. 1964; Figures 3 and 4, this volume). That these lakes and streams provided routes of dispersal for fishes and other aquatic organisms has been convincingly demonstrated for the Great Basin and adjacent arid regions (Hubbs et al. 1974; Hubbs and Miller 1948). The term pluvial as used here (see definition by Morrison 1968) has a much greater time span (Plio-Pleistocene time) than admitted by Hubbs and Miller (1948:22), who generally restricted "Pluvial" to lakes of Wisconsinan age.

The time-stratigraphic correlation between glacial advances and the expansion of pluvial lakes is now firmly demonstrated for the northwestern part of the Basin and Range Province (Morrison 1968; G. I. Smith 1979). In Mexico, to the south, where glacial records are weak or lacking, such correlations have not been possible or apparently do not exist. It is reasonable to assume, however, that the more southerly Chihuahuan pluvial lakes were associated with the same kinds of climatic event that produced their less ephemeral northern counterparts.

More than 120 pluvial lakes are known from the Great Basin and Mohave deserts alone (most recently mapped by G. R. Smith 1978, and Mifflin and Wheat 1979), and at least 85 others occurred in the Sonoran and Chihuahuan deserts and neighboring regions (Figures 3 and 4). Those pluvial lakes that are of major concern here are numbered 2 to 15 (Figure 3) and 1 to 7 (Figure 4). With their associated streams (of which only some are shown on these maps), they include (Figure 3) the Owens River-Death Valley region of southeastern California and southwestern Nevada (2); Lakes LeConte (3) and Pattie (4) and annectent sloughs and delta (Figure 6) of the Colorado River in southeastern California and northeastern Baja California Norte; Lake Otero (8) of the Tularosa basin, New Mexico; Lake Palomas (10) of northern Chihuahua, Mexico, within what has been termed the Guzman complex (Miller 1978); Lakes Encinillas and

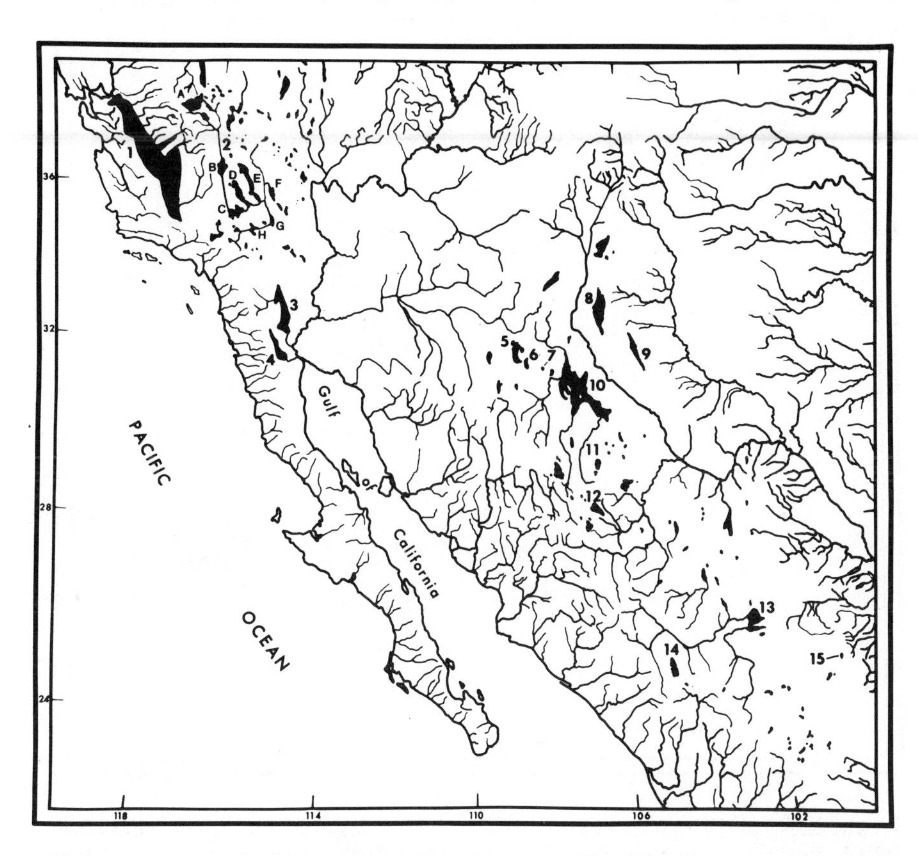

Figure 3. Pluvial lakes and waterways of Wisconsinan time in the Mohave, Sonoran, and Chihuahuan Deserts and contiguous regions. 1. Lake Tulare (from Feth 1961); 2. Owens-Death Valley drainages: A = Lake Russell; B = Lake Owens; C = Lake Searles; D = Lake Panamint; E = Lake Manly; F = Lake Pahrump; G = Lake Mohave; H = Lake Manix. 3. Lake LeConte. 4. Lake Pattie. 5. Lake Animas. 6. Lake Playas. 7. Lake Hachita. 8. Lake Otero. 9. Lake King. 10. Lake Palomas. 11. Lakes Sauz and Encinillas. 12. Lake Bustillos. 13. Lake Mayran. 14. Lake Santiaguillo. 15. Lake Potosi.

Sauz (11) just north of Chihuahua City; Lake Bustillos (12) west of Chihuahua City; Lake Mayran (13) near Torreon, Durango; Lake Santiaguillo (14) north of Durango City and technically outside the Chihuahuan

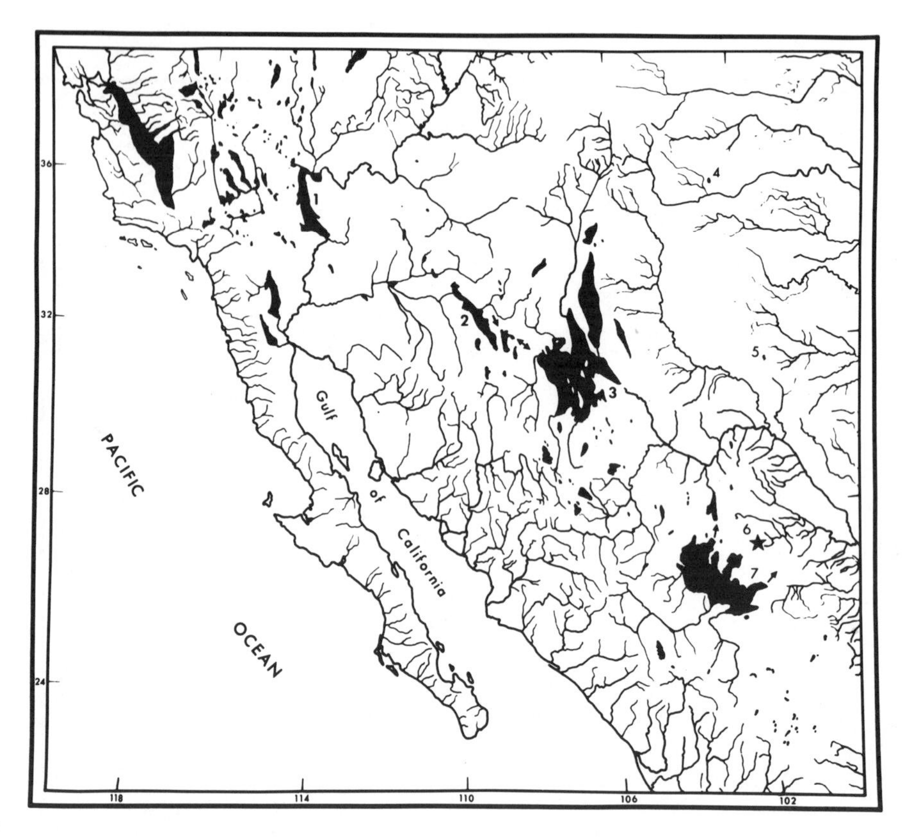

Figure 4. Pluvial lakes and some connectives at maximum Pleistocene expansion in the Mohave, Sonoran, and Chihuahuan deserts and contiguous regions. 1. Lake Hualapai and associated bodies (see text). 2. Lake Morrison. 3. "Lake" Cabeza de Vaca. 4. Lake Rita Blanca. 5. Lake Lomax. 6. Lake or lakes in Cuatro Cienegas Basin. 7. Lake Tlahualilo. See Paleohydrology for explanation.

Desert; and Lake Potosi (15) on the high, arid plateau south of Saltillo, Coahuila.

Virtually all these lakes (except those east of 105° and north of 27°) are believed to have contained species of Cyprinodon because pupfishes are known historically from their remnant waters. Pluvial lakes probably existed as long ago as 2 my because major continental glaciation in the Missouri Valley region occurred before emplacement of Pearlette-type B ash (for which there are fission-track age estimates in the 2 to 2.2 my range). Pearlette-B ash also occurs in the southern High Plains (Llano Estacado) and Hueco Bolson (near El Paso). Fluvial and lacustrine deposits in these areas indicate the existence of pluvial conditions about 2 my ago in the Southwest (John W. Hawley, personal communication 1980). One reason that pluvial lakes often have not been identified or mapped in the southern part of the Basin and Range Province (especially in southerly latitudes of the Chihuahuan Desert) is that when a lake is small, shallow, does not endure, or shows rapid changes in level evidence of its existence such as prominent strandlines fail to form; hence the former lake may be difficult to detect without studying its deposits. In the central part of the Great Basin a relatively modest increase in precipitation and decrease in evaporation rate would provide the conditions needed for restoration of late Pleistocene lakes (Mifflin and Wheat 1979; Snyder and Langbein 1962). In the more southerly deserts, especially the Chihuahuan, higher Pleistocene temperatures and rarity of mountain glaciation were partly compensated for by the greater precipitation that resulted from the "Mexican monsoon" (midsummer rains correlated with an influx of moist Caribbean air from the Gulf of Mexico), which produces today, and presumably did in the Pleistocene, far more rainfall in the Chihuahuan and eastern Sonoran deserts than in deserts to the west and northwest (Martin 1963: 3-4; Mosino and Garcia 1974, Figure 15 and 18; Wells 1979; see also Table 1).

The Pleistocene lakes and streams of the Owens River-Death Valley region have been mapped and discussed (Feth 1964; Hubbs and Miller 1948; Miller 1946; Morrison 1965; G. R. Smith 1978) but are treated at some length here because the continuity described is open to serious question.

The hypothesis of an integrated Death Valley drainage system during Wisconsinan time, through which waters from Owens Valley along the eastern flank of the Sierra Nevada overflowed into a succession of progressively lower lakes to terminate in Death Valley, was proposed by a geologist 65 years ago (Gale 1915). The hypothesis also assumed that the Amargosa and Mohave rivers, entering from the east and south, contributed contemporaneously to Lake Manly, the terminal body of water in Death Valley (Figure 3E). This idea was supported by me (Miller 1946, 1948) and in the summary by Morrison (1965). Countering the statements by Hunt and Mabey (1966) and Hunt (1975) that overflow of Lake Panamint into Death Valley was meager or nonexistent are the studies by Hooke (1972), R. S. U. Smith (1978), and G. I. Smith (1979). Although the idea of an integrated "Death Valley System" now appears to be too simplistic, this most recent geological field work reveals at least one overflow of Lake Panamint through Wingate Pass in Pleistocene (probably Tahoe or pre-Tahoe) time (G. I. Smith, personal communication), based on shorelines well above the pass level in Panamint Valley and on deltalike deposits at the foot of the pass in Death Valley. The precipitous nature of this outlet stream down Wingate Wash, however, argues strongly against its use by Cyprinodon, a fish not adapted to swim against strong currents. Radiocarbon dates (Hubbs et al. 1963, 1965) of pluvial lakes in the lower course of the Mohave River demonstrate that this stream reached Death Valley well after Tahoe time.

Lacustrine conditions in the Searles basin started about 3.2 my BP, with two prominent early lake stages, between 3.2 and 2.6 my and 1.3 to 1.0 my BP. Presumably within the last 3 my the Bouse Embayment extended westward into the Danby-Cadiz (but perhaps not the Bristol) basins (Figure 5). Evidence of such a connection is obtained from foraminifera in cores from Danby and Cadiz playas that show close relationships to those in the Bouse Formation (P. B. Smith 1960).

Hewett (1954 and personal communication) suggested (as did Dwight W. Taylor, personal communication) that after the proto-Owens River reached the Searles and Panamint valleys it entered the Leach Trough, the major east-west valley that follows the prominent Garlock Fault which lies just south of the Searles-Panamint-Death valleys and eventually connected with the proto-Colorado via the Danby-Cadiz basins and the Bouse Embayment. The timing was probably Pliocene, during the brackish-water period of the Bouse Embayment. Thus only euryhaline fishes were permitted westward movement

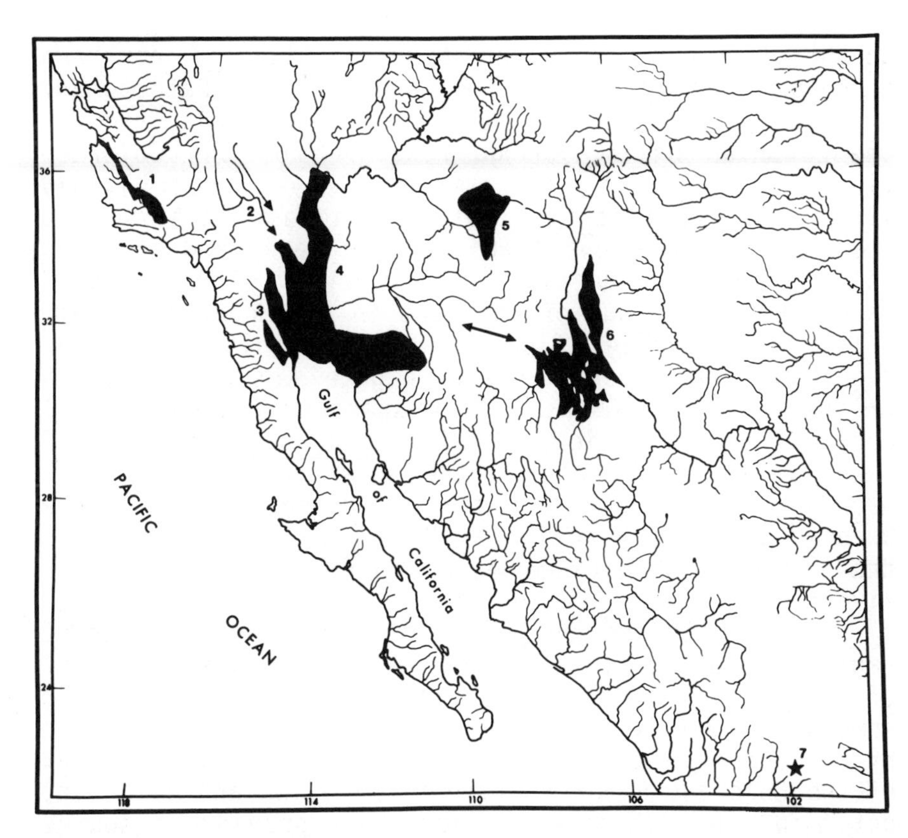

Figure 5. Mio-Pliocene drainages, lakes, and estuaries in part of the American Southwest. 1. Estuarine (brackish water) embayment. 2. Hypothetical Pliocene connection between Owens-Death Valley, southwestern Nevada, and Bouse Embayment indicated by arrows. 3-4. Embayment of Gulf of California into (3) former Lakes LeConte and Pattie basins, and (4) lower Colorado basin. 5. Middle Pliocene Lake Bidahochi. 6. "Lake" Cabeza de Vaca or progenitor. Two-headed arrow between proto-Gila River and proto-Rio Grande indicates faunal dispersal in both directions.

(G. R. Smith 1978). The Mohave River probably never acted as an effective dispersal path from Death Valley.

The present fish fauna of Owens Valley, Mohave River, and the Amargosa-Death Valley basin is of diverse origin. Cyprinodon clearly came from the southeast (Colorado River basin), *Gila* is northern (Lahontan basin), as is the Owens Valley *Catostomus*. *Rhinichthys* probably has dual origin (Owens Valley from the Lahontan basin, Amargosa River from the Colorado via stream capture with Amargosa River, see Gilluly 1929), and *Empetrichthys* has its closest relatives to the east in the pluvial White River basin and Railroad Valley, former connectives of the lower Colorado River. An undescribed genus of mollusks in Ash Meadows is comprised of a species flock that shows this same relationship, and other mollusks negate a close affinity to the west with Owens Valley species (Dwight Taylor, personal communication, 1980). *Cyprinodon radiosus* of Owens Valley is more like the ancestral form (cf. *C. macularius*), which is presumed to have invaded the Owens-Death Valley region in Pliocene time, than it is the species inhabiting Death Valley and the Amargosa River basin.

Lakes LeConte[2] and Pattie in the Colorado Desert region of southeastern California and an adjacent part of Mexico (Figure 3, Nos. 3 and 4) were described in some detail by Hubbs and Miller (1948:103-113). Their hydrographic history is intimately connected with that of the Colorado River and their formation and duration, determined by subsequent study, have provided a source of controversy that need not concern us. The recency of the last (Holocene) lake in the Salton Sink, here called Lake Cahuilla, is in a time scale close to a period of increased precipitation in the Southwest, as indicated by dendroclimatic studies (Fritts 1965).

---

[2]Called Lake Coahuila by current workers. Lake Cahuilla, however, was postpluvial (Holocene), ranging in age from about 300 to 1600 years BP (Hubbs et al. 1960:215-217). The older and deeper Lake LeConte had a shoreline elevation of about 150 ft or 45.7 m (Stanley 1966); this ancient beach line (unlike the distinct and continuous beach line of Lake Cahuilla) survives only in small fragments and available radiocarbon dates vary from about 10,000 years BP to > 35,000 and > 50,000 years BP (Hubbs et al. 1963:257; 268; 1965:89-90), thereby placing Lake LeConte in the Wisconsinan glaciation.

This sizable body of water, marked by the prominent "Ancient Beach Line," was one of the most recent of all pluvial lakes. That Cyprinodon abounded in it and in Lake Pattie, which had a similar history, seems obvious (Miller 1943).

Lakes Animas, Playas, and Hachita (Figure 3, Nos. 5-7) occupied adjacent basins in Grant County, New Mexico, and just over the Mexican border in Chihuahua. Although their now dry valleys contain no fish life, these lakes (Axtell 1978:499-500; Feth 1964; Fleischhauer 1978; Hubbs and Miller 1948:114, 116, 155, 165-166; Meinzer 1922: Figure 1) may have allowed the passage of aquatic biota between the Gila River and the Guzman complex (Lake Palomas). More plausibly, however, this connection was earlier and involved fluvial transfers (see below).

Lake Hachita has been consistently misplaced and/or misdrawn (Axtell 1978; Feth 1961, 1964; Meinzer 1922). As indicated by Brand (1937), clearly shown by Hawley (1969, 1975), Hawley et al. (1976), Morrison (1969) and the Hachita topographic sheet (USGS 1:62,500, 1918 ed.), and also noted by Hubbs and Miller (1948:116), Hachita Valley has a central draw that drains southeastward into the play of Laguna de los Moscos in adjacent Chihuahua, Mexico. That playa, part of what Brand (1937:69) called the Los Moscos plain, has been known on rare occasions to drain into Rio Casas Grandes, thus confirming Darton's (1933:166) statement that the Hachita basin drained into Laguna de Guzman, the historic terminus of Rio Casas Grandes.

The Tularosa Basin in south-central New Mexico (famous for White Sands National Monument) held a pluvial lake, Lake Otero (Figure 3, No. 8), which had a depth of about 15 m and covered about 460 km^2; its maximum expansion at an elevation of 1198 m (Hawley, personal communication, 1980) was probably correlated with the Tahoka Pluvial of the Great Plains (Kottlowski et al. 1965; Reeves 1973, 1976; Wendorf and Hester 1975), which is estimated to have occurred 15,000 to 22,500 years BP. Although this lake evidently did not discharge in Wisconsinan time, it was earlier tributary to the Lake Cabeza de Vaca complex (Figure 4, No. 5), hence part of the early Pleistocene proto-Rio Grande basin; this confirms the view that the fish now inhabiting the Tularosa basin gained access to it from the south (Miller and Echelle 1975).

Pluvial Lake King (Figure 3, No. 9) lay in a depression south and east of the Tularosa basin in the Trans-Pecos region of Texas that now contains saline lakes. Neither this basin nor the Sacramento River, which probably connected with Lake King from the north during the Tahoka Pluvial, are known to contain fish life (Hubbs and Miller 1948:117). As pointed out to me by John Hawley, it is possible that this basin connected with the Rio Grande-Conchos system by passes south of Van Horn which are much lower than the present passes that connect with the Pecos River. This pluvial lake is herewith named Lake King for the outstanding research investigations of Philip B. King, who was the first to map the lake (King 1948:Plate 22). [Lake King is not shown on the map of maximum Pleistocene lakes (Figure 4) because its outline for that stage has not been determined.]

In a brief discussion of pluvial lakes in Mexico, Flint (1971:451) wrote:

> Data on pluvial lakes in the Americas south of the United States are meager. Although such lakes are said (Jaeger, 1926) to have been abundant throughout the Mexican plateau, details are known only for the vicinity of Mexico City.

The Valley of Mexico, in which Mexico City lies, is outside the area treated herein, but a recent account of the biogeography of that region (with a discussion of Pleistocene hydrology) is given by Barbour (1973). In mapping the pluvial lakes of the Chihuahuan Desert region I assumed that each existing endorheic basin with a definite playa held a lake, however small or transient (see Hubbs and Miller 1948:145-146). This explains the many small bodies of water that appear on the Mexican portion of Figure 3 (most of which have been purposefully enlarged to be visible).

A large lake of Wisconsinan age received the waters of the streams of the Guzman complex: the

Mimbres River in New Mexico and the Casas Grandes, Santa Maria, and del Carmen in northern Chihuahua, Mexico. This lake (Figure 3, No. 10) was named Lake Palomas by Reeves (1969); its outline is based approximately on a contour of 1210 m. Lake Palomas covered about 7700 km^2 in northwestern Chihuahua and a tip of southwestern New Mexico during post-Kansan time; it had no outlet. Abandoned shorelines, spits, beaches, and lacustrine deposits provide evidence of the lake (Reeves 1969:Figures 4, 6 and 7). Either during its existence or that of an earlier, larger precursor complex (see below) the three streams entering Lake Palomas from the Sierra Madre Occidental had more extensive headwaters than now; thus, aided by direct precipitation and a presumably lower evapotranspiration rate, the lake level endured long enough to leave recognizable markers.

The Rio Sauz Bolson, which lies just north of Chihuahua City, held two shallow pluvial lakes, herein named Lakes Sauz and Encinillas (Figure 3, No. 11) after the small settlements in the endorheic basin. The two remnant playas, separated by a large alluvial fan, occur south of and near the lowest point in the bolson. Lake Encinillas, approximately 20 km long, was the larger; its outline and that of Lake Sauz (about 10 km long) are based on the 1500 m contour on Map NH 13-7 (Series F 501) Buenaventura, 1:250,000, Army Map Service, 1964.

Pluvial Lake Bustillos (Figure 3, No. 12) lay south of the Sauz Bolson in the enclosed Bustillos basin just north of Cuauhtemoc, Chihuahua. It and Lakes Sauz and Encinillas, just discussed, technically fall outside the Chihuahuan Desert, as interpreted by botanists and most zoologists (including Miller 1978: Figure 1). I have included these two basins here (and also that of Santiaguillo, north of Durango City; see below) because they lie on or near the desert margin and their fish faunas are historically related to those of the Chihuahuan Desert proper. To exclude them on floristic or climatologic grounds would make this consideration of pupfish evolution incomplete. The outline of Lake Bustillos follows approximately the 2000 m contour (AMS Map NH 13-10, 1:250,000, Chihuahua, 1965) on the assumption that its Wisconsinan level reached at least that elevation. Although there is no evidence of the discharge of Lake Bustillos in Wisconsinan time, the fish fauna of the basin (a cyprinid of the genus *Gila* and a pupfish, *Cyprinodon* n. sp., (Figure 2, No. 19), indicates a waterway connection to the northwest into the Rio Santa Maria basin (the present lowest point on the divide appears to be less than 2050 m).

In the flat, arid region known as the Parras Basin (Tamayo and West 1964:115), near Torreon, Coahuila, the Rios Nazas and Aguanaval combined at their mouths to form Pluvial Lake Mayran (Figure 3, No. 13), which had no outlet. Its maximum extent may have been even greater than mapped, for these authors indicated that Rio Nazas at one time emptied into Laguna de Tlahualilo to the northwest of historic Laguna de Mayran. The town of Parras de la Fuente contains a stream that was tributary to Lake Mayran and holds a remnant of its fish fauna (three cyprinids, a cyprinodontid, and a goodeid; see below).

The basin of Laguna de Santiaguillo, which has a surface area of 1790 km^2 (Tamayo and West, 1964:121) and whose center is about 85 km north of Durango City and well west of the Chihuahuan Desert proper, contained a pluvial lake that lay between the headwaters of the Pacific-flowing Rio Mezquital and the interior Rio Nazas drainage. Pluvial Lake Santiaguillo (Figure 3, No. 14), drawn at approximately 1950 m elevation, was one of the highest pluvial lakes in northern Mexico. Although it is separated by only a low divide from the upper Rio Mezquital, and one might therefore expect its fish fauna to be the same as or similar to that of the latter stream, there is evidence of multiple invasions that involved dual drainage connections. An older invasion, or relict of a former widespread hydrographic connection, came from the east and is represented by the atherinid fish *Chirostoma mezquital* Meek, whose ancestors probably originated in the Rio Lerma basin far to the southeast. The more recent invader, *Cyprinodon nazas* Miller (1976), is closely related to other populatios of this species in the Nazas-Aguanaval drainage system rather than to the Rio Mezquital representative, *C. meeki* Miller (1976).

Evidently *C. nazas* gained access to the Santiaguillo basin relatively recently. Albritton (1958) has described how the present upper end of Rio Mezquital, which he called "...the southern headwaters of the Pleistocene Rio Grande," was once connected with Rio Nazas, probably in pre-Wisconsinan time. After this capture the resultant loss of water to the Nazas (and to Pleistocene Lake Tlahualilo; see below), increasing aridity, and higher evaporation rates considerably reduced the size of Pleistocene lakes in the Parras Basin. Agricultural practices during the last half-century have eliminated altogether the historic remnant of Lake Mayran.

In the high (2000 m), arid Mesa Central south of Saltillo, Coahuila, labeled El Salado (Tamayo and West 1964:Figure 1), scant surface water exists. In the endorheic basin of La Hediondilla, around the small settlement of El Potosi, Nuevo Leon, there is evidence of a former marsh and lake, herein named Lake Potosi (Figure 3, No. 15). Dissected marsh deposits indicate the former lake, but its extent and depth have not been determined (Miller and Walters 1972:9); it is among the many smaller pluvial lakes on the map which have been purposefully enlarged to make them more readily visible. The only fishes in the remnant waters point to derivation from the northwest, although the basin now slopes to the southeast. The connection was almost surely pre-Wisconsinan and probably involved more than one invasion.

During early Pleistocene time other pluvial lakes that occurred in the lower Colorado River, the upper Gila River, the upper Rio Grande, and the lower Rio Nazas (Figure 4, Nos. 1 to 3 and 6) were important to the dispersal of fish faunas. The lake in the lower Colorado (No. 1), taken largely from G. R. Smith (1978) with some modification provided by Ivo Lucchitta (unpublished ms. 1979), represents Lake Hualapai in part (at the west end of Grand Canyon). Lake Morrison (No. 2, Feth 1961, named by Axtell 1978 for geologist Roger B. Morrison) has been mapped as occupying the Gila River above Safford, Arizona. However, I am informed by John W. Hawley that field data to support the existence of such a deep lake are lacking. According to

Cooley and Wilson (1968:101), the Gila River did not attain its present course until the early part of the Pleistocene. Lake Cabeza de Vaca (No. 3, modified slightly from Axtell 1978 and named by Strain 1966:10 "in honor of the first white man to enter its basin") consisted of a complex of interconnected lakes and streams that occupied bolson floors (Hawley 1975:Figure 2); current elevations do not exceed 1219 m. These waters were fed by the proto-Rio Grande and drained in mid-Pleistocene time into the present course of the Rio Grande, most likely as a result of headward erosion of a stream in the Hueco Basin, east of El Paso, or by the meandering of the proto-Rio Grande in the Mesilla Basin northwest of El Paso (Strain 1966). Although the Great Plains region of the United States was dotted with Pleistocene lakes, only two are considered here. Lake Lomax (Frye and Leonard 1968), one of the largest, was an early Wisconsinan lake (No. 5); Lake Rita Blanca (No. 4), very small, was an early Pleistocene lake (Anderson and Kirkland 1969). I know of no evidence that _Cyprinodon_ inhabited any of the Pleistocene lakes east of the Pecos River, but it is logical to assume that they did because the Llano Estacado probably supported them in historic time. The Cuatro Cienegas Bolson of northern Coahuila may have held an early Pleistocene lake or lakes (No. 6; Minckley 1969). A large lake, here named Lake Tlahualilo (No. 7) after the recent playa northwest of Tlahualilo (Tamayo and West 1964:115), was fed by the proto-Rio Nazas drainage (including Rio Aguanaval).

In Arizona, New Mexico, and Mexico these major Pleistocene lakes and annectent streams served as dispersal routes for aquatic and semiaquatic organisms, which allowed transfer across the continental divide between the Cabeza de Vaca complex and the upper Gila River, movement of upper (proto) Rio Grande fishes into the Chihuahuan Desert, and invasion of the proto-Rio Nazas system by the Rio Grande biota via an outlet (or outlets) from Lake Tlahualilo to the east or north (see arrows, Figure 4, No. 7). That some of the fishes inhabiting remnant tributaries of Lake Tlahualilo came from the northwest (via proto-Rio Grande-Colorado connections; Miller 1978:377-378) rather than from the later, through-flowing Rio Grande or Rio Conchos suggests past connections between the upper Rio Conchos and the Tlahualilo basins, presumably south and east of Jimenez, Chihuahua. In Pliocene through mid-Pleistocene time a plausible link existed between the ancestral upper Gila River drainage and the Chihuahuan (and Sonoran) drainage basins.

> An 11.7 my valley basalt flow (U. of Arizona K-Ar date) near Hachita demonstrates that stream valleys (and lake basins) in that area have occupied near present positions since middle to late Miocene time (Hawley, personal communication 1980).

During late Miocene and early Pliocene time, approximately 9 to 5.5 my BP (Berggren and Van Couvering 1974), an extensive estuary (Figure 5, No. 3) in what is now the lower Colorado River area was formed largely by the Bouse Embayment (Blair and Armstrong 1979:Figure 8; Metzger et al. 1973) after the opening of the Gulf of California in mid-Miocene time, perhaps 18 my BP (Gastil and Jensky, 1973). In the Lake Mead area outcroppings of Hualapai limestone, which have been analyzed for chert and diatoms, indicate deposition under marine or brackish-water conditions (Blair and Armstrong 1979) and suggest a northward extension of this estuary. At that time the present course of the Colorado River (i.e., the connection between the upper and lower rivers) had not become established, although a proto-Colorado River apparently flowed through what is now Peach Springs Wash (Hunt 1969: Figure 79) to the Gulf of California. No large river discharged from Grand Canyon until middle or late Pliocene time (Hunt 1969:113; or until 3.3 my BP, Lucchitta 1972) and no appreciable deepening of the Grand Canyon has taken place during the last million years (Leopold 1969:131). Blair and Armstrong (1979:12) described the proto-Gulf of California:

> ...a shallow, marshy, marine estuary existed in late Miocene time, extending up the area of the lower Colorado River to the base of

Figure 6. View northward up the delta of the Colorado River (before 1935). Desert pupfish abounded here until the 1950s (from Godfrey and Sykes, The Colorado Delta, Carnegie Institution of Washington and American Geographical Society of New York, 1937; Figure 54).

> the Colorado Plateau where the river now emerges from the Grand Canyon. The elevation at this point was then at sea level, some 420 m lower than it is now.

We can perhaps visualize how the delta of the precursor of the Colorado River, as it entered this estuary, might have appeared by examining the pristine delta (Figure 6) of this historical river before it was destroyed and which provided habitat for a variety of aquatic organisms, including *Cyprinodon*.

At some time during the formation of the proto-Gulf of California and its use by the Colorado River or its antecedent, perhaps around mid-Pliocene time, the Owens River-Death Valley drainage connected (as described earlier) with the Bouse Embayment of the developing Colorado; this, in turn, had connections eastward with the proto-Rio Grande and Chihuahuan Desert regions (Figure 5, Nos. 2 to 4 and 6). Hewett (1954:18), a long-time student of Mohave Desert geology, speculated on the Owens-Death Valley connection:

> To the writer, it seems probable that before the waters of Owens River entered and successively filled Owens Lake, China Lake, Searles Lake, Panamint Lake and finally Death Valley, they entered the Leach trough, the major valley that follows the Garlock fault, and thence joined the Death Valley trough north of the Avawatz Mountains. From this point they probably flowed southward through Silver Lake, Soda Lake, and Bristol Lake to join the Colorado River estuary.

Metzger et al. (1973:G34-35) have also discussed the surface-water connection between the Death Valley region and the Colorado River. That it was ancient (pre-Pleistocene) is indicated by the total erasure of any remnant channel.

There is no question that the genus *Cyprinodon* gained access to the Owens-Death Valley regions from the east or south (rather than from the north via the Great Basin or from the west via the Pacific coast); hence this biogeographic information provides a powerful tool for inferring past hydrological events. In fact, biologists have been among the first to point out former waterways that geologists have subsequently or independently confirmed (Gilbert 1893; Hayes and Campbell 1900; Hubbs et al. 1974; Hubbs and Miller 1948; Meek 1903; Miller 1946; G. R. Smith 1978; van der Schalie 1945, 1973). Confident paleohydrologic and geomorphic reconstructions may not be possible, however, without detailed supporting geologic evidence from stratigraphy, sedimentology, and structural geology.

DEGREE OF DIFFERENTIATION AND BARRIER CHRONOLOGY

Species of *Cyprinodon* that are morphologically and

behaviorally distinct have often proved to be interfertile when tested in no-choice experiments (Cokendolpher 1980 and references cited). This is also true to a similarly remarkable extent for the poeciliid fishes (mollies) of the subgenus *Mollienesia* (genus *Poecilia*), wherein artificial crosses are easily made (again in no-choice experiments) between such obviously distinct allopatric species as *Poecilia velifera* and *P. sphenops*. Even trihybrids or tetrahybrids are readily constructed, as in *Cyprinodon*. The ability to hybridize under artificial conditions or when sympatry in the field has been caused by human interference does not provide conclusive evidence that allopatric taxa are not good biological species. Critical tests to determine whether *Cyprinodon* possesses behavioral traits that serve as premating isolating mechanisms have not been made.

Biochemical genetics, as measured by gel electrophoresis, is proving to be helpful in yielding clues to relatedness, but so far this tool has detected little genetic divergence in the genus *Cyprinodon*. For example, similarity indices are .84 to .97 for Death Valley species and .98 to .99 for four species in Laguna Chichancanab, Yucatan (J. M. Humphries, personal communication 1980; Humphries and Miller 1981; Turner 1973, 1974). Furthermore, chromosomal divergence also seems to be minimal; the diploid number is invariably 48 (arm number 50, or 64 if submetacentrics are counted as two) in the 16 species karyotyped so far (Miller and Echelle 1975; M. L. Smith and Miller 1980; Stevenson 1981; Turner and Liu 1977).

Wherever species of *Cyprinodon* have hybridized by human interference (Miller 1968; Minckley 1969; Stevenson and Buchanan 1973) or naturally (Humphries and Miller 1981) the integrity of the gene pools of the respective species has been maintained and this indicates selection against hybridization. In the example of the Cuatro Cienegas Bolson the intermediate or hybrid habitat (*sensu* Anderson 1949) was subsequently destroyed along with the hybrid swarm (Minckley 1978).

Lack of postmating genetic barriers to reproductive isolation, biochemical similarity of species, and possession of a seemingly stable karyotype all suggest that *Cyprinodon* has a rather "conservative" genome. Could it be that this genome has enabled pupfishes to "escape from specialization," despite their frequent but reversible adaptation to extreme habitats?

The extensive interspecific genetic compatibility in *Cyprinodon* raises the question whether the allopatric morphotypes that have often been regarded as belonging only to the Wisconsinan or postpluvial age are really valid species. Failure, however, to detect allozymic or obvious chromosomal differences between allopatric populations does not prove lack of speciation. Hirshfield et al. (1980) found that a population of *Cyprinodon nevadensis mionectes* inhabiting a constant-temperature spring and one of *C. n. amargosae* living in a variable habitat, when tested for F and F generations reared under identical conditions, had significantly different ranges of temperature and oxygen tolerances, which indicated genetic divergence since their isolation, presumably in postpluvial time.

We are currently left with the usual approach to a determination of the validity of allopatric species: if the amount of morphological and other divergence is at the general level found in the few examples of sympatry (or similar to that in related genera that commonly occur together), we may judge the isolated populations to be worthy of taxonomic recognition at the species level. In this connection it would be of interest to know the relative ages of the many allopatric species of *Cyprinodon*. An attempt to determine time of differentiation since isolation follows.

Cladistic inferences and phenetic similarity, with some experimental data and inferred geological events, provide a reasonable basis for the assignment of many species of *Cyprinodon* to different chronologies in relation to barriers, although firm ages for differentiation are difficult to obtain. Means of timing events that bear on evolutionary history are greatly strengthened by consistent results from a multiple of disciplines (pluvial lake history, isotopes, palynology, radiogenic datings, and marker fossils). These correlative data to determine when various pupfishes became isolated are mostly lacking. Even when we know that a large lake (such as Lake Manly) existed in Wisconsinan time and was fed by the then continuous Amargosa River, we cannot be absolutely certain that the lake did not contain a different species of pupfish (i.e., *C. salinus*) than the river and its tributary springs (i.e., *C. nevadensis*) (see also Miller 1961b: 552). We may feel strongly that during the existence of Lake Manly it and the Amargosa River, at least its lower end, contained a single species but we do not know that it did. This further complicates the estimation of time since the now-isolated allopatric pupfishes became differentiated.

The following species are judged to be relatively old for the reasons given:

Cyprinodon radiosus. Early Pleistocene or late Pliocene. Although the Pluvial Owens River filled a series of successively lower lakes probably to Death Valley during Wisconsinan time (and perhaps during earlier pluvial stages), it evidently was not an effective dispersal route to the west from Lake Manly; hence the pupfishes of the Amargosa River basin and Death Valley were isolated from Owens Valley and its lakes, at least throughout the Pleistocene. The ancestor of the Owens pupfish reached Owens Valley during the much earlier, hypothetical connection between that valley and the proto-Colorado River, when a connection coexisted with the Death Valley region, perhaps in late Pliocene time (see Paleohydrology: Figure 5). In support of this idea is the observation that the Owens pupfish resembles the lower Colorado species, *C. macularius*, in more features than it does the other species of the Death Valley region (Miller 1948:87-93); this reflects an early isolation and differentiaton from the ancestral Death Valley form, or more rapid evolution in the Amargosa-Death Valley drainage, or both. How long this early connection between the Owens and Death Valley troughs existed is unknown and perhaps undecipherable.

Cyprinodon nevadensis. Early or mid-Pleistocene. If we assume that the invasion westward from the proto-Colorado River of an ancestral pupfish occurred in the latter part of Pliocene time and that the stream connection to Owens Valley (via Garlock Fault) was disrupted by or before mid-Pleistocene time, the differentiation of the basic Death Valley-Amargosa River form, *C. nevadensis*, presumably began in the early Pleistocene. The evolution of its offshoots, *C. salinus* and *C. diabolis*, however, is plausibly a much later development (see below).

Cyprinodon macularius. Late Pliocene, from an ancestor in the Chihuahuan Desert region. This species, which gained its present distribution by derivation from an eastern ancestral form, presumably got into the lower Colorado basin well before Pleistocene time (see Figure 5, double-headed arrow). Some form of *Cyprinodon* had to be in the lower Colorado at the time of connection with the Death Valley trough, plausibly in mid- to late Pliocene. As explained in the Paleohydrology section, the course of the Recent Gila River was not established until after mid-Pleistocene time.

Cyprinodon elegans. Early Pleistocene. This is a distinctive species in morphology and coloration (Echelle 1974 and personal observation), being readily distinguishable at sight from other pupfishes. Unlike them, it breeds in swift water (Itzkowitz 1969). Available genetic data also suggest that it is at least partially isolated (males sterile) from *C. variegatus* and *C. pecosensis*. In an analysis of field-caught hybrids between *C. elegans* and the exotic *C. variegatus*, Stevenson and Buchanan (1973) found an aberrant sex ratio (80 percent gravid females) and little or no gonadal development in the 20 percent males. The only surviving male of a hybrid between *pecosensis* and *elegans* was sterile, although F hybrids of both sexes that resulted from the hybrid between *pecosensis* and *variegatus* were fertile unto the F and F generations (Drewry 1967, cited by Stevenson and Buchanan). For these reasons I place *elegans* among the earlier differentiated species of pupfish.

Cyprinodon bifasciatus. Early Pleistocene. This species is remarkable among pupfishes in that it lives an open, active, competitive life and evidently is unique in its lek behavior (Arnold 1972). It occurs with and is not at all intimidated by large, predatory fishes (e.g., cichlids, sunfish, and largemouth bass), toward which it often shows aggressive behavior; the young are darterlike in behavior and appearance and occur on the bottom in foraging schools of 10 to many hundreds of individuals (Minckley 1969). Body form, coloration, weak sexual dimorphism, and breeding colors are unlike other pupfishes, for not even juveniles possess an ocellus on the dorsal fin, a mark almost always present in juveniles and females of Cyprinodon. Unlike other species of Cyprinodon tested or observed to date (save one other undescribed Chihuahuan Desert species), this pupfish is highly sensitive to thermal changes; it is essentially restricted to warm waters (25 to 33°C) of headwater springs and limnocrenes in the Cuatro Cienegas Bolson, Coahuila (Miller 1968; Minckley 1969, 1978). I hypothesized a pre-Pleistocene origin for this species in the paper cited; however, selection pressure in its multispecies habitat could have accelerated the rate of evolution.

Cyprinodon beltrani, C. maya, C. labiosus, and C. simus (Chichancanab species flock). Early to mid-Pleistocene. Although living well outside the Chihuahuan Desert, this remarkable species flock (Humphries and Miller 1981), confined to a single lake in Yucatan, Mexico, deserves mention because it is unique for the genus in constituting a group of endemic sympatric species with completely overlapping habitats. Morphological diversity (including feeding habits unique for Cyprinodon) is greater here than for all other pupfish species combined. Except to say that Laguna Chichancanab is probably post-Pliocene, as the Yucatan Peninsula gradually emerged from the Gulf of Mexico during Pliocene time (de Cserna 1975), no reliable information is available on its age.

The following arid-land species are judged to be younger than the group just discussed, although their origin is believed to antedate the last glaciation:

Cyprinodon n. sp. (Whitefin pupfish). Pre-Wisconsinan. The distinctiveness (particularly the conspicuous, white dorsal fin) and distribution of this pupfish in two well-separated drainages (Rio Yaqui and Rio Santa Maria) suggest a fairly old time of differentiation. No dates are available for a connection between the upper Yaqui (Rio Papigochic) and the Santa Maria basins, but the assumption that it was pre-Wisconsinan is reasonable. Meek (1903:775) stated that the upper Rio Papigochic no doubt once had its outlet through Laguna Bustillos (his Lago de Castillos) into Rio Conchos. The fish fauna of Laguna Bustillos, at least at present, does not support this hypothesis (see Paleohydrology). That there was a connection somewhere between the drainages of Rio Conchos and Rio Yaqui is, however, supported by the distribution of fishes other than those currently inhabiting the Bustillos Basin (Miller 1978).

Cyprinodon meeki. Pre-Wisconsinan. As explained by Albritton (1958; see Paleohydrology), the upper Rio Mezquital near Durango City, to which this species is presently restricted, was disrupted from the proto-Rio Nazas well before Wisconsinan time (maybe in the Illinoian).

Cyprinodon nazas. Pre-Wisconsinan. This species, with one exception (basin of Laguna Santiaguillo), is restricted to the endorheic Rio Nazas basin (including the Rio Aguanaval). Its closest relative is probably _C. eximius_, which inhabits the Rio Conchos basin next to the north. No one has yet suggested a direct surface-water connection between these two rivers, but that is plausible. Because in early to mid-Pleistocene time, when Pluvial Lake Tlahualilo (Figure 4, No. 7) overflowed, there was no throughflowing Rio Grande, the ancestral pupfish that presumably gave rise to both species did not inhabit a common drainage system. This suggests, then, a shorter means of connection between these two systems at some time in the Pleistocene.

Cyprinodon alvarezi. Pre-Wisconsinan. This species, also probably closest to _C. eximius_, is the only pupfish isolated on the elevated, southern extension of the Basin and Range Province (at El Potosi, Nuevo Leon;

Figure 2, No. 33). Any connection between that endorheic plateau and the Nazas and Conchos basins was surely earlier than the last glaciation.

Cyprinodon tularosa. Pre-Wisconsinan. There is no evidence that Lake Otero (Figure 3, No. 8) overflowed during Wisconsinan time. During the existence of the Lake Cabeza de Vaca complex (see Paleohydrology; Figure 4, No. 3), however, it was part of that early to mid-Pleistocene system. Whether it had a connection to the Rio Grande between mid- and late Pleistocene time is not definitely known but may be assumed.

Cyprinodon n. sp. (Palomas pupfish). Pre-Wisconsinan. This species is confined to the drainages that once fed Pluvial Lake Palomas (Figure 3, No. 10). It and two other species (*C. fontinalis* and *C.* n. sp.) were presumably once sympatric because all three occur in the Palomas basin. Assuming that no Wisconsinan outlet existed for that lake (and none has been suggested), the isolation of this form must antedate the late Pleistocene. Careful comparisons of the now disrupted populations are under study.

Cyprinodon macrolepis. Pre-Wisconsinan. Although this distinctive species lives in a spring (probably the largest in the Chihuahuan Desert) that is tributary to Rio Conchos, home of *C. eximius*, no overlap in ranges has been noted. A large collection of *C. eximius* was made in 1901 by Meek at Jimenez, Chihuahua, not far from the mouth of the ditch that exits from El Ojo de la Hacienda Dolores (type locality of *C. macrolepis*) only 12.5 km to the south-southwest. This hot-spring species may have lived in its present habitat since mid-Pleistocene time, although I know of no study to determine the age of the spring.

Cyprinodon atrorus. Pre-Wisconsinan. If the closest relative of this species is *C. eximius*, as stated by Miller (1968), its isolation from that species may predate Wisconsinan glaciation. It seems more likely that it originated from south and west of the Cuatro Cienegas Bolson rather than by way of a Rio Salado-Rio

Grande connection, despite the occurrence of *C. eximius* today in Devil's River, Texas (which I believe represents a late downstream migrant from Rio Conchos).

Cyprinodon eximius. Pre-Wisconsinan. This is one of the most widely distributed of the Chihuahuan Desert pupfishes. It occurs in Rio Conchos (presently tributary to Rio Grande, the basin of Pluvial Lakes Encinillas and Sauz (Figure 3, No. 11), Alamito Creek in Presidio County, Texas, Rio Alamo, Chihuahua, across from Alamito Creek, Tornillo Creek, Brewster County, Texas, and Devil's River in Val Verde County. The last four streams are all independent tributaries to the Rio Grande, the first three above and the last one below Big Bend National Park. It is reasonable to believe that this wide range was not attained all in Wisconsinan time; there is no clear-cut evidence, for example, that Pluvial Lakes Encinillas and Sauz were connected with Rio Conchos during late Pleistocene time and it has been suggested that Rio Conchos was ponded in endorheic basins in its lower course in late Tertiary or Quaternary time (King and Adkins 1946).

Cyprinodon latifasciatus. Pre-Wisconsinan. This distinctive pupfish (Miller 1964), unfortunately extinct for perhaps 50 years, is so unlike *C. nazas*, the geographically nearest species, that it probably had been in the Parras Valley for a long time -- possibly since mid-Pleistocene. This valley also contained the endemic cyprinid genus *Stypodon* and at one time a goodeid (identifiable now only as *Characodon lateralis* on the basis of inadequate material). Miller assumed that the closest relative of this pupfish is *C. eximius*.

Cyprinodon fontinalis. Pre-Wisconsinan. This spring-inhabiting desert pupfish of the Pluvial Lake Palomas basin (Figure 2; Figure 3, No. 10) is most similar to *C. macrolepis* of a warm spring in the Rio Conchos drainage, next to the south. If so, it has been isolated from that congener since before the last glaciation, for Lake Palomas had no outlet. The two other species of pupfish in the basin of this pluvial lake, both undescribed, are more distant relatives, as inferred by morphological studies.

The following species are judged to be of Wisconsinan or postpluvial age:

Cyprinodon bovinus. Wisconsinan. This species is similar to *C. pecosensis*, which also inhabits the Pecos River drainage, and to *C. rubrofluviatilis* of adjacent river systems to the east (Echelle and Echelle 1978: Figures 1, 4). All three are closely allied to *C. variegatus*, which normally does not (or did not) ascend the Rio Grande farther upstream than Zapata County, Texas (Robinson 1959). That each is distinct from *C. variegatus* is evident from morphological comparisons, but whether we are really dealing with three separate species or only one may be debated. *C. pecosensis* like *C. rubrofluviatilis* and, to a lesser extent, *C. bovinus* lives in fluctuating environments. During pluvial periods of Wisconsinan time probably all three now allopatric species were in contact. The recency of their isolation seems to be well supported by the many similarities shared by these disrupted populations.

Cyprinodon diabolis. Late Wisconsinan. The age of the water-filled, fault-formed, semicave that constitutes the unique habitat of this remarkable species has not been determined. Perhaps it formed 20,000 years BP or as recently as 10,000 years BP (Soltz and Naiman 1978: 35). Devils Hole is the highest and first isolated of all the springs in Ash Meadows. That it overflowed in the past seems obvious, but no firm date for or unequivocal evidence of surface discharge has been discerned. The differentiation of this species from a common ancestor with *C. nevadensis* in adjacent Ash Meadows could have been due to rapid selection on or drift in a small founder population isolated in a habitat unlike that of any other species of pupfish (Miller, 1977). Factors contributing to such rapid speciation are a small population (at times, *C. diabolis* is reduced to about 200 individuals), the completeness of isolation, longevity, and the number of generations per year. In 32.8 to 33.9°C water more than one generation can be produced annually, but the bulk of the population probably lives only one year.

Cyprinodon salinus. Late Wisconsinan. As mentioned earlier, we cannot be sure that during the existence of Lake Manly, which persisted perhaps to 12,000 years BP, there were not already two well-differentiated species of pupfish (C. nevadensis and C. salinus) in Death Valley. In fact, the origin of C. salinus conceivably could go back to what Hubbs and Miller referred to as prepluvial time, in this case to late Pliocene. The few rather extreme characteristics (e.g., high scale counts) might suggest such long isolation, but again differentiation could have been relatively rapid when a small population of the ancestral Lake Manly form was cut off by the desiccation of that body of water. The case seems equivocal.

The pupfish so thoroughly described as C. milleri (LaBounty and Deacon 1972), on the other hand, was probably not isolated from its congener, C. salinus, much more than 2000 years ago, when a Holocene lake 9 m deep covered enough of the floor of Death Valley to unite Cottonball Marsh and Salt Creek, their respective habitats. On the basis of radiocarbon dating of pollen seeds, Mehringer (in LaBounty and Deacon 1972:777) suggested dates of about 3600 and 300 to 400 years BP as "probably times of significant connection of aquatic habitats on the floor of Death Valley." (See also Fritts 1965.) Current reanalysis of C. salinus and C. milleri suggests the reduction of the latter to subspecific status as C. salinus milleri; for example, the median ridge on the outer face of the jaw teeth occurs in both species.

SURVIVAL STATUS OF PUPFISHES

There can be no question that some pupfishes, like many other organisms, have declined markedly because of the perturbations promulgated directly or indirectly by man during the last three to five decades. However, accurate information regarding the status of fish species thought to be threatened is often scanty and difficult to obtain. For this reason some pupfishes (e.g., the Devil's River population of Cyprinodon eximius, C. bovinus, and C. radiosus) were prematurely reported as extinct (Davis 1980; Echelle and Miller 1974; Miller 1948). As far as we know now only two species of *Cyprinodon* have become extinct, *C. latifasciatus* of Mexico and *C.* n. sp. of Arizona (Miller 1964; Minckley 1973:192-194). Heroic efforts to save two other species, the Devils Hole pupfish (*C. diabolis*) and Owens pupfish (*C. radiosus*), probably prevented their demise (Deacon and Deacon 1979; Miller 1977; Pister 1974).

The chief threats to pupfish survival are habitat destruction or adverse biotic interaction that results from pollution, mining of groundwater, draining of marshes (cienegas), dam construction, introduction of exotic species, diseases or parasites, overgrazing, and deforestation. Interference by man with the delicate ecological balances that are so typical of the fragile desert environment has produced a marked biotic change in the American Southwest (Deacon 1979, and references cited; Miller 1961a). Because populations of distinctive taxa are often restricted to single springs and their outflows, man-induced changes may rapidly affect the total genome and result in sudden extinction. This happened to two pupfish subspecies, exterminated since 1950; namely the highly localized *Cyprinodon nevadensis calidae* and *C. n. shoshone* of the Amargosa River valley in eastern California.

Currently endangered species of *Cyprinodon* are *C. bovinus*, *C. diabolis*, *C. elegans*, *C. macularius*, *C. nevadensis pectoralis*, and *C. radiosus* (Deacon 1979; Miller 1979). Destruction of its original spring and subsequent hybridization (ca. 1975) in a newly discovered habitat has threatened *C. bovinus* with extinction (Hubbs 1980). A thriving population, however, (as of May 5, 1980), from Leon Creek is being cultured at the Dexter National Fish Hatchery of the U. S. Fish and Wildlife Service. The story of *C. diabolis*, saved ultimately by the U. S. Supreme Court, has been recounted in detail (Deacon and Deacon 1979; Miller 1977). *Cyprinodon elegans* has been extinct at its type locality, Comanche Springs, since 1956 (these enormous springs disappeared because of excessive groundwater pumping) but survives in the Toyahvale-Balmorhea area in Reeves County, Texas (Echelle 1975), although under stress by hybridization (Stevenson and Buchanan 1973).

A stock is thriving at the Dexter National Fish Hatchery (as of May 5, 1980) and a refuge for the species is in operation at Balmorhea State Park by the Texas Parks and Wildlife Department. So many populations of *Cyprinodon macularius* have become exterminated in the last 50 years that this species is now completely extirpated from the Gila River basin, Arizona, and from virtually the entire Colorado River delta, where it was once abundant (see summary by Miller 1979). Two refuges have been established for it at Anza-Borego State Park in eastern San Diego County, California, where it is doing well (Frances N. Clark, personal communication, April 15, 1980). *Cyprinodon nevadensis pectoralis*, of Ash Meadows, Nevada, is regarded as endangered by the U. S. Department of the Interior. A refuge is operated by the Bureau of Land Management (Miller 1979). *Cyprinodon radiosus*, in Owens Valley, survives in four refuges but requires frequent monitoring to exclude exotic species (both fishes and crayfish) or to control excessive growth of vegetation (Miller 1979).

Pupfishes regarded as threatened include the Potosi pupfish, *Cyprinodon alvarezi*, whose Mexican habitat is potentially under threat by agricultural use of groundwater. The main spring habitat is now occupied in part by largemouth bass (M. L. Smith 1980). Many populations of the Palomas pupfish, *Cyprinodon* n. sp., have been exterminated from the lower Rio del Carmen and tributary springs in Chihuahua (observations by M. L. Smith and R. R. Miller, April 1980); the Carmen stock apparently now survives only in Rio Santa Clara, its headwater region. In the Rio Casas Grandes basin the same species has been extirpated from two springs near Las Palomas, where it was abundant in 1950, and is elsewhere under stress from exotic mosquitofish. *Cyprinodon* n. sp., confined to Laguna Bustillos, Chihuahua, is now known for certain in only one arroyo in that bolson. The current status of *Cyprinodon meeki* is uncertain but it was difficult to find in the upper Rio Mezquital drainage near Durango in 1976 (observatins by J. M. Humphries and M. L. Smith). The Ash Meadows pupfish, *C. nevadensis mionectes*, has suffered the loss of many spring populations and an overall decline of unknown proportions since

1969; a spring refuge has been purchased for this subspecies by the Nature Conservancy and a management plan is in preparation. The White Sands pupfish, _Cyprinodon tularosa_, is mentioned only because it is rare, confined to three known populations in the Tularosa Basin, New Mexico. In 1978 there was no evidence of a decline in numbers. The species lives in the U. S. Army White Sands Missile Range, which is difficult of access. A proposed desalinization plant at Alamogordo, however, may pose a serious problem (Paul R. Turner, personal communication 1980).

CONCLUSIONS AND SUMMARY

Cyprinodon is a unique genus of fish. Endowed with a "conservative" primitive karyotype and with high genetic similarity indices for morphologically divergent species (Turner 1974; J. M. Humphries, personal communication 1980), pupfishes can maintain viable populations in highly stressed environments in which other fishes would perish or can adjust rapidly to changed conditions of water chemistry, oxygen, temperature, and food sources. Successful spawning may occur over a temperature range of 13 to 43.8°C. Minimal genetic divergence may occur in spite of considerable morphological differentiation, especially in cases of rapid or "explosive" speciation. Although most pupfishes appear to have evolved over time spans of 300,000 to more than 1,000,000 years, some could have differentiated in the last 10,000 years or less. The genus is at least as old as the Miocene.

Pupfishes occur in three different deserts of the American Southwest: the Mohave, Sonoran, and Chihuahuan; the latter supports the greatest number of species. In Plio-Pleistocene times these deserts were relatively well watered and except for the interglacial periods had more equable climates than now. The genus attained its present wide distribution in arid regions during the expansion of lakes, drainage systems, and estuaries that occurred during late Cenozoic time. The "Death Valley System" is shown to have been based on too simplistic a model and was not a continuously integrated drainage as previously conceived. Pluvial lakes are described in reconnaissance fashion ("mapped") for the first time for the entire Chihuahuan Desert and their role in the dispersal and evolution of _Cyprinodon_ is presented. Much more field work is needed by multidisciplinary biologist-geologist teams in that region before the history of these lakes can be deciphered and objective data on isolation of populations relative to their differentiation can be obtained.

Two species and two subspecies of another species have been exterminated, many local populations have been eliminated, and the habitats of a number of pupfishes are threatened. A recent greater awareness of the need to maintain biotic diversity is helping _Cyprinodon_, but further efforts to educate land and water developers to use water without destroying resources is still urgently needed (Pister 1979 and this volume).

The "species complexes" of _Cyprinodon_ have not yet been objectively defined and a thorough revision of the entire genus is needed before explanations concerning the mode of evolution of these remarkable fishes can be invoked. Comparative osteology, largely neglected, may prove to be helpful in defining groups of related species.

ACKNOWLEDGMENTS

I remain indebted to the late Carl L. Hubbs for initiating my studies of desert fishes, especially _Cyprinodon_, and for his enthusiastic support and wise counsel over much of my career. Gerald R. Smith provided provoking stimulation of concepts of differentiation and rate of evolution and much improved the text. G. I. Smith and especially J. W. Hawley gave constructive criticism and provided valuable references that greatly improved the section on paleohydrology. An anonymous reviewer also improved the text. M. L. Smith, J. M. Humphries, and B. Chernoff contributed fruitful discussions on speciational aspects. D. W. Taylor and J. E. Deacon helped to retire the concept of a "Death

Valley System." My wife Frances served as typist for rough drafts, proofreader, critic, and helped over the years in the field work; in short she made possible the completion of this study. Others too numerous to mention have assisted in the field. The National Science Foundation, New York Zoological Society, and U. S. Fish and Wildlife Service, in particular, made possible the field work in the Chihuahuan Desert that was critical to the formulation of ideas in regard to paleohydrology and the evolution of *Cyprinodon* in that region. The maps were drafted under my supervision by M. L. Smith. Linda Krakker typed the final copy.

REFERENCES

Albritton, C. C., Jr. 1958. Quaternary stratigraphy of the Guadiana Valley, Durango, Mexico. Bulletin of the Geological Society of America 69:1197-1216.

Anderson, E. 1949. Introgressive Hybridization. Wiley, New York.

Anderson, R. Y. and D. W. Kirkland. 1969. Paleoecology of an early Pleistocene lake on the High Plains of Texas. Geological Society of America Memoir 113:1-215.

Arnold, E. T. 1972. Behavioral ecology of two pupfishes (Cyprinodontidae, genus *Cyprinodon*) from northern Mexico. Ph.D. dissertation, Arizona State University, Tempe.

Axelrod, D. I. 1979. Age and origin of Sonoran Desert vegetation. Occasional Papers of the California Academy of Sciences 132:1-74.

Axtell, R. W. 1978. Ancient playas and their influence on the Recent herpetofauna of the northern Chihuahuan Desert. *In*: R. H. Wauer and D. H. Riskind (Eds.), Symposium on the Biological Resources of the Chihuahuan Desert Region, United States and Mexico. United States National Park Service Transactions and Proceedings Series No. 3 (1977), pp. 493-512.

Barbour, C. D. 1973. A biogeographical history of *Chirostoma* (Pisces: Atherinidae): a species flock from the Mexican Plateau. Copeia 1973:533-556.

Bayly, I. A. E. 1972. Salinity tolerance and osmotic behavior of animals in athalassic saline and marine hypersaline waters. Annual Review of Ecology and Systematics 3:233-268.

Beadle, L. C. 1943. Osmotic regulation and the fauna of inland waters. Biological Reviews 18:172-183.

Berggren, W. A. and J. A. Van Couvering. 1974. The Late Neogene. Palaeogeography, Palaeoclimatology, and Palaeoecology 16:1-228.

Blair, W. N. and A. K. Armstrong. 1979. Hualapai limestone member of the Muddy Creek formation: the youngest deposit predating the Grand Canyon, southeastern Nevada and northwestern Arizona. United States Geological Survey Professional Papers 1111:1-14.

Brand, D. B. 1937. The natural landscape of northwestern Chihuahua. University of New Mexico Bulletin, Geological Series 5:1-74.

Cokendolpher, J. C. 1980. Hybridization experiments with the genus _Cyprinodon_ (Teleostei: Cyprinodontidae). Copeia 1980:173-176.

Cooley, M. E. and A. Wilson. 1968. Canyon cutting in the Colorado River system. _In_: R. W. Fairbridge (Ed.), The Encyclopedia of Geomorphology, Rheinhold, New York, pp. 98-102.

Darton, N. H. 1933. Guidebook of the Western United States. Part F. The Southern Pacific Lines. New Orleans to Los Angeles. United States Geological Survey Bulletin 845:1-304.

Davis, J. R. 1980. Rediscovery, distribution, and populational status of _Cyprinodon eximius_ (Cyprinodontidae) in Devil's River, Texas. Southwestern Naturalist 25:81-88.

Deacon, J. E. 1979. Endangered and threatened fishes of the West. _In_: The endangered species: a symposium. Great Basin Naturalist Memoir No. 3, pp. 41-64.

Deacon, J. E. and M. S. Deacon. 1979. Research on endangered fishes in the national parks with special emphasis on the Devils Hole pupfish. _In_: Proceedings of the First Conference on Scientific Research in the National Parks, United States National Park Service Transactions and Proceedings Series No. 5, pp. 9-19.

Deacon, J. E. and W. L. Minckley. 1974. Desert fishes. In: G. W. Brown, Jr. (Ed.), Desert Biology, Vol. II. Academic, New York, pp. 385-488.

de Cserna, Z. 1975. Mexico. In: R. W. Fairbridge (Ed.), The Encyclopedia of World Regional Geology, Part I: Western Hemisphere. Dowden, Hutchinson, and Ross, Stroudsburg, PA, pp. 384-360.

Echelle, A. A. 1975. A multivariate analysis of variation in an endangered fish, Cyprinodon elegans, with an assessment of population status. Texas Journal of Science 26:529-538.

Echelle, A. A. and A. F. Echelle. 1978. The Pecos pupfish, Cyprinodon pecosensis n. sp. (Cyprinodontidae), with comments on its evolutionary origin. Copeia 1978:569-582.

Echelle, A. A., C. Hubbs and A. F. Echelle. 1972. Developmental rates and tolerances of the Red River pupfish, Cyprinodon rubrofluviatilis. Southwestern Naturalist 17:55-60.

Echelle, A. A. and R. R. Miller. 1974. Rediscovery and redescription of the Leon Springs pupfish, Cyprinodon bovinus, from Pecos County, Texas. Southwestern Naturalist 19:179-190.

Eigenmann, C. H. 1894. Notes on some South American fishes. Annals of the New York Academy of Science, 7:625-637.

Fenneman, N. M. 1931. Physiography of Western United States. McGraw-Hill, New York.

Feth, J. H. 1961. A new map of western conterminous United States showing the maximum known or inferred extent of Pleistocene lakes. United States Geological Survey Professional Papers 424B:110-112.

Feth, J. H. 1964. Review and annotated bibliography of ancient lake deposits (Precambrian to Pleistocene) in the western states. United States Geological Survey Bulletin 1080:1-119.

Fleischhauer, H. L., Jr. 1978. Summary of the late Quaternary geology of Lake Animas, Hidalgo County, New Mexico. New Mexico Geological Society Guidebook, 29th Field Conference, Land of Cochise, pp. 283-284.

Flint, R. F. 1971. Glacial and Quaternary Geology. Wiley, New York.

Fritts, H. C. 1965. Dendrochronology. In: H. E. Wright, Jr. and E. G. Frey (Eds.), The Quaternary of the United States. Princeton University Press, Princeton, NJ, pp. 871-879.

Frye, J. C. and A. B. Leonard. 1968. Late-Pleistocene Lake Lomax in western Texas. In: R. B. Morrison and H. E. Wright, Jr. (Eds.), Means of Correlation of Quaternary Successions. Proceedings of the VIIth Congress of the International Association for Quaternary Research, 8. University of Utah Press, Salt Lake City, pp. 519-534.

Gale, H. S. 1915. Salines in the Owens, Searles, and Panamint basins, southeastern California. Bulletin United States Geological Survey 580-L:251-323.

Gastil, R. G. and W. Jensky. 1973. Evidence for strike-slip displacement beneath the Trans-Mexican volcanic belt. In: R. L. Kovack and A. Nur (Eds.), Proceedings of the Conference on Tectonic Problems of the San Andreas Fault System. School of Earth Sciences, Stanford University, Stanford, CA, pp. 171-180.

Gilbert, C. H. 1893. Report on fishes of the Death Valley expedition collected in southern California and Nevada in 1891, with descriptions of new species. North American Fauna 7:229-234.

Gilluly, J. 1929. Possible desert-basin integration in Utah. Journal of Geology 37:672-682.

Hawley, J. W. 1969. Notes on the geomorphology and late Cenozoic geology of northwestern Chihuahua. In: New Mexico Geological Society Guidebook, 20th Field Conference, The Bordor Region, pp. 131-142.

Hawley, J. W. 1975. Quaternary history of Dona Ana County region, south-central New Mexico. In: New Mexico Geological Society, Guidebook 26th Field Conference, Las Cruces Country, pp. 139-150.

Hawley, J. W., G. O. Bachman and K. Manley. 1976. Quaternary stratigraphy in the Basin and Range and Great Plains provinces, New Mexico and western Texas. In: W. C. Mahaney (Ed.), Quaternary Stratigraphy of North America. Dowden, Hutchinson, and Ross, Stroudsburg, PA, pp. 235-274.

Hayes, C. W. and M. R. Campbell. 1900. The relation of biology to physiography. Science, new series 12:131-133.

Hewett, D. F. 1954. General geology of the Mojave Desert region, California. _In_: R. H. Jahns (Ed.), Geology of Southern California. California Division of Mines Bulletin 170, Ch. II, Pt. 1, pp. 5-20.

Hirshfield, M. F., C. R. Feldmeth and D. L. Soltz. 1980. Genetic differences in physiological tolerances of Amargosa pupfish (_Cyprinodon nevadensis_) populations. Science 207:999-1001.

Hooke, R. LeB. 1972. Geomorphic evidence for late-Wisconsinan and Holocene tectonic deformation, Death Valley, California. Bulletin of the Geological Society of America 83:2073-2098.

Hubbs, C. L., G. S. Bien, and H. E. Suess. 1960. La Jolla natural radiocarbon measurements. American Journal of Science Radiocarbon Supplement 2:197-223.

Hubbs, C. L., G. S. Bien, and H. E. Suess. 1963. La Jolla natural radiocarbon measurements III. Radiocarbon 5:254-272.

Hubbs, C. L., G. S. Bien, and H. E. Suess. 1965. La Jolla natural radiocarbon measurements IV. Radiocarbon 7:66-117.

Hubbs, C. L. and R. R. Miller. 1948. Correlation between fish distribution and hydrographic history in the desert basins of western United States. Bulletin of the University of Utah 38(20):18-166.

Hubbs, C. L., R. R. Miller and L. C. Hubbs. 1974. Hydrographic history and relict fishes of the north-central Great Basin. Memoirs of the California Academy of Sciences 7:1-259.

Hubbs, Clark. 1980. The solution to the _Cyprinodon bovinus_ problem: eradication of a pupfish genome. Proceedings of the Desert Fishes Council 1978:9-18.

Humphries, J. M. and R. R. Miller. 1981. A remarkable species flock of _Cyprinodon_ from Lake Chichancanab, Yucatan, Mexico. Copeia 1981:52-64.

Hunt, C. B. 1969. Geologic history of the Colorado River. In: The Colorado River region and John Wesley Powell. United States Geological Survey Professional Papers 669, pp. 59-130.

Hunt, C. B. 1975. Death Valley: Geology, Ecology, Archaeology. University of California Press, Berkeley.

Hunt, C. B. and D. R. Mabey. 1966. Stratigraphy and structure, Death Valley, California. United States Geological Survey Professional Papers 494A:A1-A162.

Itzkowitz, M. 1969. Observations on the breeding behavior of *Cyprinodon elegans* in swift water. Texas Journal of Science 21:229-231.

King, P. B. 1948. Geology of the southern Guadalupe Mountains, Texas. United States Geological Survey Professional Papers 215:1-183.

King, R. E. and W. S. Adkins. 1946. Geology of a part of the lower Conchos Valley, Chihuahua, Mexico. Bulletin of the Geological Society of America 57: 275-294.

Kottlowski, F. E., M. E. Cooley, and R. V. Ruhe. 1965. Quaternary geology of the Southwest. In: H. E. Wright and D. G. Frey (Eds.), The Quaternary of the United States. Princeton University Press, Princeton, NJ pp. 287-298.

LaBounty, J. F. and J. E. Deacon. 1972. *Cyprinodon milleri*, a new species of pupfish (family Cyprinodontidae) from Death Valley, California. Copeia 1972:769-780.

Leopold, L. B. 1969. The rapids and the pools--Grand Canyon. In: The Colorado River Region and John Wesley Powell. United States Geological Survey Professional Papers 669, pp. 131-145.

Lotan, R. 1971. Osmotic adjustment in the euryhaline teleost *Aphanius dispar* (Cyprinodontidae). Zeitschrift Vergleichende Physiologische 75:383-387.

Lowe, C. H. 1955. The eastern limit of the Sonoran Desert in the United States with additions to the known herpetofauna of New Mexico. Ecology 36:343-345.

Lowe, C. H., D. S. Hinds and E. A. Halpern. 1967. Experimental catastrophic selection and tolerances to low oxygen concentration in native Arizona freshwater fishes. Ecology 48:1013-1017.

Lucchitta, I. 1972. Early history of the Colorado River in the Basin and Range Province. Bulletin of the Geological Society of America 83:1933-1947.

Martin, P. S. 1963. The last 10,000 years. University of Arizona Press pp. 1-87.

Meek, S. E. 1903. Distribution of the fresh-water fishes of Mexico. American Naturalist 37: 771-784.

Mehringer, P. J., Jr. 1977. Great Basin late Quaternary environments and chronology. In: D. D. Fowler (Ed.), Models and Great Basin prehistory: A symposium. Desert Research Institute Publications in Social Science, University of Nevada, Reno 12:113-117.

Meinzer, O. E. 1922. Map of the Pleistocene lakes of the Basin and Range Province and its significance. Bulletin of the Geological Society of America 33: 541-552.

Metzger, D. G., O. J. Loeltz and B. Irelna. 1973. Geohydrology of the Parker-Blythe-Cibola area, Arizona and California. United States Geological Survey Professional Papers 486-G:G1-G130.

Mifflin, M. D. and M. M. Wheat. 1979. Pluvial lakes and estimated pluvial climates of Nevada. Nevada Bureau of Mines and Geology, Bulletin 94:1-57.

Miller, R. R. 1943. The status of Cyprinodon macularius and Cyprinodon nevadensis, two desert fishes of western North America. Occasional Papers Museum of Zoology, University of Michigan 473:125.

Miller, R. R. 1945. Four new species of fossil cyprinodontid fishes from eastern California. Journal of the Washington Academy of Sciences 35: 315-321.

Miller, R. R. 1946. Correlation between fish distribution and Pleistocene hydrography in eastern California and southwestern Nevada, with a map of the Pleistocene waters. Journal of Geology 54:43-53.

Miller, R. R. 1948. The cyprinodont fishes of the Death Valley system of eastern California and southwestern Nevada. Miscellaneous Publications Museum of Zoology, University of Michigan 68:1-155.

Miller, R. R. 1950. Speciation in fishes of the genera Cyprinodon and Empetrichthys inhabiting the Death Valley region. Evolution 4:155-162.

Miller, R. R. 1961a. Man and the changing fish fauna of the American Southwest. Papers of the Michigan Academy of Science, Arts, and Letters 46:365-404.

Miller, R. R. 1961b. Speciation rates in some freshwater fishes of Western North America. In: W. F. Blair (Ed.), Vertebrate Speciation. University of Texas Press, Austin, pp. 537-560.

Miller, R. R. 1964. Redescription and illustration of Cyprinodon latifasciatus, an extinct cyprindontid fish from Coahuila, Mexico. Southwestern Naturalist 9:62-67.

Miller, R. R. 1968. Two new fishes of the genus Cyprinodon from the Cuatro Cienegas basin, Coahuila, Mexico. Occasional Papers Museum of Zoology, University of Michigan 659:1-15.

Miller, R. R. 1976. Four new pupfishes of the genus Cyprinodon from Mexico. Bulletin of the Southern California Academy of Sciences 75:68-75.

Miller, R. R. 1977. The desert pupfish fights for survival. In: N. Sitwell (Ed.), The World of Wildlife. Hamlyn Publications Group, London, pp. 102-109.

Miller, R. R. 1978. Composition and derivation of the native fish fauna of the Chihuahuan Desert region. In: R. H. Wauer and D. H. Riskind (Eds.), Symposium on the Biological Resources of the Chihuahuan Desert Region, United States and Mexico. United States National Park Service Transactions and Proceedings Series No. 3 (1977), pp. 365-381.

Miller, R. R. 1979. Freshwater fishes. Red Data Book, 4: Pisces. Rev. ed. International Union for Conservation of Nature and Natural Resources, Morges, Switzerland (1977).

Miller, R. R. and A. A. Echelle. 1975. Cyprinodon tularosa, a new cyprinodontid fish from the Tularosa basin, New Mexico. Southwestern Naturalist 19:365-377.

Miller, R. R. and V. Walters. 1972. A new genus of cyprinodontid fish from Nuevo Leon, Mexico. Natural History Museum of Los Angeles County, Contributions in Science 233:1-13.

Minckley, W. L. 1969. Environments of the bolson of Cuatro Cienegas, Coahuila, Mexico. Science Series, University of Texas at El Paso 2:1-65.

Minckley, W. L. 1973. Fishes of Arizona. Arizona Department of Fish and Game, Phoenix.

Minckley, W. L. 1978. Endemic fishes of the Cuatro Cienegas basin, northern Coahuila, Mexico. In: R. H. Wauer and D. H. Riskind (Eds.), Symposium on the Biological Resources of the Chihuahua Desert Region, United States and Mexico. United States National Park Service Transactions and Proceedings Series No. 3 (1977), pp. 383-404.

Morrison, R. B. 1965. Quaternary geology of the Great Basin. In: H. E. Wright and D. G. Frey (Eds.), The Quaternary of the United States. Princeton University Press, Princeton, NJ, pp. 265-285.

Morrison, R. B. 1968. Pluvial lakes. In: R. W. Fairbridge (Ed.), The Encyclopedia of Geomorphology. Encyclopedia Earth Science Series 3. Rheinhold, New York, pp. 1-4.

Morrison, R. B. 1969. Photointerpretive mapping from space photographs of Quaternary geomorphic features and soil associations in northern Chihuahua and adjoining New Mexico and Texas. In: New Mexico Geological Society, Guidebook 20th Field Conference, The Border Region, pp. 116-129.

Mosino, A. P. and E. Garcia. 1974. The climate of Mexico. In: R. A. Bryson and F. K. Hare (Eds.), Climates of North America. World Survey of Climatology, Vol. 11. Elsevier, Amsterdam, pp. 345-404.

Moyle, P. B. 1976. Inland Fishes of California. University of California Press, Berkeley.

Pister, E. P. 1974. Desert fishes and their habitats. Transactions of the American Fisheries Society 103:531-540.

Pister, E. P. 1979. Endangered species: costs and benefits. In: The endangered species: a symposium. Great Basin Naturalist Memoir 3, pp. 151-158.

Raisz, E. 1964. Landforms of Mexico. Cambridge, Massachusetts (map, with text, 1:3,000,000). 2nd rev. ed. Private publisher.

Raven, P. H. and D. I. Axelrod. 1975. History of the flora and fauna of Latin America. American Scientist 63:420-429.

Reeves, C. C. Jr. 1969. Pluvial Lake Palomas, northwestern Chihuahua, Mexico. In: D. A. Cordoba, S. A. Wengerd, and J. Shomaker (Eds.), Guidebook of the Border Region. New Mexico Geologic Society Field Conference Guidebook 20, pp. 143-154.

Reeves, C. C. Jr. 1973. The full-glacial climate of the southern High Plains, west Texas. Journal of Geology 81:693-704.

Reeves, C. C. Jr. 1976. Quaternary stratigraphy and geologic history of southern High Plains, Texas and New Mexico. In: W. C. Mahaney (Ed.), Quaternary stratigraphy of North America, Dowden, Hutchinson, and Ross, Stroudsburg, PA, pp. 213-234.

Robinson, D. T. 1959. The ichthyofauna of the lower Rio Grande, Texas and Mexico. Copeia 1959:253-256.

Rzedowski, J. 1973. Geographical relationships of the flora of Mexican dry regions. In: A. Graham (Ed.), Vegetation and vegetational history of northern Latin America. Elsevier, Amsterdam, pp. 61-72.

Schumm, S. A. 1965. Quaternary paleohydrology. In: H. E. Wright and D. G. Frey (Eds.), The Quaternary of the United States. Princeton University Press, Princeton, NJ, pp. 783-794.

Shreve, F. 1942. The desert vegetation of North America. Botanical Review 8:195-246.

Simpson, D. G. and G. Gunter. 1956. Notes on habitats, systematic characters and life histories of Texas salt water Cyprinodontes. Tulane Studies in Zoology 4:115-134.

Smith, G. I. 1979. Subsurface stratigraphy and geochemistry of late Quaternary evaporites, California. United States Geological Survey Professional Papers 1043:1-130.

Smith, G. R. 1978. Biogeography of intermountain fishes. In: Intermountain Biogeography: A Symposium. Great Basin Naturalist Memoirs 2:17-42.

Smith, M. L. 1980. The status of Mequpsilon aporus and Cyprinodon alvarezi at El Potosi, Nuevo Leon. Proceedings of the Desert Fishes Council 1978:24-29.

Smith, M. L. and B. Chernoff. 1981. Breeding populations of cyprinodontoid fishes in a thermal stream. Copeia, 1981 (in press).

Smith, M. L. and R. R. Miller. 1980. Systematics and variation of a new cyprinodontid fish, Cyprinodon fontinalis, from Chihuahua, Mexico. Proceedings of the Biological Society of Washington. 93:405-416.

Smith, P. B. 1960. Fossil foraminifera from the southeastern California deserts. United States Geological Survey Professional Papers 400B:B278-279.

Smith, R. S. U. 1978. Pluvial history of Panamint Valley, California. Friends of the Pleistocene, Pacific Cell, Guidebook. Geology Department, University of Houston, Texas, pp. 1-36.

Snyder, C. T., G. Hardman and F. F. Zdenek. 1964. Pleistocene lakes in the Great Basin. United States Geological Survey Miscellaneous Geological Investigations, Map I-416.

Snyder, C. T. and W. B. Langbein. 1962. The Pleistocene lake in Spring Valley, Nevada, and its climatic implications. Journal of Geophysical Research 67:2385-2394.

Soltz, D. L. and R. J. Naiman. 1978. The Natural History of Native Fishes in the Death Valley System. Natural History Museum of Los Angeles County, Science Series 30:1-76.

Stanley, G. M. 1966. Deformation of Pleistocene Lake Cahuilla shoreline, Salton Sea basin, California (Abstr.). Geological Society of America Special Papers 87:165.

Stevenson, M. M. 1981. Karyomorphology of several species of Cyprinodon. Copeia 1981 (in press).

Stevenson, M. M. and T. M. Buchanan. 1973. An analysis of hybridization between the cyprinodont fishes Cyprinodon variegatus and C. elegans. Copeia 1973:682-692.

Strain, W. S. 1966. Blancan mammalian fauna and Pleistocene formations, Hudspeth County, Texas. Bulletin of the Texas Memorial Museum 10:1-55.

Strain, W. S. 1970. Late Cenozoic bolson integration in the Chihuahua tectonic belt. In: K. Seewald and D. Sundeen (Eds.), The geologic framework of the Chihuahuan tectonic belt. West Texas Geological Society Publication 71-59, pp. 167-173.

Sykes, G. 1937. The Colorado delta. Carnegie Institution of Washington Publications 460:1-193.

Tamayo, J. L. and R. C. West. 1964. The hydrography of Middle America. In: R. C. West (Ed.), Handbook of Middle American Indians, vol. 1. University of Texas Press, Austin, pp. 84-121.

Thornbury, W. D. 1965. Regional Geomorphology of the United States. Wiley, New York.

Turner, B. J. 1964. An introduction to the fishes of the genus Nothobranchius. African Wild Life 18: 117-124.

Turner, B. J. 1973. Genetic divergence of Death Valley pupfish populations: species-specific esterases. Comparative Biochemistry and Physiology 46B:53-70.

Turner, B. J. 1974. Genetic divergence of Death Valley pupfish species: biochemical versus morphological evidence. Evolution 28:281-294.

Turner, B. J. and R. K. Liu. 1977. Extensive interspecific genetic compatibility in the New World killifish genus Cyprinodon. Copeia 1977:259-269.

van der Schalie, H. 1945. The value of mussel distribution in tracing stream confluence. Papers of the Michigan Academy of Science, Arts, and Letters 30:355-373.

van der Schalie, H. 1973. The mollusks of the Duck River drainage in central Tennessee. Sterckiana 52:45-55.

Van Devender, T. R. and W. G. Spaulding. 1979. Development of vegetation and climate in the southwestern United States. Science 204:701-710.

Weitzman, S. H. 1966. Review of South American characid fishes of subtribe Nannostomina. Proceedings of the United States National Museum 119:1-56.

Wells, P. V. 1978. Post-glacial origin of the present Chihuahuan Desert less than 11,500 years ago. In: R. H. Wauer and D. H. Riskind (Eds.), Symposium on the Biological Resources of the Chihuahuan Desert Region, United States and Mexico. United States National Park Service Transactions and Proceedings Series No. 3, 67-73 (1977).

Wells, P. V. 1979. An equable glaciopluvial in the West: pleniglacial evidence of increased precipitation on a gradient from the Great Basin to the Sonoran and Chihuahuan deserts. Quaternary Research 12:311-325.

Wendorf, F. and J. J. Hester, Eds. 1975. Late Pleistocene environments of the southern High Plains. Southern Methodist University, Fort Burgwin Research Center Publication 9:1-290.

4 Systematic and Zoogeographical Interpretation of Great Basin Trouts

ROBERT J. BEHNKE
Colorado State University, Fort Collins

ABSTRACT

The systematics of trout native to the Great Basin is not well known. Six endemic subspecies of cutthroat trout (*Salmo clarki*) are native to the Bonneville, Lahontan, and Alvord Basins, each of which has lacustrine and fluvial specialized forms. The redband trout, an interior group of rainbow trout (*Salmo gairdneri*), became established in several desert basins of Oregon. All of the native trouts of the Great Basin are rare as pure populations and some forms are probably extinct.

INTRODUCTION

There is little published information on the

trout native to the several desert basins that form the Great Basin of western North America. Trout received only brief mention in the epic works on Great Basin fishes by Hubbs et al. (1974), Hubbs and Miller (1948), and Smith (1978).

Trout native to the Great Basin are not pre-Pleistocene relicts comparable in age to the sucker genus *Chasmistes*. The widespread pliocene western North American trout (*Rhabdofario*) evidently was completely replaced by ancestors of the present species of western North American *Salmo* (subgenus *Parasalmo*) during upper Pliocene-Pleistocene times. Although endemism is expressed at no more than the subspecific level, distinctly different times of invasion and degrees and lengths of isolation can be correlated with divergences in cutthroat trout (*Salmo clarki*) and rainbow trout (*S. gairdneri*) evolutionary lines (Figure 1).

The cutthroat trout was the first to invade the present Great Basin waters, but only after two major divergences occurred in the phylogeny of this species. Cutthroat trout persisted only in those basins isolated from recent contact with coastal drainages (Bonneville, Lahontan, and Alvord basins). In a series of six desiccating basins in southern Oregon and in the upper Klamath Lake Basin I assume that the original cutthroat trout inhabitant was completely replaced by the later, invading primitive form of rainbow trout that I have called the redband trout. This phenomenon of replacement of interior forms of cutthroat trout by the rainbow trout, except in segments of the basin isolated by barrier falls, is characteristic of the Columbia River Basin east of the Cascade Range.

Various degrees of differentiation can be noted in the cutthroat trout and rainbow trout of the Great Basin in comparison to "sister groups" of these species in the Columbia River Basin. A more ancient influence of selection in large, pluvial lakes is evident in the increased numbers of gillrakers, in the spotting pattern, and in the proclivity for predation on other fishes. After the final desiccation of the major pluvial lakes about 8000 years ago (Benson 1978; Broecker and Kaufman 1965; Smith 1978) selective pressures favored adaptations to survive in harsh and

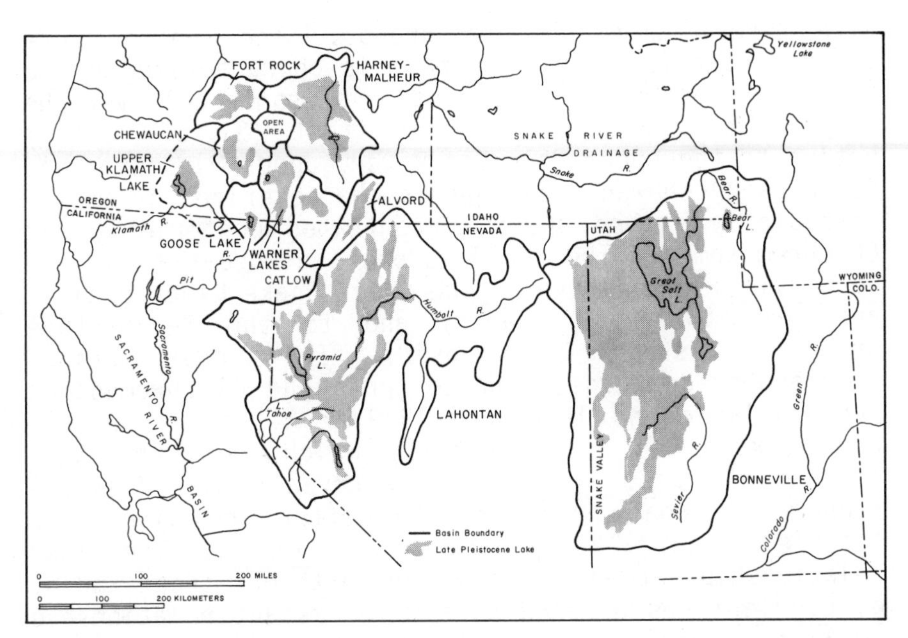

FIGURE 1. Segments of the Great Basin known to have native trouts with an indication of the approximate maximum extent of the late Pleistocene lakes in each basin. A derivative of the "Yellowstone" cutthroat trout was first to invade the Great Basin. This ancestral cutthroat probably inhabited all of the present separate basins but was replaced in all basins with more recent contact to the Columbia River basin by the redband trout. Native cutthroat trout persisted only in the Lahontan, Alvord, and Bonneville basins.

unstable environments. This is particularly apparent in the Humboldt River cutthroat trout of the Lahontan Basin, the Bear River cutthroat trout of the Bonneville Basin, and in the redband trout of the Oregon desert basins.

CUTTHROAT TROUT

I tentatively recognize 15 subspecies of cutthroat trout. The diversity within *S. clarki* can be associated with three major phyletic groups. The coastal cutthroat trout (S. clarki clarki) is characterized by 68 to 70 chromosomes and a profusion of spots all over the body. The interior cutthroat trout (S. c. lewisi) native to the upper Columbia River, upper Missouri, and South Saskatchewan drainages is characterized by small, irregularly shaped spots concentrated posteriorly and 66 chromosomes. The "Yellowstone" cutthroat trout (S. c. bouvieri) is native to the upper Snake River drainage (Columbia River Basin) and to the Yellowstone drainage (Missouri River Basin). It is characterized by pronounced, large, rounded spots and 64 chromosomes (Loudenslager and Thorgaard 1979). All other subspecies (except S. c. alpestris, which is probably a synonym of S. c. lewisi) were derived from a "Yellowstone" cutthroat trout ancestor after the divergence that led to S. c. bouvieri and S. c. lewisi. The close relationship between Great Basin cutthroat trout and S. c. bouvieri, as opposed to a more distant relationship to S. c. lewisi, is also supported by electrophoretic data (Loudenslager and Gall 1980).

The most recent information on the late Pleistocene chronology of the Lahontan Basin is given by Benson (1978). Pluvial Lake Lahontan at maximum elevation was approximately the size of Lake Erie. A large lacustrine environment, although flucutating greatly in size, was present in the basin from about 75,000 years BP to about 10,000 BP. The presence of the specialized lacustrine sucker genus Chasmistes in Pyramid Lake (also in Klamath Lake and Utah Lake) indicates that a persistant, but geographically disrupted, lacustrine "track" of migration into the Great Basin has existed at times, at least since the Pliocene. The degree of divergence expressed between the endemic Great Basin species of Chasmistes suggests that they have been isolated in the Lahontan, Bonneville, and Klamath basins since the Pliocene or early Pleistocene. If this assumption is correct, then some lacustrine environments, suitable for the survival of Chasmistes, have had continuity since the Pliocene. This, in turn, raises the question of the fate of the salmonid species which lived with Chasmistes in Pliocene Lake Idaho (Smith 1975). Why are there no living descendents of

Pliocene *Rhabdofario*, interior species of *Oncorhynchus*, and other Pliocene fossil salmonid genera in the Great Basin or anywhere else? I suggest that their "tracks" of continuity were "derailed and dead-ended" by the invasion of species of the subgenus *Parasalmo*, particularly the cutthroat trout, which completely replaced the earlier species in the Pleistocene.

Lahontan Basin

The Lahontan cutthroat trout (*S. c. henshawi*) exhibits the greatest morphological divergence of any subspecies of *S. clarki* (Figure 2). This degree of

FIGURE 2. *Salmo clarki henshawi*, showing ventral spotting pattern diagnosis for this subspecies. This specimen is derived from Summit Lake, Nevada, but was stocked and reared in Pyramid Lake.

divergence suggests a long period of isolation and selective pressures as the only large predatory fish in pluvial Lake Lahontan. Based on electrophoretic data, Loudenslager and Gall (1980) found *S. c. henshawi* to be the most divergent member of the "Yellowstone" cutthroat trout group of subspecies. *S. c. henshawi* evolved into an effective predator on the abundant forage fishes of Lake Lahontan. Almost certainly the most abundant and available pelagic forage fish would have been *Gila pectinifer* in Lake Lahontan, as it is today in Pyramid Lake, Nevada, the remnant sump of the pluvial lake. *G. pectinifer* attains a relatively large size to about 400 mm. To make full use of the available prey *S. c. henshawi* acquired the genetic basis to attain the greatest size of any cutthroat trout and perhaps greater than any form of trout in the subgenus *Parasalmo* (Behnke and Zarn 1976; Hickman and Behnke 1979). Although well adapted to large lacustrine environments, *S. c. henshawi* was highly vulnerable to replacement by introduced brook, brown, and rainbow trout in Lahontan basin streams and persisted as pure populations only in a few small isolated headwaters. Native lacustrine populations of *S. c. henshawi* persist in Summit Lake, Nevada, and in Independence Lake, California. The Walker Lake, Nevada, population is maintained by a hatchery brood stock. Fish from Summit Lake are propagated for stocking Pyramid Lake to maintain a sport fishery. *S. c. henshawi* was originally listed as endangered under the Endangered Species Act but its status was changed to threatened, mainly to legalize sport fishing in Pyramid Lake.

The Paiute trout, *S. c. seleniris*, named from Silver King Creek, a tributary of the Carson River drainage in California, represents an isolated population of Lahontan cutthroat trout with few or no spots on the body. The genetic relationship between *seleniris* and Carson River drainage *henshawi* is closer than the relationship between Carson and Truckee drainage *henshawi*. Being formally recognized as a unique taxon has benefited the Paiute trout by its inclusion as an endangered species (later changed to threatened) under the Endangered Species Act. If the Paiute trout had lacked formal taxonomic recognition and was considered only as an unspotted form of *S. c. henshawi*, it is not likely that the active restoration efforts made in recent years to save the Silver King Creek population from extinction (due to rainbow trout hybridization) would have occured.

A fluvial adapted cutthroat trout, differentiated from *S. c. henshawi*, is native to the Humboldt River drainage of the Lahontan Basin in Nevada (see Figure 2). The Humboldt cutthroat is an unnamed subspecies but its diagnostic characters have been given by Behnke and Zarn (1976).

The major differences between *henshawi* and the Humboldt cutthroat trout evolved before the final desiccation of Lake Lahontan. This assumption is based on the fact that all *henshawi* populations isolated in the Truckee, Carson, and Walker river drainages since the last desiccation of Lake Lahontan are uniform in their characters and all Humboldt cutthroat populations isolated in distinct segments of the Humboldt drainage share the same consistent differences from *henshawi*.

Ecologically, the Humboldt cutthroat trout is different from *henshawi*. It appears to have evolved an adaptive strategy to survive in harsh and unstable small-stream environments. Several populations still persist in small, intermittent streams in arid watersheds. In the Ruby Mountains, which contain the best trout streams in the Humboldt drainage, the native trout has been almost entirely replaced by the nonnative. Vigg and Koch (1980) found the lethal maximum temperature for *S. c. henshawi* to be 23°C. I have observed the Humboldt cutthroat trout in Frazier Creek, Elko County, Nevada, in water of 25.5°C. Although these high temperatures are moderated at night, it is likely that the ability to function at higher temperatures is one of the evolutionary adaptations acquired by the Humboldt cutthroat trout.

This evolutionary pattern of divergence of an ancestral cutthroat trout into lacustrine and fluvial specialized forms is also paralleled in the Bonneville and Alvord basins.

Other Species of Trout Described from the Lahontan Basin. J. O. Snyder named *Salmo regalis* the royal silver trout of Lake Tahoe, *S. smaragdus*, the emerald trout of Pyramid Lake, and *S. aquilarum*, the Eagle Lake trout (Snyder 1914, 1917). Examination of the type specimens and evaluation of available evidence has led me to believe that *S. regalis* and *S. smaragdus* are not native trouts but were based on introductions of rainbow trout (Behnke 1972a). Eagle Lake, California, is an isolated segment of the Lahontan Basin which contains a typical Lahontan fish fauna except for its trout. The Eagle Lake trout is obviously more closely related to rainbow trout than it is to the Lahontan cutthroat trout. Hubbs and Miller (1948) suggested that the Eagle Lake trout resulted from hybridization between introduced rainbow trout and a native population of *S. c. henshawi*. I find no evidence, however, of cutthroat-rainbow hybridization in specimens of Eagle Lake trout and its chromosome complement of 2N = 58 also argues against a hybrid influence. I believe the original Lahontan cutthroat trout became extinct in Eagle Lake in a warmer, drier period during the last several thousand years, as described by Benson (1978) for the Lahontan Basin. Later, a headwater transfer from the Pit River drainage brought in a form of the rainbow trout that became established.

Bonneville Basin

The cutthroat trout native to the Bonneville Basin is probably a relatively recent invader that gained entrance into the basin when the Bear River changed its connections from the Snake River to the Bonneville Basin about 30,000 years BP (Broecker and Kaufman 1965). The timing of this entry route is supported by morphological and biochemical similarities between the Bonneville cutthroat trout *S. c. utah* and *S. c. bouvieri* of the Snake River drainage.

After the Bear River became tributary to pluvial Lake Bonneville the lake reached a maximum size comparable to Lake Michigan about 16,000 to 18,000 BP when it overflowed the basin and caused catastrophic flooding in the Snake River (Broecker and Kaufman 1965). Malde (1968), however, believed that the overflow from Lake Bonneville occurred at least 30,000 years ago.

Although the Bonneville cutthroat trout is not well differentiated from its progenitor (*S. c. bouvieri*), three evolutionary groups of Bonneville trout can be recognized by morphological and ecological differences. The cutthroat trout native to the Bear

River division of the basin is a fluvial adapted form that persists in harsh, highly fluctuating stream environments which parallel the evolutionary direction of the Humboldt cutthroat trout in the Lahontan Basin. The morphological differences between the Bear River drainage cutthroat trout and *S. c. utah* of the rest of the Bonneville Basin (higher numbers of scales and pyloric caeca) are slight and not deserving of subspecies separation in a taxonomic scheme, but the ecological differences seem to be pronounced. I made a survey of *S. c. utah* in parts of the Bear River drainage in Wyoming for the U.S. Bureau of Lake Management in 1976. In the Thomas Fork and Smith Fork drainages near Cokeville, Wyoming, I was amazed to find that the native cutthroat trout was dominant in turbid streams that carry heavy sediment loads. The environment appeared to be marginal for trout. From a general observation of the Thomas Fork and Smith Fork rivers I would have assumed that only the Europen brown trout (*S. trutta*) could establish reproducing populations in those locations. I found brown trout in only two cold, clear tributary streams (Pine Creek and Hobble Creek). Evidently the Bear River form of the Bonneville cutthroat trout has evolved adaptations superior to *S. trutta* to cope with these unstable and fluctuating stream environments.

Outside the Bear River drainage I have found no single example in which *S. c. utah* has persisted in coexistence with any nonnative trout. As with *S. c. henshawi* of the Lahontan Basin, the form of *S. c. utah* with the evolutionary programming derived from pluvial Lake Bonneville appears to be ill-adapted to life in small streams and has been replaced by nonnative trouts except on a few isolated headwaters.

A divergent group of *S. c. utah* is native to the Snake Valley region of the basin. The Snake Valley cutthroat trout differs from other *S. c. utah* in numbers of gillrakers and basibranchial teeth and body contours. It also differs by possessing a unique allele for a LDH enzyme (Klar and Stalnaker 1979).

At maximum elevation of Lake Bonneville Snake Valley was an arm of the lake, but with slight declines in lake levels it became isolated from the rest of the basin. The cyclical fluctuations of Lake Bonneville provided opportunities for isolation and incipient divergence of the cutthroat trout in Snake Valley.

The Snake Valley cutthroat trout first came to my and R. R. Miller's attention in an introduced population in Pine Creek, Spring Valley, Nevada. We both considered it subspecifically distinct from *S. c. utah* and it is mentioned as an undescribed subspecies in Hubbs et al. (1974). Since then new populations of Snake Valley cutthroat trout and *S. c. utah* were discovered which reduced the degree of distinction between them (Hickman and Duff 1978).

Alvord Basin

Except for a brief description by Behnke and Zarn (1976), there is virtually no published information on the native trout of the Alvord Basin. The Alvord Basin lies north of the Lahontan Basin and is contiguous with the Owyhee drainage of the Snake River on its northern and eastern borders. It is separated from the Guano-Catlow and Malheur basins to the west by the Steens Mountains. The Alvord Basin has had long and relatively complete isolation from surrounding basins. Besides the cutthroat trout, only one species of fish, the endemic *Gila alvordensis*, has been described from the basin (Hubbs and Miller 1948, 1972). An isolated derivative of *G. alvordensis*, the Borax Lake chub, is currently being called a new species and has been granted emergency listing under the Endangered Species Act (Endangered Species Tech. Bull. June 1980).

Cutthroat trout were first collected from several streams in the Alvord Basin by Carl Hubbs in 1934. Hubbs and Miller (1948) mentioned that these trout showed local peculiarities correlated with their isolation. When I examined these specimens I realized that they represented two distinct subspecies, one of which was collected in Thousand Creek, Nevada, and in Trout Creek, Oregon (Figure 3a). The other subspecies occurs in Willow and Whitehorse creeks, Oregon. Thousand Creek and Trout Creek were tributary to pluvial Lake Alvord. Willow and Whitehorse creeks drain out onto a high desert plateau east of the Alvord sump and above

FIGURE 3. (a) Unnamed subspecies of cutthroat trout native to Alvord sump basin. These specimens (University of Michigan Museum of Zoology 130493) were collected in 1934 by Carl Hubbs and family from Trout Creek, Harney County, Oregon. (b) Same form of cutthroat trout collected by Hubbs from Virgin Creek, Humboldt County, Nevada (UMMZ 130532). This is the lacustrine specialized subspecies of Alvord basin cutthroat trout. Evidently its lacustrine evolutionary heritage from pluvial Lake Alvord made it ill-adapted to compete with nonnative trouts in the remnant streams of the basin. It has not been found in the Virgin or Trout Creek drainages since 1934 and is presumed extinct.

the pluvial lake level. I have never found *Gila alvordensis* in Willow and Whitehorse creeks; this suggests that these streams were not direct tributaries to pluvial Lake Alvord.

The evolutionary parallels in lacustrine and fluvial specialized forms and their relative vulnerability to replacement by nonnative trouts are similar for Alvord Basin cutthroat trout, as discussed for the Lahontan and Bonneville basins. The more lacustrine specialized subspecies (higher numbers of gillrakers), which Hubbs collected in Thousand and Trout creeks in 1934, were gone when I made collections in the area in 1972 (Figure 3b). Only rainbow trout with no indication of cutthroat trout hybrid influence were found up to the very headwaters of Trout Creek and in tributaries to Virgin Creek (Thousand Creek drainage). A sample in the Oregon State University collection made in 1970 at the same site in Thousand Creek where Hubbs collected cutthroat trout in 1934 also represents typical rainbow trout.

In Willow and Whitehorse Creeks I found the same cutthroat trout (no detectable hybrid influence) in 1972 that occurred there in 1934 (Figure 4a). It may be that no nonnative trout have ever been stocked in Willow and Whitehorse creeks, but I find this difficult to believe. More probably, in the harsh environmental regime of desert basin streams, natural selection has endowed the more fluvial specialized subspecies of Alvord trout with ecological adaptations that make it superior to nonnative trout in such environments, paralleling the evolution of the Humboldt cutthroat trout of the Lahontan Basin and the Bear River cutthroat trout of the Bonneville Basin.

The origin of the Alvord cutthroat trout is obscure. Based on the degree of differentiation from the nearest related species, the ancestor of the Alvord chub invaded the basin long before the cutthroat trout. Hubbs and Miller's (1948) suggestion of a Pliocene origin for the chub seems plausible. The trout's progenitor may have entered the basin from a headwater transfer from the Owyhee drainage (although cutthroat trout are not known from the Owyhee in historical times). More probable is an origin from the Lahontan

FIGURE 4. (a) The fluvial specialized form of Alvord cutthroat trout from Whitehorse Creek, Malheur County, Oregon. This unnamed subspecies appears to be well-adapted to the harsh, unstable environment of desert basin streams. It is still the only fish found in Willow and Whitehorse Creeks where Carl Hubbs first collected this trout in 1934. (b) Unnamed subspecies of cutthroat trout native to Humboldt River drainage of the Lahontan basin. This cutthroat trout represents the fluvial specialized form of Lahontan cutthroat trout. Specimens from Seguna Creek, Elko County, Nevada.

Basin via Summit Lake into the Thousand-Virgin Creek drainage. The origin of trout in Willow and Whitehorse creeks probably occurred in a headwater transfer from Trout Creek into Willow Creek in the Trout Creek Mountains, where the headwaters of the two streams are adjacent. I assume that this transfer occurred during an early stage of the trout's speciation in pluvial Lake Alvord because the Willow and Whitehorse Creek trout average 21 gillrakers (slight lacustrine specialization) and the trout native to the Alvord sump drainage average almost 24 (advanced lacustrine specialization).

RAINBOW OR REDBAND TROUT

I recognize two major evolutionary groups of rainbow trout: the coastal rainbow and the interior rainbow or redband trout[1]. The redband trout is distributed in parts of the Sacramento River Basin, the Upper Klamath Lake Basin, the Columbia and Fraser River basins east of the Cascade Range, and the Oregon desert basins. The Sacramento Basin redband trout (including the golden trout *S. aguabonita*) is differentiated from the Columbia and Fraser redband trout by fewer meristic elements and brighter coloration. Evidently derivatives of the Columbia and Sacramento redband trout mixed in the Oregon basins.

Oregon Desert Basins

Native trout occurred in the following desiccating basins of southern Oregon [names established by Hubbs and Miller (1948)]: Malheur (or Harney), Catlow Valley, Warner Lakes, Chewaucan, and Fort Rock. Trout are also native to the Goose Lake Basin, which is a semidisrupted part of the Sacramento Basin, and to the Upper Klamath Lake Basin, which once had connections to Great

---

[1]The interior form of rainbow trout is not presently recognized taxonomically but considered a part of *Salmo gairdneri*.

Basin (Hubbs and Miller 1948; Smith 1978). The correct classification of the native trout of these basins has long been a mystery. Cope (1883) believed that they were cutthroat trout (which he called S. purpuratus). Snyder (1908) published taxonomic accounts of several collections made in 1904 and also concluded that they were cutthroat trout (S. clarki). Snyder, however, was not certain whether the trout he collected were native or introduced. Hubbs and Miller (1948), following Snyder, assumed that S. clarki is the native trout of these basins but they believed that hybridization with introduced rainbow trout caused the demise of the indigenous trout.

Many redband trout opulations bear a strong phenotypic resemblance to cutthroat trout (a faint cutthroat mark, bright coloration, and relatively large spots). It is understandable therefore why Cope and Snyder considered the native trout of the Oregon basins cutthroat trout. It was the redband trout of the middle Columbia River basin that caused Gilbert and Evermann (1894) to conclude that a complete transition occurred in the Columbia Basin between the interior forms of cutthroat trout and coastal rainbow trout.

I examined old museum specimens of the Oregon desert basins trout in the Stanford University collection (now part of California Academy of Sciences collection) and in the collection of the U. S. National Museum to document the taxonomic characters of the native trout before the introducton of nonnative trout became widespread. I then made collections from the same or nearby sites of the earlier collections to evaluate what is extant of the native genotypes. Although I have not found a population I consider absolutely pure, I found surprising similarities between the present trout populations in the Oregon basins and the original native trout. Most of these waters have been exposed to long and continued stocking of hatchery trout. In the harsh and fluctuating environments of desert basins streams the predominance of the native genotypes in the Oregon basins is similar to the persistence of Bear River and Humboldt River cutthroat trout. The practical fishery management value of the genetic resource of the Oregon desert basins' redband trout has been recognized by the Oregon Fish and Game Department and a propagation program initiated for the Catlow Valley native trout.

Invasion of the Oregon basins by redband trout occurred via the Deshutes River into the Fort Rock Basin and via the Snake drainage into the Malheur Basin. The Catlow Valley trout was derived from the Malheur Basin. Redband trout from the Sacramento River system invaded the Goose Lake Basin from the Pit River and transfers from the Goose Lake Basin established native trout in the Warner Lakes Basin. Native trout of Goose Lake and Warner Lakes were probably established in these basins before the Columbia River invasions occurred in the other basins. The evidence lies in the higher number of gillrakers in the native trout of the Goose Lake and Warner Lakes basins, compared with the other Oregon basins (average of 23 gill-rakers versus about 21). The Chewaucan basin trout appear to be derived from the Goose Lake Basin and the Fort Rock Basin (transfers from the Upper Kalmath Basin may also be involved). The duration of isolation of trout in the Oregon basins was not sufficient to evolve trenchant morphological divergences from "sister" groups in the Columbia and Sacramento basins or from one another. In general the lacustrine evolutionary influence of the pluvial lakes can be noted in the increased numbers of gillrakers in the native trout of the Oregon desert basins in comparison with their sister groups.

In the Upper Klamath Lake Basin I detected two groups of native trout, a lacustrine specialized form from Upper Klamath Lake and a resident stream form in small tributaries of the drainage. The two forms differ in meristic elements. Also, anadromous steelhead trout once used the Upper Klamath watershed for spawning and rearing before dams blocked access. Thus three groups of the rainbow trout evolutionary line once occurred sympatrically with reproductive isolation in the Upper Klamath Lake drainage. The innate reproductive homing behavior of salmonid fishes can maintain reproduction isolation between populations with only slight genetic divergence if selective pressure that favors ecological divergence and specialization are present (Behnke 1972a). Girard (1856) named *Salmo newberryi* for the lacustrine form of Klamath Lake trout (based on my examination of the holotype, USNM 578). This is the earliest published name for any form of the redband trout group and can serve as a starting point for any future revision of *S. gairdneri*.

The occurrence of the bull trout *Salvelinus confluentus* in the Upper Klamath Lake Basin and in the McCloud River of the Sacramento Basin (Cavender 1978) indicates a pattern of dispersal from the Columbia River basin through the Oregon desert basins during the Pleistocene. The absence of bull trout in the Oregon basins in historical times is probably due to the lack of suitable cold-water habitat in these basins after the desiccation of the pluvial lakes.

GENERAL DISCUSSION AND CONCLUSIONS

In addition to the rainbow and cutthroat trout species, the subgenus *Parasalmo* includes *Salmo apache*, endemic to the headwater of the Gila River Basin and the headwaters of the Little Colorado River Basin in Arizona, *S. gilae*, endemic to the Gila River Basin of Arizona and New Mexico (Miller 1950, 1972), and *S. chrysogaster*, endemic to three river systems in Mexico which drain to the Gulf of California (Needham and Gard 1964). Table 1 compares important meristic characters of the groups of the subgenus *Parasalmo*.

The precise relationships of the latter three species are uncertain. *S. apache* and *S. gilae* appear to be closely related to each other. Only the spotting pattern can differentiate consistently between them. I believe that the genetic differentiation between *S. apache* and *S. gilae* is comparable to the differentiation between subspecies of the "Yellowstone" group of cutthroat trout and that *S. apache* will eventually be considered more correctly as a subspecies of *S. gilae*. The question, however, of the connection of the most recent phylogenetic branching point in the ancestry of *S. gilae* and *S. apache*, in relation to *S. chrysogaster*, *S. gairdneri*, and *S. clarki*, is not known.

The origins of *S. gilae*, *S. apache*, and *S. chrysogaster* appear to be associated with the Gulf of

TABLE 1. Some Meristic Characters of Great Basin Trout with Comparisons to other Groups of the Subgenus *Parasalmo*[a]

	Gillrakers	Scales, Lateral Series	Pyloric Caeca	Comments
Lahontan Basin				
A. Lacustrine specialized, *S. c. henshawi*	21-28(23-25)	150-180(160-170)	40-80(50-65)	Hereditary basis for large size; spots evenly distributed on body and ventral region.
B. Fluvial speicalized, unnamed subspecies Humboldt River drainage	18-24(21)	125-155(135-145)	40-70(50-60)	Spotting pattern with fewer spots seldom on ventral region.
Bonneville Basin *S. c. utah*				
A. Bear River drainage fluvial specialized (except for Bear Lake population)	16-20(17-19)	150-190(160-175)	35-60(ca. 45)	All forms of Bonneville cutthroat trout have large, roundish spots, more-or-less evenly distributed over the body but seldom on the ventral region. Spots more profuse and basibranchial teeth are better developed in the Snake Valley form.
B. Snake Valley form	18-24(21)	135-170(145-150)	25-50(ca. 35)	
C. Remainder of basin	17-21(18-20)	140-180(150-160)	25-50(ca. 35)	

Alvord Basin				
A. Lacustrine specialized unnamed subspecies, Alvord sump drainage (presumed extinct)	20-26(23-24)	125-150(135)	35-50(42)	Spots are sparse. Basibranchial teeth poorly developed. Branchiostegal rays 8-9.
B. Fluvial specialized unnamed subspecies, Willow and Whitehorse Creeks	18-23(21)	130-165(145-150)	40-60(45-48)	Spots similar to Humboldt subspecies of Lahontan Basin. Branchiostegal rays 9-11.
Yellowstone cutthroat trout *S. c. bouvier*[b]	17-21(18-20)	150-200(165-180)	25-50(35-40)	Shares same karyotype (2N = 64) with all Great Basin cutthroat trout.
Oregon Lake Basin Redband Trout				
Harney-Malheur basin	20-24(21-22)	130-160(145-150)	30-50(37-43)	High vertebrate counts (63-66). No basibranchial teeth.
Catlow Valley	19-23(21)	130-150(140)	30-45(37)	Derived from Harney-Malheur basin. No basibranchial teeth. Vertebrae 62-65.
Fort Rock Basin	19-22(20-21)	135-160(142-147)	35-55(40-45)	Vestigial basibranchial teeth. Vertebrae 62-65.
Chewaucan Basin	20-24(22)	135-158(145)	30-55(40-45)	Vestigial basibranchial teeth. Vertebrae 62-65.

(Continued)

Table 1. (Continued)

	Gillrakers	Scales, Lateral Series	Pyloric Caeca	Comments
Warner Lakes Basin	21-25(23)	135-155(145)	30-50(40-45)	Basibranchial teeth absent. Vertebrae 61-64.
Goose Lake Basin	21-25(23)	135-155(145)	30-50(40-45)	Two of 71 specimens examined have basibranchial teeth. Vertebrae 61-64.
Upper Klamath Lake Basin				
A. Lacustrine specialized (data from only four specimens)	20-23	142-148	45-58	Vertebrae 63-65. Branchiostegal rays 12-14(13).
B. Fluvial specialized	17-21(19)	130-150(135-140)	35-55(46-48)	Vertebrae 62-65. Branchiostegal rays 9-13(11). One of 37 specimens has a basibranchial tooth.
Columbia R. Basin Redband Trout	16-22(18-19)	130-170(135-155)	30-55(40-45)	Vertebrae counts high; 63-67 (64-65). Coloration subdued.

Sacramento R. Basin Redband Trout	15-22(17-20)	135-200+(140-175)	25-50(30-40)	Includes golden trout *S. aguabonita*. Probably more than one ancestral form responsible for the great variability found within the basins. Vertebrae 58-65 (60-63). Coloration brighter than in Columbia River redband trout. Especially brilliant colors in Kern River Basin (golden trout). Basibranchial teeth vestigial in populations native to Kern River, Little Kern River and headwater McCloud River.
Coastal Rainbow Trout (west of Cascade Range)	16-22(18-19)	120-140(125-130)	40-75(50-60)	Spots smaller, more profuse; parr marks roundish with poorly developed supplementary rows, fading in adult, in comparison with redband trout. Yellow-orange tints on the body, "cutthroat" mark and white or yellow tips on dorsal, pelvic and anal fins, characteristic of redband trout are absent or much reduced in coastal rainbow trout.

(Continued)

Table 1. (Continued)

	Gillrakers	Scales, Lateral Series	Pyloric Caeca	Comments
Arizona Trout				
S. gilae	18-21(19)	135-165(150-155)	24-45(32-35) except Spruce Creek population with mean of 48.	Gila and Apache trout are closely related to each other. Only the spotting pattern is consistently different. They both differ from other members of the subgenus in their deeper bodies and longer fins. Vertebrae 58-62. Basibranchial teeth vestigial.
S. apache	18-21(19)	135-170(150-155)	20-45(27-33)	
Mexican Trout				
S. chrysogaster	15-20(17-18)	120-160(126-142)	10-30(21-23)	Vertebrae 56-59. Branchiostegal rays 8-10. Basibranchial teeth absent.

[a]Typical modal values in parentheses.
[b]Derived from common ancestor with all Great Basin cutthroat trout.

California, lower Colorado River area, from one or more ancestral species displaced to the south during an early glacial period. An interpretation can be made from this distribution that the ancestors of the *S. gilae*-*S. apache* and *S. chrysogaster* group of *Parasalmo* were the first members of the subgenus in North America and were replaced in the north by the later, invading cutthroat and rainbow trout. An alternative hypothesis places the origin of *Parasalmo* in the Gulf of California region.

The sequence of speciation and distribution of cutthroat and rainbow trout in the western United States can now be more clearly interpreted. As mentioned, three distinct phylogenies can be detected in cutthroat trout: the coastal cutthroat *S. c. clarki*, the upper Columbia River or "westslope" cutthroat *S. c. lewisi*, and the "Yellowstone" cutthroat *S. c. bouvieri* (and its derived subspecies).

The distribution of *S. c. lewisi* above major barrier falls in the Kootenai, Pend Oreille, and Spokane river drainages and the distribution of *S. c. bouvieri* above Shoshone Falls on the Snake River indicates that the cutthroat trout was in North America, moved inland and diverged into three distinct groups before any form of rainbow trout appeared on the scene. The invasion of the Columbia River Basin by the redband trout resulted in the virtual elimination of cutthroat trout from the Cascade Range eastward to the barrier falls that blocked further penetration. Cutthroat trout persisted only in those segments of the Great Basin (Lahontan, Bonneville, and Alvord basins) that were isolated from invasion by the redband trout. I assume that cutthroat trout were established in the Oregon desert basins during an earlier pluvial period but were replaced by the redband trout when the lakes in the Fort Rock and Malheur basins cut outlets to the Columbia Basin in the last pluvial period. The timing of the original invasion of the western United States by cutthroat trout (or their immediate ancestor) is not known. I know of no valid evidence that *S. clarki* or its most recent progenitor existed before the Pleistocene. Much is still to be learned however, of late Cenozoic salmonid fish fossils. It would not be surprising if my current ideas in regard to the timing of speciation and distribution events are revised in light of future discoveries.

In any event I do not believe that native Great Basin trouts are ancient (Pliocene-early Pleistocene) relicts of a more primitive and broad track of distribution. Rather they had their origins in separate dispersals of the "Yellowstone" phylogeny of cutthroat trout into the Lahontan, Bonneville, and Alvord basins and the Columbia and Sacramento river forms of the redband trout in the Oregon basins during the last glacial epoch.

Based on the degree of morphological divergence from *S. c. bouvieri* and on the degree of divergence into lacustrine and fluvial specialized forms in the basin, the cutthroat trout of the Lahontan Basin has been isolated longer than any other Great Basin trout. The penetration of the ancestor of the Lahontan cutthroat trout into the basin had probably occurred by the beginning of the last glacial epoch. The fluctuations in pluvial lake level promoted isolation and divergence of the original stock into a lacustrine specialized form (*S. c. henshawi*) and a fluvial specialized form (undescribed subspecies native to Humboldt River drainage). This phenomenon of intrabasin divergence into lacustrine and fluvial specializations also occurred among the cutthroat trout of the Alvord and Bonneville basins. Adaptations acquired by natural selection in the pluvial lakes made the lacustrine specialized forms of cutthroat trout poorly adapted to survive in competition with nonnative trout. I know of no example in which the lacustrine specialized cutthroat trout of the Lahontan, Alvord, or Bonneville basin has successfully coexisted in a stream environment with nonnative trout. On the other hand, the fluvial specialized cutthroat trout of the Humboldt River drainage of the Lahontan Basin, the Bear River drainage of the Bonneville Basin, and in Willow and Whitehouse creeks of the Alvord Basin continue to be the dominant fish in arid region streams characterized by a harsh envionment with highly fluctuating temperatures, flows, and sediment loads.

In the Great Basin redband trout I found evidence of lacustrine and fluvial specialized forms in the Upper Klamath Lake Basin. Based on four museum specimens (the holotype of *S. newberryi* plus three specimens collected by Captain Bendire in 1883), I found 20 to 23 gillrakers and 12 to 14 branchiostegal rays. In 33 specimens from three samples of the river form of the Klamath redband trout collected in 1968 and 1970 I found 17 to 21 gillrakers and 9 to 13 (typically 11) branchiostegal rays. Evidence of divergence and reproductive isolation between lacustrine and fluvial specialized forms of redband trout in the Klamath Basin is not so conclusive as it is among the Great Basin cutthroat trout, but I believe it is nonetheless real. Upper Klamath Lake and its large network of tributary streams has maintained the necessary environments to promote and reinforce incipient speciation into lacustrine and fluvial forms up to the present time. With environmental degradation and massive stocking of nonnative trout into the basin, the present status of the lacustrine specialized trout of Upper Klamath Lake is not known.

I have long promoted my opinions among state and federal agencies concerned with fisheries management that the preservation of the remnant genetic diversity still extant in the native trout of western North America can be of great practical value (Behnke 1972a, b). For too long, I believe, "fisheries management" programs in most western states have been preoccupied with the ever-increasing production of massive numbers and biomass of domesticated hatchery trout, particularly the hatchery rainbow trout derived mainly from the coastal rainbow trout. State and federally funded fisheries programs of the past, in relation to the present status of the native western trouts and in particular the native trouts of the Great Basin, can be characterized as ranging from benign neglect to outright extermination.

In recent years, under the stimulation of the Endangered Species Act and an increasingly popular demand for "wild trout" fisheries as opposed to the stocking of hatchery trout, most of the natural resource agencies engaged in fisheries management have become more aware, concerned, and protective of the native trout resource. The point I have tried to make to management personnel is that the "evolutionary programming" acquired by many forms of western trout endow them with attributes such as functioning at high temperatures, survival in harsh and unstable environments, and predatory feeding habits. I have caught the desert basin redband trout by angling in water of 28.3°C. They fought well and had considerable metabolic reserve or scope for activity at this temperature. These hereditary attributes can be of great value for innovative and creative fisheries management programs. Often significant divergence can occur in life history, behavioral, and ecological traits without significant divergence in morphological characters; for example the cutthroat trout *S. c. utah*, native to the Bear River drainage of the Bonneville Basin, compared with *S. c. utah* in the rest of the basin. Thus specific attributes important to fisheries management may be local adaptations with little or no correlation with taxonomic units (Hickman and Behnke 1979).

In Willow Creek Reservoir in the Lahontan Basin, Elko County, Nevada, a turbid, fluctuating irrigation reservoir with abundant nongame fishes, trout stocked from a hatchery exhibit poor survival and negative growth if they do survive. The native Humboldt cutthroat trout enters the reservoir from tributary streams and attains weights to more than 3 kg (Patrick Coffin, Nevada Fish and Game Department biologist, personal communication).

I observed the native redband trout at three Mile Creek in the Catlow Valley Basin, Oregon, where, during high flows, some of these trout have access to a shallow, warm, irrigation reservoir with a dense population of chub, *Gila bicolor*. After two years of reservoir life the native redband trout averages 1.7 kg (Kunkel 1976). The practical fisheries management value of this genetic resource is obvious and the Oregon Department of Fish and Wildlife is currently propagating the Catlow Valley redband trout (Kunkel and Hosford 1978). The typical fisheries management solution to a "rough" or "trash" fish problem has been to eradicate all fish chemically from a body of water and stock the barren water with domesticated hatchery trout. Native species such as the Lahontan and Bonneville basin cutthroat trout and redband trout of the Oregon basins, in which predator-prey relationships between the native trout and native cyprinids evolved during pluvial periods, offer a management option to convert a nongame fish liability into a forage fish asset.

Over the years millions of hatchery rainbow trout and nonnative cutthroat trout were stocked in Bear Lake, Utah-Idaho, of the Bonneville Basin. Survival to reproduction by these nonnative hatchery trout has been essentially nil. It was finally realized that only the offspring of the native cutthroat trout, which existed in minimal numbers because of blockage and destruction of tributary spawning streams, could survive and exhibit good growth in Bear Lake with its unique endemic fish fauna. Stocking Bear Lake with the young of the native cutthroat trout is now bearing excellent results (Nielson and Archer 1978 and Nielson personal communication). The average size of spawning trout is about 3 kg and trout to almost 9 kg have been taken.

These fisheries management examples clearly demonstrate the practical value inherent in the genetic resources of native trout. A more progressive outlook by management agencies concerning the preservation and management of native trout is gradually evolving in the West. There is much yet to be done, however, before an era of more "enlightened" fisheries management is achieved in regard to the appropriate recognition and management of all native fishes. The years of ingrained biological ignorance, which resulted from an overwhelming reliance on massive production of domesticated hatchery trout to solve all management problems, will take some time to overcome.

The native trout of the Great Basin offer the opportunity to apply evolutionary principles to modern fisheries management. By so doing these trout which now occur as only a trace of their former numbers would have their distribution and abundance greatly increased. For some, however, such as the cutthroat trout that once inhabited the Alvord Lake drainage and evolved in pluvial Lake Alvord, it is too late. They are gone forever. In 1934 Carl Hubbs was the first and last person to record their existence. The influence of man during the last 100 years which has produced environmental degradation and the introduction of nonnative fishes has probably caused more extinctions and drastic rearrangement of Great Basin fishes than occurred naturally during the preceding 100,000 years of great cyclical fluctuations of the climate.

REFERENCES

Behnke, R. J. 1972a. The salmonid fishes of recently glaciated lakes. Journal of the Fisheries Research Board Canada 19:639-671.

Behnke, R. J. 1972b. The rationale of preserving genetic diversity. Proceedings of the Annual Meeting Western Association State Game and Fish Commissioners 52:559-561.

Behnke, R. J. and M. Zarn. 1976. Biology and management of threatened and endangered western trouts. USDA Forest Service, General Technical Report RM-28:45 p.

Benson, L. V. 1978. Fluctuations in the level of pluvial Lake Lahontan during the last 40,000 years. Quaternary Research 9:300-318.

Broecker, W. S. and A. Kaufman. 1965. Radiocarbon chronology of Lake Lahontan and Lake Bonneville II, Great Basin. Geological Society of America Bulletin 76:537-566.

Cavender, T. M. 1978. Taxonomy and distribution of bull trout, <u>Salvelinus</u> <u>confluentus</u> (Suckley), from the American Northwest. California Fish and Game 64:139-174.

Cope, E. D. 1883. On the fishes of the eastern part of the Great Basin, and of the Idaho Pliocene Lake. Proceedings of the Academy of Natural Sciences of Philadelphia 1883:134-166.

Gilbert, D. H. and B. W. Evermann. 1894. A report on investigations in the Columbia River basin, with descriptions of four new species of fishes. U. S. Fish Commission Bulletin 14:169-207.

Girard, C. F. 1856. Notice upon the species of the genus *Salmo* of authors observed chiefly in Oregon and California. Proceedings of the Academy of Natural Sciences of Philadelphia 8:217-220.

Hickman, T. J. and R. J. Behnke. 1979. Probable discovery of the original Pyramid Lake cutthroat trout. Progressive Fish Culturist 41:135-137.

Hickman, T. J. and D. A. Duff. 1978. Current status of cutthroat trout subspecies in the western Bonneville Basin. Great Basin Naturalist 38:193-202.

Hubbs, C. L. and R. R. Miller. 1948. Correlation between fish distribution and hydrographic history in the desert basins of western United States. University of Utah Bulletin 38(20):17-166.

Hubbs, C. L. and R. R. Miller. 1972. Diagnosis of new cyprinid fishes in isolated waters in the Great Basin of western North America. Transactions of the San Diego Society of Natural History 17:101-106.

Hubbs, C. L., R. R. Miller and L. C. Hubbs. 1974. Hydrographic history and relict fishes of the North-central Great Basin. California Academy of Sciences Memoirs 7:1-259.

Klar, G. T. and C. B. Stalnaker. 1979. Electrophoretic variation in muscle lactate dehydrogenase in Snake Valley cutthroat trout. Comparative Biochemistry and Physiology 64B:391-394.

Kunkel, C. M. 1976. Biology and production of the red-band trout (*Salmo* sp.) in four southeastern Oregon streams. M. S. thesis, Oregon State University, Corvallis.

Kunkel, C. M. and W. Hosford. 1978. The native red-band trout-one alternative to stocking hatchery rainbows in southeastern Oregon. *In*: J. R. Moring (Ed.), Proceedings of the wild trout-catchable trout symposium. Oregon Department of Fish and Wildlife, Portland, pp. 41-51.

Loudenslager, E. J. and G. A. E. Gall. 1980. Geographic patterns of protein variation and subspeciation in the cutthroat trout, *Salmo clarki*. Systematic Zoology 29:27-42.

Loudenslager, E. J. and G. N. Thorgaard. 1979. Karyotypic and evolutionary relationships of the Yellowstone (Salmo clarki bouvieri) and westslope (S. c. lewisi) cutthroat trout. Journal of the Fisheries Research Board of Canada 36:630-635.

Malde, H. E. 1968. The catastrophic late Pleistocene Bonneville flood in the Snake River Plain, Idaho. U. S. Geological Survey Professional Paper 596:1-52.

Miller, R. R. 1950. Notes on the cutthroat and rainbow trouts with a description of a new species from the Gila River, New Mexico. Occasional Papers of the Museum of Zoology, University of Michigan. 529:1-42.

Miller, R. R. 1972. Classification of the native trouts of Arizona with the description of a new species, Salmo apache. Copeia 1972:401-422.

Needham, P. R. and R. Gard. 1964. A new trout from central Mexico: Salmo chrysogaster, the Mexican golden trout. Copeia 1964:169-173.

Nielsen, B. R. and D. L. Archer. 1978. The Bear Lake cutthroat trout enhancement program. Utah Division of Wildlife Resources Publication 78-6:1-47.

Smith, G. R. 1975. Fishes of the Pliocene Glenns Ferry Formation, southwest Idaho. University of Michigan, Museum of Paleontology, Papers on Paleontology 14:1-68.

Smith, G. R. 1978. Biogeography of intermountain fishes. In: Intermountain Biogeography: A Symposium. Great Basin Naturalist Memoirs, No. 3, pp. 17-42.

Snyder, J. O. 1908. Relationships of the fish fauna of the lakes of southeastern Oregon. U.S. Bureau of Fisheries Bulletin 27:69-102.

Snyder, J. O. 1914. A new species of trout from Lake Tahoe. U.S. Bureau of Fisheries Bulletin 32:23-28.

Snyder, J. O. 1917. The fishes of the Lahontan system of Nevada and northeastern California. U.S. Bureau of Fisheries Bulletin 35:31-86.

Viggs, S. C. and D. L. Koch. 1980. Upper lethal temperature range of Lahontan cutthroat trout in waters of different ionic concentrations. Transactions of the American Fisheries Society 109:336-339.

5 Effects of Habitat Size on Species Richness and Adult Body Sizes of Desert Fishes

G. R. SMITH
University of Michigan, Ann Arbor

ABSTRACT

A comparison of samples from deserts in the United States and the Paraguayan Chaco shows that species richness, life history patterns, and evolution of desert fishes are strongly influenced by habitat size and drainage directions with respect to habitat stability. In the intermountain (United States) desert, streams head in relatively stable habitats and flow onto deserts where conditions are frequently more variable and habitat may disappear. Chaco streams originate in low, flat headwaters where wet-season precipitation collects in temporary swamps and drains to a large stable river with a rich fauna. Species richness ranges from 0 to more than 40 species per sample in the Chaco, depending on habitat size. Annual extinction is high locally, but recolonization rate is also high and long-term extinction rates are probably low. This pattern contrasts with that in the intermountain desert where the annual cycles are part of a longer postpluvial desiccation and extinction cycle. Here barriers to recolonization are imposed by basin and range topography and species numbers range from 0 to 11 per sample.

The pluvial fluctuations in the Great Basin might have cyclicly disrupted evolutionary trends before they produced species-level adaptations to deserts or pluvial great lakes. Chaco aridity is not so old as in the intermountain desert, but colonization from tropical, humid forest streams has introduced fishes with diverse adaptations for air breathing and for surviving dry seasons as aestivating adults. Selection has produced annual life-history adaptations that include survival of eggs in desiccated substrate.

Body size is also related to habitat size in most nonbenthic freshwater fishes. Many intermountain minnows, suckers, and trouts are selected locally for large size by increased adult survival and consequent late reproduction in large habitats. Annual fluctuations that reduce habitats seasonally, thus causing heavy mortality, lead to persistence of phenotypes that reproduce early at the expense of later growth. Fossil evidence suggests that late Cenozoic species richness and body size in intermountain fishes were generally correlated with habitat size. Most desert fishes belong to widespread groups apparently adaptable to a broad size range of habitats.

INTRODUCTION

Latitudinal variation in species richness has been shown by Fischer (1960) to be strongly influenced by long- and short-term stability of environment. Pianka (1966) and Emery (1980) summarized the influence of environmental stability as well as spatial heterogeneity, productivity, and aspects of community structure on species richness. Barbour and Brown (1972) showed that surface area and latitude account for 90 percent of the variation in numbers of species of fishes in lakes.

The number of species of fishes at a freshwater stream locality depends on the volume, temporal stability, and spatial heterogeneity of habitats (Evans and Noble 1979; Gorman and Karr 1978; Sheldon 1968). The species richness in a drainage is influenced by the same factors and also by the nature of habitat connections that allow colonization among localities and drainages (Horwitz 1978; MacArthur and Wilson 1967). Volume and connectedness affect species richness indirectly by their effect on habitat diversity and colonization rate.

Desert fishes provide an opportunity to examine the relative effects of these ecologic and biogeographic factors because fish habitats in arid lands are subject to extreme reduction and fluctuation. The deserts of western North America offer an additional dimension for study: parts of the evolutionary and ecological history can be inferred from paleontological and geological evidence.

This study tests several propositions about the interaction of habitat volume, climate, geography, and time in the regulation of fish species richness, life histories, and body size in arid lands:

1. Habitat volume is a dominant determinant of species richness in fluvial freshwater fishes. It is controlled by interaction of precipitation, evaporation, groundwater geology, and geographical relief.
2. Fluctuations in volume cause local extinctions, but colonization enables species numbers to approach equilibrium in proportion to the volume and persistence of habitats that connect the localities. This is a form of the extinction-colonization hypotheses of MacArthur and Wilson (1967). It and the preceding postulate must be considered in the context of the species-area relationship (Levins and Heatwole 1963; MacArthur and Wilson 1967).
3. Two contrasting life-history responses to fluctuating aquatic environments in arid regions are reduction of adult body size because of increased early investment in reproduction and increased adult body size because of uncertain offspring survival relative to adult survival.

COMPARISONS

The fish fauna studied here is that of the cold deserts of the US intermountain west. [The warm-desert fishes are treated by R. R. Miller and M. L. Smith in this volume.] For comparison, observations of fishes and habitats of the Paraguayan Chaco are also presented. The Chaco of Paraguay is one-fourth the size of the intermountain cold desert; each is a part of a larger arid region. The Chaco is warm; the intermountain cold desert is defined to exclude its warmer extension to the south. The Chaco is flat and connected to a great fluvial system; the intermountain cold desert is traversed by rows of north-south trending mountain ranges and shows corresponding isolation among its basins. The dry seasons of the two regions are similar in length and precipitation but the wet seasons in the Chaco receive more rain. The species richness in the two areas is similar in the dry but different in the wettest sections.

The Intermountain Cold Desert

For the purposes of this study this region is defined as those lowland areas of the Basin and Range Province of the western United States in which the annual rainfall is less than 40 cm, the mean January temperature is less than 0°C, and the mean annual number of days with a minimum temperature as low as 0°C ranges from 60 to about 210. Included are the lowlands of California east of the Sierras and south to Death Valley, Nevada north of Las Vegas, northern Arizona, northwest New Mexico, western Colorado, Utah, southwest Wyoming, the Snake River Plain of Idaho, southeast Oregon, and the strip of central Oregon and Washington east of the Cascades (Figure 1). The elevation generally ranges from 600 to 2400 m, with extremes from -85 to 4000 m (but areas above 2400 m are excluded from most of this analysis). Most of the area is between 35° and 45° N latitude.

From the Climatic Atlas of the United States (US Department of Commerce 1968, pp. 1-23) we can characterize the dominant climatic feature of the region as

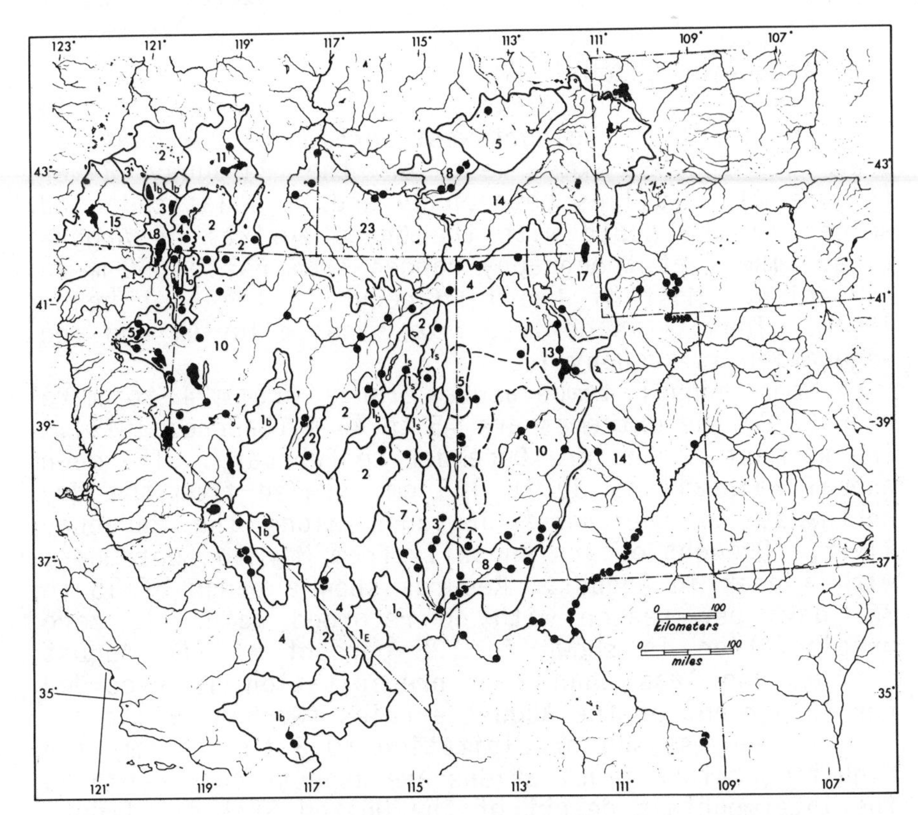

FIGURE 1. The intermountain cold desert (sampling area marked with round symbols). Numerals indicate numbers of native in each basin.

its extreme daily temprature fluctuation; the variation is 10 to 20°C (usually 14 to 17°C) in the spring and 14 to 25°C (usually 17 to 22°C) in the summer. The normal daily ranges are consistently the highest in the United States. The mean annual lake evaporation ranges from 86 cm on the Snake River Plain to 218 cm in Death Valley and usually between 100 and 150 cm over most of the region. Summer precipitation is low, usually 0.5 to 1.5 cm/month (in contrast to the monsoonal deserts to the southeast, which have about four times that amount). The winter precipitation is slightly higher but still usually less than 2.5 cm month^{-1}. The mean annual runoff is generally less than 2.5 cm except in the uplands (Langbein et al. 1949).

The Paraguayan Chaco

The Chaco is the northwest portion of Paraguay, an area of 246,925 km^2 west of the Paraguay River (Figure 2). It lies between 19 and 25°S latitude and merges to the south with the Chaco Austral of Argentina. The flat topography ranges from 64 to 350 m in elevation. Most of the fish occur in the east below 100 m. Topsoil is derived from sandy alluvium and the vegetation is xerophytic.

Mean annual temperatures range from 24 to more than 26°C; the extremes are -5 to 44°C (Sanchez 1973). The warmest months are December to February. The mean annual rainfall is 30 to 145 cm, graded from the dry northwest to the moist lowlands along the Paraguay River. Drought occurs annually from May to September, with a peak in August. Records show a range of 15 to 367 days per season with no recorded rainfall; zero precipitation is shown for 38 percent of the August entries and less than 2 cm precipitation is recorded for 75 percent of the August entries (Gorham 1973).

Low dry-season precipitation (0.5 to 6.5 cm) and high temperature fluctuations are aspects comparable to the intermountain desert of the United States. Important differences in the Chaco are the higher winter temperatures, the greater summer precipitation, and the summer connections of all habitats to the Paraguay system because of the flat topography. The last distinction has the most important impact on fish distribution and species richness.

GEOLOGIC AND CLIMATIC HISTORY

Intermountain Cold Desert

Two distinct geologic provinces make up this region: the Basin and Range (and associated Snake River Plain), and the Colorado Plateau. These are partly separated by the Wasatch Mountains in central Utah. The Rocky Mountains form the eastern boundary, the Idaho batholith a part of the northern boundary, and the Sierras and Cascades make up the western boundary

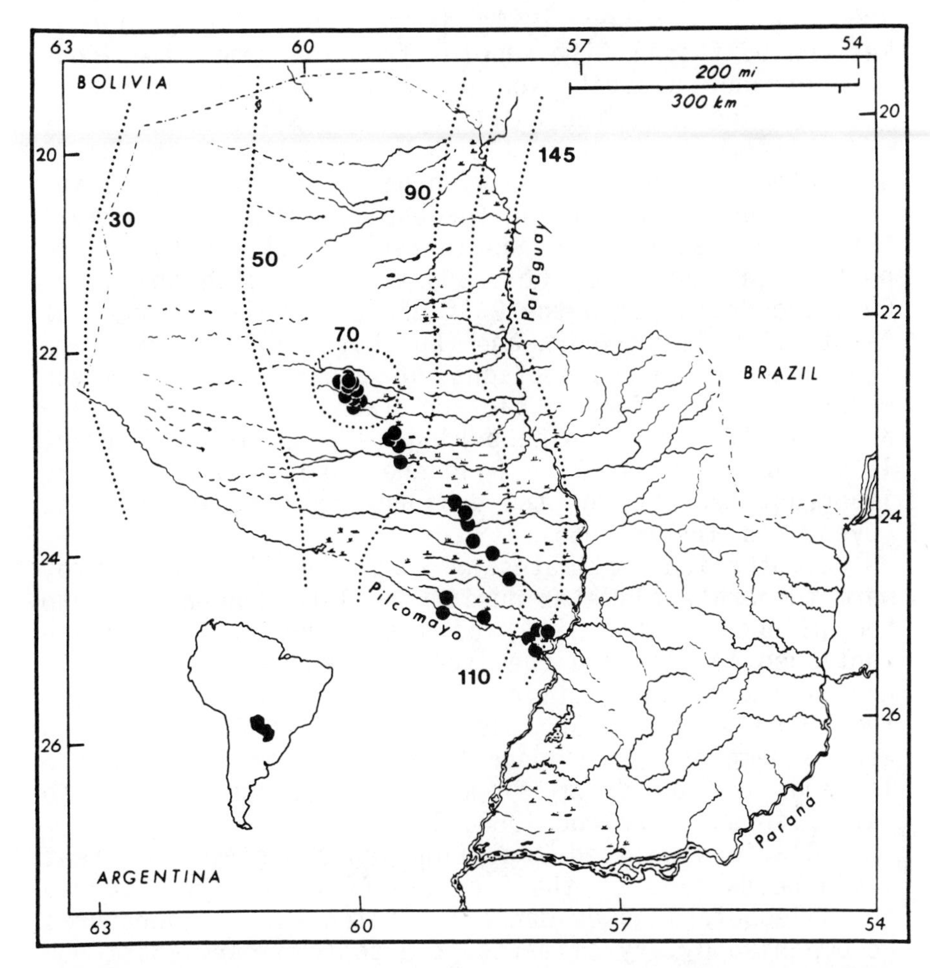

FIGURE 2. Map of Paraguay. Round symbols indicate sampling localities in the Chaco (west of the Paraguay River). Dotted lines show rainfall isoclines (in cm precipation).

and the all-important barrier to the moisture of Pacific air masses. The present topography of the Great Basin and Snake River Plain began with rifting and extensional faulting 14 to 17 million years ago (Zoback and Thompson 1978). A significant shift toward a cooler, drier climate occurred at about the same time according to paleobotanical evidence (Axelrod 1950,

1956, 1968; Chaney 1944; Dorf 1930; Smiley 1963). Western US fossil fish records from the same time indicate broader distribution patterns than in any subsequent period (Smith and Miller in press).

Differentiation of the Colorado Plateau and Basin and Range Provinces began with basin and range faulting accompanied by uplift and (probably) increased precipitation in western Colorado (Larson et al. 1975). Connection of the upper Colorado drainage with the lower Colorado drainage to the Gulf of California occurred at least 3.3 million years ago (Lucchitta 1972). The time and location of connections of upper Colorado River waters to the lower Colorado (above the Lake Meade area) in the intervening period is problematical (Hunt 1969; Lucchitta 1972). Pliocene lakes, however, were important features of the drainage in northern Arizona (Uyeno and Miller 1965).

Subtropical climates were gradually replaced by warm temperate climates during the late Miocene through the Pliocene (Dorf 1959) as basin and range faulting continued to increase the isolation and internal drainage of the Great Basin and parts of the Colorado system. The elements of intermountain desert vegetation were assembled by response to aridity, at least in the lowlands and on drier slopes, through the Miocene to late Pliocene (Axelrod 1950, 1956).

Glaciers existed 3 million years ago in the Great Basin headwaters of the Sierras (Curry 1966). Because the paleobotanical evidence indicates warm temperatures accompanied by dry climates, the contemporary existence of widespread lake deposits and onset of montane glaciation seem to require much lower local evaporation rates; specifically, cool summers. Geological evidence (Longwell 1928; Mannion 1962) also indicates dry climates. Even during the subsequent Pleistocene pluvial episodes the precipitation was less than twice that of today (Snyder and Langbein 1962). Many studies of pluvial cycles in relation to lower temperature and higher precipitation are reviewed by Morrison (1965) and refined by Brakenridge (1978). In general, it is clear that lake waters were not derived from glaciers; it seems likely that precipitation was modestly increased and that low summer temperatures and lower evaporation were important factors. Northern occurrences of cold-intolerant fossil reptiles also support the conclusion that climates were more equable than today until late in the Pleistocene (Hibbard 1960; Van Devender and Spaulding 1979; Van Devender and Wiseman 1977).

A climatic mechanism offered by Bryson and Wendland (1967) involves the exclusion of arctic air from middle latitudes by continental ice sheets and local adiabatic warming of drainage winds off the ice. Morrison (1965), on the basis of evidence of weathering profiles, erosion, and surficial deposits in relation to pluvial and glacial evidence, suggests that the typical cycle began with cool-dry conditions at the beginning of an interpluvial-interglacial and progressed through warm-dry, warm-wet, cold-wet (beginning of pluvial-glacial), cool-moist, then back to cool-dry. Examination of O^{18} curves (Emiliani 1972) has suggested that glacial episodes were several times as long as interglacials. Paleolacustrine data assembled by Morrison (1965) seem to suggest that pluvials were often shorter than interpluvials. Certainly pluvials were frequently interrupted by desiccation episodes and stable lacustrine habitats were of relatively short duration, possibly less than 20,000 years (Emiliani 1972). Pliocene lake deposits are more continuous than Pleistocene, which indicates longer duration of stable habitats, at least on the Snake River Plain (Glenns Ferry Formation, Malde and Powers 1962) and in parts of the western and northern Great Basin.

In summary, despite evidence of remarkable freshwater lakes, arid conditions have been at least as frequent: for fishes the history has been marked by scores of oscillations between pluvial- and aridity-dominated habitats.

The main effects of this instability on intermountain fish distributions have been severe extinctions and barriers to colonization (Smith 1978). Species richness is unusually low in small drainages in relation to large drainages (i.e., the species:area curve is steep) which indicates (1) higher extinction of species that were restricted to small populations during interpluvials and (2) low colonization because of relief- and aridity-induced barriers to dispersal among basins. Interpluvial extinctions were more severe in the south than in the north. Northern and peripheral drainages have more species than southern and central basins and species distributions are much broader in the northern parts of the region; this indicates freer dispersal. These conditions hold despite the fact that because of the relationship of temperature to elevation and latitude mountain barriers were (effectively) higher in the north and lower in the south (Smith 1978).

The Chaco

The first unmistakable evidence of desert and semidesert conditions in southern South America occurs in the Pliocene (Solbrig 1976). At this time the final uplift of the Bolivian Cordillera Central completed the topographic relief that began in the Miocene (Harrington 1962). Miocene climates represented continuing trends toward lower temperature and precipitation, but the climate was more humid than today. Grasslands were increasing, forests decreasing (Solbrig 1976). With the establishment of the vast alluvial plain in the rainshadow of the Andes by the end of the Pliocene the present setting was nearly complete. At least three major montane glacial stages constituted Pleistocene interruptions, but the effect on fishes in the Chaco was probably minor because of the distance and difference in elevation. Examination of numerous exposures along the southeastern trans-Chaco highway in August 1979 revealed no lacustrine beds. In all probability the present pattern of annual flood cycles that swell the Paraguay tributaries in the wet season, then dry them in the winter, also characterized the Pleistocene. Long-term variations in temperature and habitat persistence no doubt occurred but probably did not cause long-term isolation and extinction of fish species as in the Great Basin. From the standpoint of the fishes the most important aspects of the Chaco environment are continuous connections to a central habitat system and the seasonal existence of widespread, shallow, nutrient-rich swamps.

Fossil Fishes

About 25 Miocene, Pliocene, and Pleistocene fish localities in the intermountain cold desert offer an opportunity to compare past environments and diversity with the present. High diversity evolved only in a few intermountain lacustrine environments, the richest being the Miocene and Pliocene lakes on the Snake River Plain (Kimmel 1975; Smith 1975). Historical data permit the evaluation of long-term ecology and evolution of species richness in the intermountain area.

CLIMATE, GEOGRAPHY, AND FISH SPECIES RICHNESS

Variables that influence fish species richness in streams are temperature, oxygen concentration, salinity, habitat volume, and heterogeneity, as well as (indirectly) precipitation, evaporation, elevation, stream order, and discharge. The extremes and predictability are probably more significant than averages (Fischer 1960).

Temperature

An unexpected result in the early stages of this study was the failure to find a major correlation between temperature (or its variance) and species number. The effect of elevation and temperature in the Basin and Range Province is not reflected in patterns of species richness. A major effect of temperature variation on fish species richness is probably demonstrable along a latitudinal gradient in the Mississippi basin but does not appear in the intermountain deserts because the effect of topographic relief on temperature, precipitation, and evaporation causes a strong negative correlation between temperature and habitat permanence (discussed below). In the Chaco temperature is higher in the north and species richness increases to the east. Scanty data on temperature extremes suggest a negative correlation between range of temperature and species richness.

There is no doubt that temperature is an important factor that affects the growth, metabolism, and distribution of fishes. Failure to find a positive temperature effect on species richness in the intermountain desert might reflect the history of extinction by desiccation in the south and substantial colonization by coolwater forms from the north (Smith 1978). The major effect of temperature on species richness is the local elimination of narrowly adapted species by extreme variations in temperature. The evolutionary significance of this phenomenon has been investigated by Brown and Feldmeth (1971) and Hirshfield et al. (1980).

Salinity

Salinity is important to desert fishes because of the high evaporative accumulation of dissolved salt. High salinity environments in both the Chaco and the intermountain deserts show reduced faunas, which usually consist mainly of cyprinodontids (e.g., Miller 1948). Extreme effects of salinity are seen in the fishless waters of the Great Salt Lake (Utah) and the middle Rio Pilcomayo of the Chaco.

Cyprinodontids are the most salinity-tolerant desert fishes (Renfro and Hill 1971). Their derivation from estuarine sister groups suggests that they are technically "preadapted" rather than "desert-adapted" to salinity (M. L. Smith, this volume).

Oxygen

One of the most discussed adaptations of arid-land fishes is the use of accessory respiratory structures to make use of atmospheric oxygen. Most of the species in the arid part of the Chaco show such adaptations (Carter and Beadle 1930, 1931). Examples include lungfish (*Lepidosiren paradoxa*), catfish (*Hoplosternum, Pterygoplichthys*), and characins (*Hoplias*). Even the cichlids when stressed readily lie at the surface to bathe their gills in oxygen-rich water. No similar adaptations, except physiological tolerances (Hirshfield et al. 1980) and behaviors for obtaining surface oxygen occur in desert fishes of North America despite their equal or greater antiquity. Furthermore, temperatures are higher in the Chaco and warm, shallow waters conducive to oxygen stress occur in all seasons: in the organic-rich swamps during the summer wet season and the drying pools of the (warm) winter. In the North American deserts drying pools and (high temperature) oxygen stress are generally restricted to the summer and moving (oxygen-enriched) waters are more constant in all seasons. In both regions, but especially in North America, recolonization after local extinction comes from oxygen-rich environments.

A possible historical reason for the absence of special desert adaptations in North American fishes, in contrast to mammals, for example (Mares 1976), is that their exposure to cool pluvial lakes at frequent geologic intervals and long-term persistence of cool running-water habitats has precluded consistent selection for such traits.

Habitat Size

The amount and persistence of fluvial habitat depends on the balance between precipitation and evaporation as well as groundwater and topographic control of discharge rate. Elevation is related to species richness only insofar as it is correlated with habitat size, temperature, and gradient. Stream order is equally indirect as a causal variable because its relation to species richness depends on its correlation with habitat volume, stability, and dispersal connections (Evans and Noble 1979; Horwitz 1978).

A test of the hypothesis that habitat size is a dominant determinant of species richness in streams might be conducted with measurements for many localities over all seasons. Presently available data (C. L. Hubbs seine collections from western United States, 1925–1945, and seine, electrofishing, and rotenone collections by the author and colleagues in western United States and Paraguay) allow examination of the effects of habitat size (estimated by habitat width) and a general index of regional water supply (precipitation and evaporation data from weather station summaries, the Climatic Atlas of the United States and Gorham

1973). Stream width is consistently related to other dimensional parameters (Leopold et al. 1964). Samples were taken in the dry season in both areas and include collections from streams, springs (US), and ponded sections of streams. Effects of springs and ponds on variations in width do not differ between the two areas. Introduced species are included in the totals because the question is ecological rather than historical. They are limited to North American samples, where except in trout streams their numbers are generally correlated with native species (Smith 1978).

Species number is significantly correlated with habitat width in both areas (Figures 3, 5). The number of species increases steeply in larger habitats, especially in the Chaco. There is considerable scatter in the data, especially in the Colorado drainage (Figure 3). The combined samples show a correlation of .58 ($p < .01$) between species number and stream width. The correlation is .71 ($p < .01$) in the Great Basin and Snake River drainage. On the other hand, the samples associated with the Colorado River drainage have a correlation coefficient of only .46 ($p < .01$). These samples include small streams near their confluence with the main river, sometimes with an inflated species number, as well as samples in the main channel of the Green and Colorado rivers, where species number is often limited by sand substrate and strong current.

Species number is not correlated ($r = .1$, $p = .3$) with precipitation in the Great Basin (plus Snake River Plain) and Colorado drainages (Figure 4). Nevertheless the graph is instructive. A few relatively high species numbers are found in the low-precipitation area of the graph, in large streams that flow out into the most arid sections of the region (upper left of Figure 4). By contrast there are 50 samples from small headwater streams in the moist upland watersheds (lower right part of Figure 4).

The role of evaporation in the interaction between precipitation, habitat size, and species number was tested by regression of species richness on a surface-water index. The index was computed as annual precipitation minus the square root of annual lake evaporation because evaporation is 2 to 10 times larger than

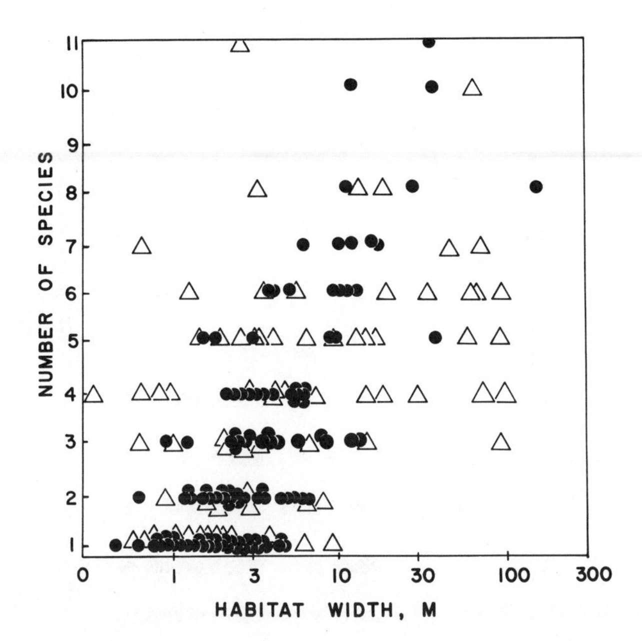

FIGURE 3. Number of species as a function of stream width for 183 localities in the intermountain cold desert. Great Basin and Snake River localities are solid circles, Colorado and adjacent localities are triangles. The overall correlation coefficient is .58 ($p < .01$).

precipitation in this region. The index is conformable with regional surface water but the relationship was not signficant ($r = -.07$, $p = .3$).

In the Chaco the number of species is far greater and the contrast between small, fishless ponds and streams and large, species-rich habitats is more pronounced (Figure 5). The correlation coefficient between species number and habitat width is .32 ($p = .04$). The correlation is lower in the flat Chaco region because drying streams and their remnant, isolated ponds do not conform to the normal close relationship among channel dimensions in flowing streams. Furthermore, the samples were taken during the period of accelerated local extinction.

Precipitation is a more forceful factor in the Chaco. The correlation coefficient between species

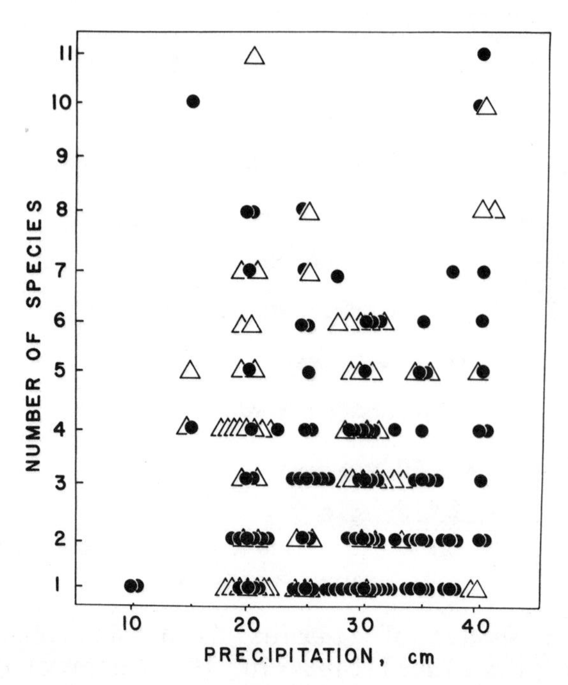

FIGURE 4. Number of species as a function of mean annual precipitation in the intermountain cold desert. Solid symbols are Great Basin and Snake River Plain localities; triangles represent localities in the Colorado and adjacent drainages.

number and precipitation is .77 ($p < .01$). A comparison of Figures 5 and 6 reveals that the smallest habitats, with few species, are in the driest areas sampled. (The slope of both Chaco regressions should probably be steeper; our samples underestimate the true species numbers in the largest waters.) When the log of habitat width is added to a stepwise multiple linear regression model, the partial correlation between precipitation and species number drops from .76 to .68, which indicates correlation of the independent variables. The coefficient of causation (multiple R^2) increases from .59 to .63. Species richness in the Chaco was also studied in relation to precipitation in the dry season, but the relationship was not significant ($r = .15$, $p = .36$).

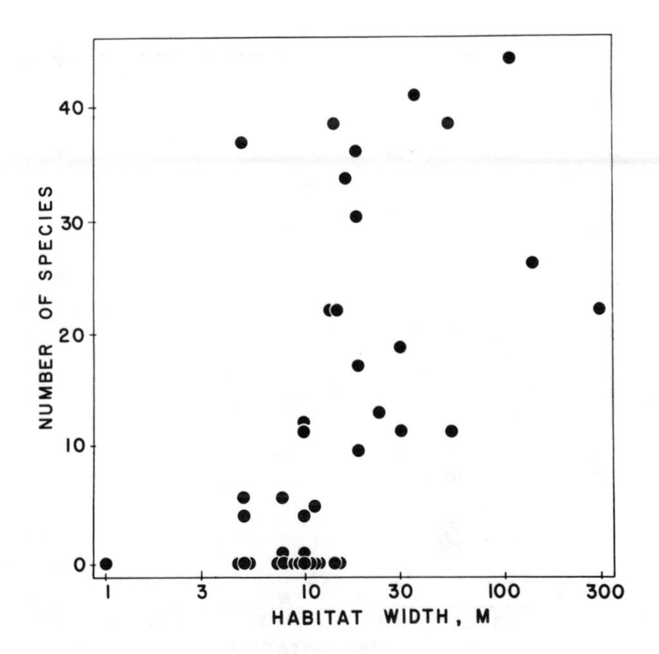

FIGURE 5. Number of species as a function of habitat width for 41 localities in the Paraguayan Chaco. The correlation coefficient is .32 (p = .04).

In summary, there is a strong relationship between habitat size and species number in both systems but the explanation is incomplete. The role of habitat size is complex in a way that involves dispersal and colonization.

Fluctuations in habitat size cause local extinctions, but recolonization occurs in proportion to the size and persistence of connecting habitats. That species richness in the intermountain region is low because of barriers to colonization is suggested by the steep species:area curve (z = .59 in Smith 1978, Figure 8). A similar line of evidence is the relative absence of rare species in truncated species-abundance curves for desert communities (Smith, unpublished data), demonstrated by Hubbell (1979) in a study of dry-forest trees.

The striking contrast in patterns of species richness between the Chaco and the intermountain samples is related to a simple but little recognized geographical

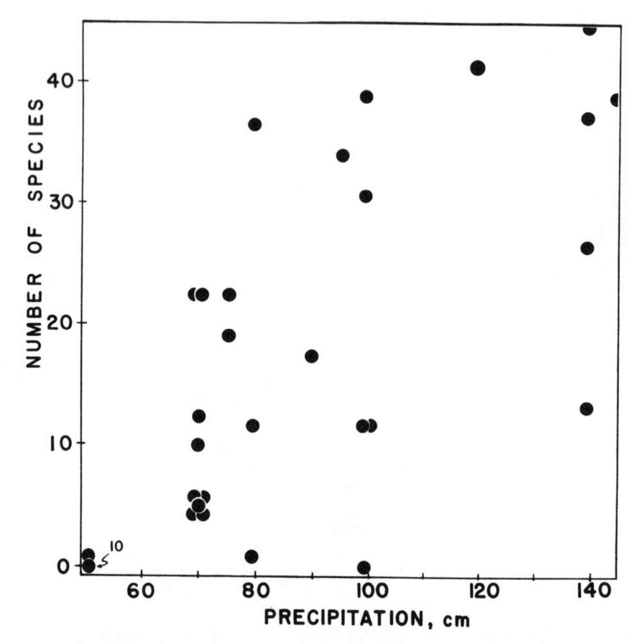

FIGURE 6. Number of species as a function of mean annual precipitation in the Paraguayan Chaco. The correlation coefficient is .77 ($p < .01$).

factor. In the intermountain deserts streams collect their waters in mountains with high precipitation and flow out onto low-precipitation alluvial plains and playas. In such cases the lower reaches are often the least stable and the least predictable with respect to their habitats and inhabitants. Local extinctions cannot be recolonized from a stable, main-trunk refuge because of the double isolation by desert and mountain barriers.

In contrast, the tributaries of the Paraguay head in flat, arid alluvial plains where the dry season reduces the apparent species number to zero. The streams then flow through increasingly wetter regions into the Paraguay River, a large stable habitat with potential for recolonizing almost any local extinction. Thus direction of flow relative to the precipitation gradient interacts with the effect of habitat volume to produce an impoverished fauna where the flow direction is wet-to-dry and a rich fauna where the direction is dry-to-wet.

Late Cenozoic Species Richness

The Late Cenozoic fossil record of fishes from the intermountain cold desert permits the examination of the effects of long-term and large-scale fluctuations in habitat size on species richness. Most of the fossil occurrences (Figure 7) are in lake deposits, whereas the comparisons emphasized above involve streams. Because lakes and streams in the same basins share most species, the comparison of species richness in and among basins is generally valid.

The hypothesis is that large habitat size promotes increased species richness by increased species packing and lower extinction rates. Instability of habitat volume causes local extinctions, whereas restriction of aquatic connections increases the likelihood of extinctions and decreases colonization. On the other hand, seasonality promotes selection for special life-history adaptations and semiisolation promotes evolutionary fixation of genetic traits. Therefore on the basis of the known geologic and climatic history we might predict high extinction rates or evolution of special adaptations and speciation, depending on the frequency of the fluctuations in volume of aquatic habitat. Seasonal fluctuations should have caused local extinctions and strong selective pressures; long-term fluctuations should have increased the breadth of the extinctions and the rate of speciation.

Two patterns, one geographic and one temporal, are shown in Figure 7 (and by comparison with Figure 1). First, there is a strong latitudinal gradient in species richness but its trend is opposite that of most such gradients. Second, the pattern of high species richness in the north (or at least in the Snake River Basin) has been characteristic of the region for all the Late Cenozoic. It has been shown that the present species richness at localities is correlated with volume. This conclusion is strengthened by the historical trends in Figure 7.

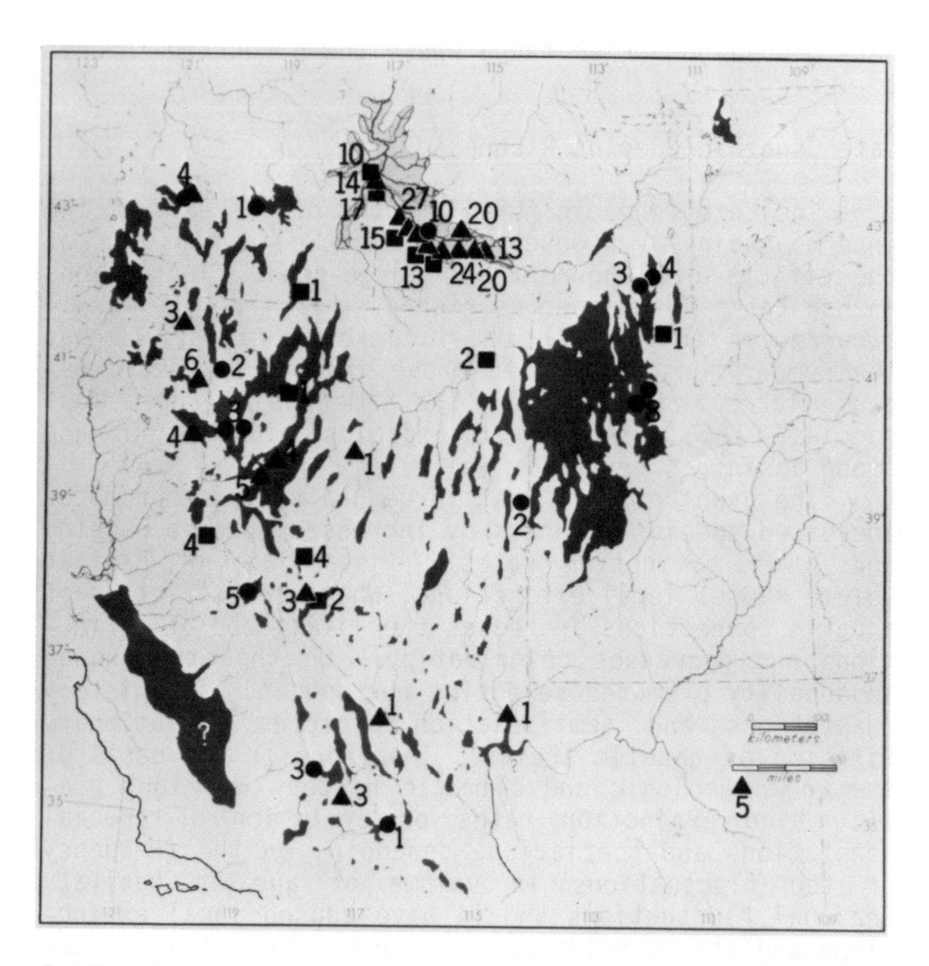

FIGURE 7. Maximum extent of Pleistocene pluvial lakes and some Pliocene lakes with estimates of species numbers in selected fossil lakes and stream systems. Data from published sources and work in progress (Table 1). Numbers are underestimates, probably by two or more species each, because small forms such as _Rhinichthys_ and _Richardsonius_ are rarely collected, although they were probably present.

Miocene localities in the Great Basin have one to four species, including trouts, minnows, and a sunfish. In the Snake River Plain we find 10 to 17 species (additional trout, salmon, _Esox_, minnows, suckers, and sculpins). The Miocene Deer Butte and Chalk Hills formations on the Snake River Plain (Kimmel 1979) are extensive, thick sections which indicate large, long-term depositional environments and probably large, long-term lacustrine and fluvial habitats. The Miocene deposits in the Great Basin are less extensive and indicate smaller, less persistent bodies of water (Table 1). (An exception is a fauna, described by P. H. McClellan from the Salt Lake Group, which is related to the Chalk Hills fauna.) The rather limited depositional systems are probably due in part to the low Miocene relief.

Pliocene localities in the Basin and Range Province (5.5 to 1.7 million years ago) are also poor in species compared with the fauna of the Snake River Plain (Table 1). Glenns Ferry lacustrine sites (Smith 1975) have up to 27 species; fluvial sites have 10 to 13 species. Again, the Basin and Range sites represent less extensive depositional systems. The Pliocene faunas differ from the Miocene by the addition of genera and species of whitefish, suckers, minnows, and sculpins. Boreal forms, such as *Prosopium*, *Myoxocephalus*, and other sculpins become conspicuous by their relative abundance in some deposits on the Snake River Plain, thus indicating cooler climates and aquatic connections to the north (compare with Axelrod 1968). At this time the number of species of fishes in the Glenns Ferry Formation reached the maximum known for Late Cenozoic western North America. Geological evidence (Kimmel 1979) reveals that the Glenns Ferry lacustrine system persisted at least one million years.

The close of the Pliocene on the Snake River Plain is marked by regression of lacustrine Glenns Ferry deposition (Kimmel 1979) and depletion of fish species richness (Smith 1975), perhaps as the Glenns Ferry lake began draining into the Columbia River system (previous drainage to the Pacific at least occasionally passed through southern Oregon or California (Kimmel 1975; Smith 1975; Taylor 1960; Wheeler and Cook 1954). Whatever the cause, the decrease in aquatic volume and species richness was roughly synchronous.

Pleistocene faunas (Table 1, Figure 7) are richer in association with large waters. Pleistocene deposits of the Bruneau Formation and the Glenns Ferry Formation

TABLE 1. Known Species Richness of Late Cenozoic Fossil Localities in the Intermountain Cold Desert[a]

State	Location	Formation	Species	Authority
		Pleistocene		
California	Afton	Manix Lake	1	2
California	China Lake	?	3	3
California	Mono Lake	?	5	3
Idaho	Grandview	Bruneau and Glenns Ferry	10	1
Idaho	Swan Lake	–	3	1
Idaho	Lake Thatcher	Main Canyon	4	4
Nevada	Duck Valley	–	2	3
Nevada	Smith Creek Cave	–	2	1
Nevada	Pyramid Lake	–	3	1,3
Oregon	Fossil Lake	Unnamed	4	3
Oregon	Harney Lake	?	1	1
Utah	Bonneville	Bonneville? Draper?	8	6
		Pliocene		
Arizona	Bidahochi	Bidahochi	5	7
California	Crowder Flat	Alturas	3	1
California	Furnace Creek	Titus Canyon	3	8
California	Honey Lake	Unnamed	4	9
California	Mohave Desert	Unnamed	1	8
California	Secret Valley	Unnamed	6	1
Oregon	Fossil Lake	Unnamed	4	3
Oregon	Adrian	Chalk Hills [Deer Butte Kimmel, 1975]	14	10
Idaho	Fossil Creek	Glenns Ferry	27	11
Idaho	Castle Creek Browns Creek Birch Creek Shoofly Poison Creek Horse Hill	Glenns Ferry	24	11
Idaho	Sand Point Bennett Spring	Glenns Ferry	20	11
Idaho	Hagerman	Glenns Ferry	13	11
Nevada	Esmeralda	Esmeralda	3	12
Nevada	Jersey Valley	Unnamed	1	13
Nevada	Logandale	?	1	1
Nevada	Mopung Hills	Upper Truckee	4	9

TABLE 1. (Continued)

State	Location	Formation	Species	Authority
		Miocene		
California	Bear Valley	Unnamed	4	1
Oregon	Trout Creek	Trout Creek	1	15
Oregon	Sand Hollow	Deer Butte	10	15
Oregon	Adrian	Chalk Hills	17	10
Idaho	Browns Creek	Poison Creek?	15	1
Idaho	Shoofly	Chalk Hills	13	10
Idaho	Horse Hill	Chalk Hills	13	10
Nevada	Esmeralda	Esmeralda	2	1,16
Nevada	Humboldt	Humboldt	2	1
Nevada	Rabbithole	?	1	17
Nevada	Stewart Spring	?	4	1
Nevada	Truckee	Truckee	5	14
Utah	Cache Valley	Salt Lake	1	18

[a]Sources: (1) unpublished data; (2) Blackwelder and Ellsworth (1936); (3) Miller and Smith (1981); (4) Bright (1967); (5) Meade and Van Devender collection; (6) Smith et al. (1968) and Madsen collection; (7) Uyeno and Miller (1965); (8) Miller (1945); (9) Taylor and Smith (1981); (10) Kimmel (1975); (11) Smith (1975); (12) LaRivers (1962); (13) Lugaski (1979); (14) Bell (1974) and subsequent collection; (15) Smith and Miller (in press); (16) LaRivers (1966); (17) LaRivers (1964); (18) Uyeno and Miller (1963).

near Grandview on the Snake River Plain, although depauperate, are the richest observed in the intermountain area. Pleistocene lakes in this area were not so permanent as the earlier lakes. Lake Bonneville was the largest and has the richest fauna outside the Snake River Plain. Nevertheless it is depauperate for its size, no doubt because of its long-term instability. Lake episodes were usually too short to permit extensive evolution and (or) interpluvials were arid enough to cause severe extinction. The species richness associated with other Pleistocene pluvial lakes indicates the same restrictions.

The negative effect of isolation on species richness is seen in the low numbers at southern localities, but not in historical trends. There is no evidence of more species in southern Miocene samples, even though distributions of Miocene species indicate fewer barriers and the paleobotanical record indicates more precipitation and moisture. Southern localities are consistently poor in species (Figure 7). The Miocene vegetation of southeastern Nevada was an oak sclerophyll woodland that implies a cool temperate climate with perhaps 760 mm annual precipitation (Axelrod 1956, 1979). The elevation was possibly 760 to 915 m, with low relief (Axelrod, 1956, 1979). The low fish species richness in the Miocene suggests seasonal aridity and perhaps isolation from the rich fauna of the Snake River Plain.

By Pliocene time some aquatic connections had occurred between the northwest and southwest Great Basin with the establishment of *Chasmistes* as far south as China Lake (Miller and Smith 1981; Taylor and Smith 1981). The biogeographic track from the southwest Great Basin northward along the Sierras, eastward across the Snake River Plain, and south into northern Utah was described by Taylor (1960) on the basis of the distribution of mollusks and *Chasmistes*. This pattern requires at least brief absences of topographic barriers, and therefore establishes the significance of ecological barriers to southern dispersal of many northern species that did not move south. The ecological barriers probably involved low, unstable, warm-water habitats. Large-scale fluctuations characterized the

Pleistocene. The pluvial lakes were surely stable habitats, optimal in almost every way. Their low species richness can be explained only by barriers between basins and failure of pluvial conditions to persist long enough to permit much species evolution.

The Recent fauna includes some taxa that apparently were evolved locally at some unknown time in the late Cenozoic; for example, the *Prosopium* species flock in Bear Lake; species differentiation in *Gila*, *Ptychocheilus*, *Richardsonius*, *Catostomus*, and *Chasmistes*; the endemic genera of minnows, *Iotichthys*, *Eremichthys*, *Relictus*, *Moapa*, and the Plagopterinae; the cyprinodontid genera, *Empetrichthys* and *Crenichthys*; and the species diversity in *Cottus* and *Cyprinodon*. Of this diversity three large groups contain some possible examples of Pleistocene isolation and speciation: the Great Basin and related *Catostomus* (subgenus *Catostomus*), the species of *Lepidomeda* (Miller and Hubbs 1960), and the species of *Cyprinodon* (Miller 1948). In each of these cases the species are similar, allopatric, and probably lacking in genetic isolating mechanisms. Data that indicate a basis for questioning the level of species evolution in *Cyprinodon* were presented by Turner (1974), Turner and Liu (1977), Cokendolpher (1980), and Stevenson (1981). The karyotypes of *Lepidomeda* were shown by Uyeno and Miller (1973) to be similar. Many species pairs of *Catostomus* in and around the Great Basin are not completely separable by known characters even though intermittent geographic separation surely dates back into the Pliocene or early Pleistocene (M. L. Smith, this volume). These observations do not detract from the importance of the differentiation in *Cyprinodon*, *Lepidomeda*, and *Catostomus*. If these genera provide examples of Pleistocene speciation, it is critical that the amount of differentiation be documented.

HABITAT VOLUME AND BODY SIZE IN INTERMOUNTAIN FISHES

It is a common observation that individuals of many species reach larger sizes in larger streams. Not only is growth potentially faster when greater input from upstream provides more ration but individuals in larger, more stable waters may live longer because they are less subject to seasonal mortality. Some fishes may also respond to restrictive living space by growing less, regardless of ration. Although it is unclear how much of the correlation between body size and habitat size is due to faster growth and how much to longevity, it is even less clear how much of the observed pattern is heritable adaptive variation in age or size of maturity and how much is opportunistic adaption to local resources.

It may be assumed that the supply of ration in a stream is proportional to the normal volume of habitat upstream (minus food extraction by competitors). The density of food and competitors is not precisely proportional to volume because the seasonal fluctuations in fish and invertebrate biomass are not coincident with each other or volume. Two additional qualifications affect productivity in desert streams. First, input of terrestrial leaf litter and insects is less than in mesic climates. Second, streams in arid lands carry more sand and consequently provide less favorable substrate for bacteria, algae, diatoms and invertebrates. Extreme cases occur when discharge is sufficiently high, in relation to channel dimensions, to move bedload; thus eliminating habitat for periphyton and benthos (e.g., in the Colorado River). The fishes that occupy these streams are subject to low ration, swift current (relatively unavoidable in the main channels of the Green and Colorado), and large seasonal fluctuations. Here, in contrast to most streams, gradients in productivity and stability may be reversed: density of benthic organisms is higher at higher elevations (and steeper gradients) because of more organic input and more stable substrate and also because rainfall, evaporation, oxygen, and groundwater supply (and its effect on temperature) are usually more stable nearer the mountains.

A consequence of the steep ecological gradient is sharp species segregation. Higher elevations require adaptations to low temperatures, short growing season, current, and low ratio of living volume to substrate. By contrast, lower, seasonally larger habitats have more species, higher temperatures, and a longer growing season but lower food density. Optimal habitat for most species occurs in intermediate reaches in the zone of transition of gradient, substrate, and temperature.

In light of these considerations, some influences on body size may be explored: (a) ration and temperature are major determinants of growth rate (Brett et al. 1969); (b) length of the growing season determines yearly increments (Gerking 1966). The first year's growth is especially important to survival and reproduction. Growth during the second and subsequent years depends on the effects of survival probability on reproductive investment (Hirshfield and Tinkle 1975; Murphy 1968; Williams 1966).

1. High adult density is unfavorable for juvenile survivorship, especially when both eat the same food (e.g., microvores) or when cannibalism is common. When these or other factors contribute to relatively low juvenile survivorship selection favors later reproduction, iteroparity, and long-term investment in large numbers of potential offspring.
2. Low adult survivorship caused by seasonal reduction in habitual size favors individuals that reproduce early in life.
3. Intermediate responses might be favored in intermediate environments or under irregular schedules of seasonal fluctuation, but divergence from intermediacy is caused by positive feedback effects within the two alternatives: In (1) longevity leads to larger size, which in turn increases longevity in the presence of factors (such as predators) that disadvantage small individuals. In (2) low adult survival favors early reproductive investment, which subtracts from future growth and leads to small size and shorter life span (Constantz 1979).

Body Size and Habitat Width

Maximum adult size was recorded for samples of several stream species in the University of Michigan Museum of Zoology. (Minimum adult size is more important but harder to measure reliably in late summer samples; thus the amount of available data is reduced. Average adult size is difficult to for the same reason. Maximum adult size is less ambiguous and was not significantly related to sample size in the examples given.)

C. L. Smith (1980) presented evidence of interaction between maximum size (per species) and community structure in fishes inhabiting patch reefs. This interaction is not demonstrable in my data. Instead maximum adult size is usually correlated with habitat dimensions (Table 2, Figure 8). The variation in body size attributable to stream width (log transformed) ranges from 8 to 5 percent (R^2) in the significant samples. In those species that display the correlation the strength of the response is proportional to adult body size: the smallest species (e.g., *Rhinichthys*) remain relatively small in large streams (where they are common); larger suckers (*Catostomus*) are larger at maturity and show a proportionally larger response to increase in habitat volume. Adults of the larger species are rarely present in small tributaries, although young are usually found in the same streams as their parents (often in shallower waters).

The large species are absent from most small isolated drainage units in the Great Basin. The fossil record suggests that the species remaining in these units are the survivors of postpluvial extinctions. Thirteen single-species survivors in drainage basins (Figure 1) are small fishes: six *Gila bicolor*, four *Relictus solitarius*, and three *Rhinichthys osculus*. This evidence indicates disproportionate extinction of large-bodied species in small habitats and small basins.

Recent samples demonstrate a plasticity that implies selection against large individuals in small habitats. *Gila bicolor* (including *G. alvordensis*) and *Gila atraria* show nearly identical responses to habitat size (Table 2, Figure 8). These species are ecologically and morphologically similar, but their lineages have been separated since the late Miocene, which suggests parallel environmental control of body size in these forms. These species are capable of maturing at small size in small creeks and springs (35 and 62 mm SL; Table 2). In large rivers and lakes, however, they

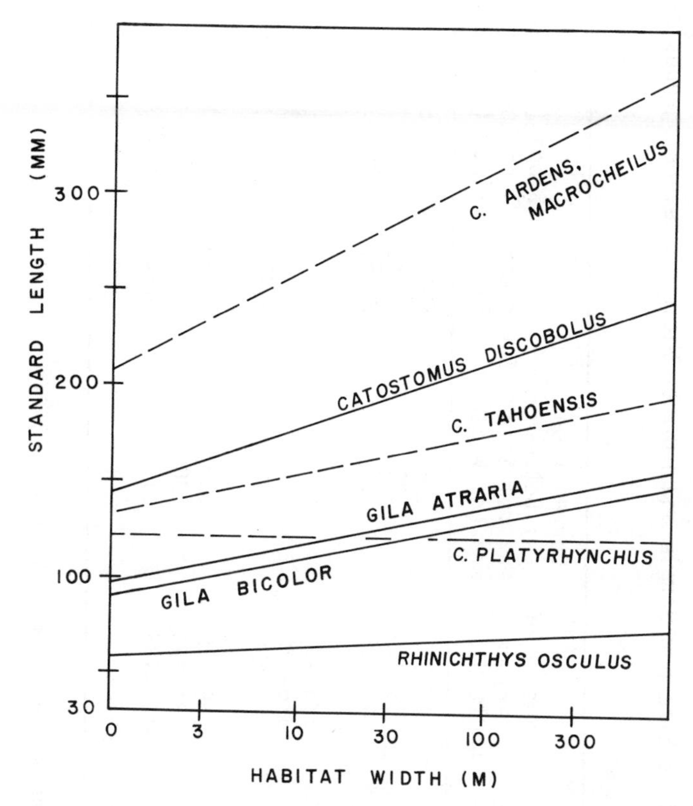

FIGURE 8. Regression of maximum adult size (in each sample) against habitat width. Solid lines are signficant; dashed lines not significant (Table 2).

mature later, at lengths of more than 150 mm. *Gila atraria* may live as long as 11 years and grow to more than 400 mm in SL in one of its largest habitats, Bear Lake, Utah and Idaho (Sigler and Miller 1963).

The smallest minnow in North America is a close relative of *Gila*; it lives in shallow waters in the Bonneville Basin. *Iotichthys phlegethontis* matures at about 25 to 30 mm SL and rarely grows larger than 50 mm SL. It is not variable in size and not widespread. *Ptychocheilus*, which includes the largest minnows in North America, inhabits coastal streams from central California to British Columbia and the Columbia and Colorado drainages. *Ptychocheilus* formerly

Table 2. Relationship Between Maximum Adult Size per Sample and Habitat Width in Several groups of Intermountain Fishes[a]

	K (Samples)	Average Maximum Size	Min–max. of Maximum Adult Size	Correlation (R) Between Maximum Size and Log Width	Intercept (SE)	Slope (SL) on Log Width (SE)	Significance
Rhinichthys osculus	192	63	35–107	.28	60(2)	5.3(2.5)	.03
Gila bicolor	121	110	35–319	.36	90(6)	21(5)	.00
Gila atraria	41	117	62–215	.39	97(9)	20(8)	.01
Ptychocheilus							
oregonensis	13	201	123–400				
lucius	6	448	376–730				
Catostomus discobolus	60	183	80–400	.32	145(17)	34(13)	.01
Catostomus							
platyrhynchus	52	120	69–178	.00	122(7)	0(8)	.97
Catostomus tahoensis	30	145	84–260	.30	132(17)	22(24)	.11
Catostomus ardens and *macrocheilus*	15	261	149–485	.19	208(80)	54(78)	.50

[a]Only adults are analyzed; measurements are in mm standard length.

inhabited the Great Basin but not in Recent time. It is excluded from small waters by ecological requirements related to size.

Because of the correlation between temperature and elevation, northern fishes (trout, whitefish, and sculpins) are the primary inhabitants of mountain headwaters in the cold desert. Trout and sculpins are small macrovores in desert streams. Trout are active midwater swimmers; sculpins are negatively buoyant, benthic predators. Trout show size modifications in different habitat sizes (about 150 to 750 mm SL); sculpins remain small, like *Rhinichthys*, which are also negatively buoyant benthic fishes. Whitefish (*Prosopium*) are medium-sized, neutrally buoyant predators on small invertebates. They appear to have a modest size response. *Rhinichthys cataractae*, like *Prosopium williamsoni*, inhabits lower montane, high-gradient streams. *Rhinichthys cataractae* is also a predator on small invertebrates; it is negatively buoyant and restricted in size.

A cross comparison of *Salmo*, *Prosopium*, *Cottus*, and *Rhinichthys cataractae* reveals an important principle concerning body size in freshwater fishes (Table 3). Benthic forms (*Rhinichthys* and *Cottus*) are smaller and less opportunistic in growth than their less benthic ecological counterparts, probably because their reduced mobility reduces their ration and increases their vulnerability to predation.

Most mountain suckers (*Catostomus*, subgenus *Pantosteus*, Tables 2 and 3) occupy the middle reaches of northern desert streams, where temperature, gradient, current, and substrate are intermediate; they are negatively buoyant benthic microvores. Most species are small and show little response to habitat size (see *platyrhynchus*, Table 2). *Catostomus* (*Pantosteus*) *discobolus*, and *columbianus*, however, inhabitants of the Colorado and Snake rivers (and some nearby systems), show a definite size response (see *discobolus*, Table 2). Southern populations (*C. plebeius*, Mexico) are small-bodied inhabitants of higher (cool) headwaters. Species of the subgenus *Catostomus* are large, benthic microvores with large swimbladders and show only a suggested size response, but the sample sizes

TABLE 3. Examples of the Relationship Between Buoyancy and Size Response[a]

Group	Food and Feeding	Buoyancy and Habit	Size and Size Response
Salmo	Macrovore; large jaws	Neutrally buoyant; midwater	Potentially large or small; opportunistic
Cottus	Macrovore; large jaws	Negatively buoyant; benthic	Small; limited variability
Prosopium	Small invertebrates; small mouth	Neutrally buoyant; midwater	Medium and large; limited variability
Rhinichthys	Small invertebrates; small mouth	Negatively buoyant, benthic	Small; limited variability
Catostomus (ss)	Microvore; benthic	Semineutrally buoyant; semibenthic	Medium to large; limited variability
Pantosteus	Microvore; benthic	Negatively buoyant; benthic	Small to medium; large spp variable

[a]Benthic fishes tend to be smaller than their less benthic ecological counterparts and their less benthic relatives.

are inadequate. Other fishes of the intermountain desert (except endemics of large lakes) are small. They inhabit small waters or are benthic.

The fossil record offers an interesting comparison. Many of the above genera (*Prosopium* and other salmonids, *Ptychocheilus*, *Catostomus*) grew much larger in Pliocene Lake Idaho (Smith 1975) than anywhere in North America today or in the Pleistocene. Benthic species such as sculpins were only slightly larger. Pleistocene Lake Bonneville (Smith et al. 1968), though large, does not appear to have had larger fishes than modern intermountain lakes, but the reasons are not known.

In summary there is a tendency for fish that occupy large habitats to respond ecophenotypically with extended growth and large size. Conversely, in small habitats these species may mature early and fail to reach large size. Benthic fishes, cyprinodontoids, and some minnows are invariably small. The explanation of this phenomenon requires consideration of reproductive life history.

Life history

It is significant that the tendency for large-volume habitats to produce large fishes may be as pronounced in the intermountain desert as it is in mesic climates. This suggests that productivity and ration do not govern the phenomenon but that life-history adaptations to length of the growing season and mortality schedules may be the primary determinants. Specifically, irrespective of ration and temperature, size seems to be determined by the effect of mortality schedules on early versus late breeding. Whether ration is abundant or minimal, low adult mortality will lead to selection for iteroparity and large size, as illustrated by *Gila*, *Ptychocheilus*, and suckers in the Colorado River. In habitats of all sizes selection will favor reproduction by one-year-olds if seasonal reduction in habitat repeatedly causes heavy mortality after a season of substantial growth. This is illustrated by stream fishes in the desert portions of the Great Basin.

If adult mortality is variable because of unpredictable variation in the severity of seasonal fluctuations, individuals that invest in early reproduction will leave more descendants after destructively dry years; those that grow larger and produce more offspring later will leave more descendants in a series of wet years. In a postpluvial transition period such as the present both genotypes could be present. This is a possible explanation of some of the growth and reproductive plasticity among and perhaps within the populations discussed.

The extreme responses to annual fluctuation in aquatic habitat, aestivation and annual reproduction with dry-season survival as eggs only, has not evolved in North American desert fishes, despite the antiquity of desert climates. In the Chaco lungfish and cyprinodonts exhibit these responses to dry seasons, even though desert climates may be less ancient in South America. It is possible that the pluvial periods have interrupted evolution of these adaptations in North American deserts. Also, because of topographic relief and its regulation of groundwater, headwaters and springs are among the more stable habitats, whereas in the Chaco headwaters are the least stable.

In limited examples adaptations to survive dry seasons in the egg stage may release cyprinodonts from the small size and early reproduction syndrome that characterizes them elsewhere. Wherever the dry season is highly predictable and the wet season favorable for rapid growth, early reproduction (more than one generation per season) may not be so successful as will larger numbers of eggs immediately before desiccation. In such conditions cyprinodonts in the Chaco may reach lengths of 75 mm SL, which is somewhat larger than they become in habitats that favor earliest reproduction.

The other major adaptation to dry seasons is migration. Avoiding desiccation by roundtrip alternation of habitats is a common adaptation of inhabitants of large river systems with headwaters in arid lands (Lowe-McConnell 1964). In intermountain desert fishes migration is not well studied, but it seems to be related to choice spawning sites rather than desiccation avoidance. Because of the relationship between swimming performance and body size, migration is limited to large, nonbenthic species that live in large habitats.

Emigration is a special problem. In fluctuating environments survival value could accrue to genotypes that show colonization behavior. This is the opposite of migration in that success depends on the probability of existence of an unoccupied environment more favorable for reproduction than the individual's birthplace. This probability is not insignificant in environments like the intermountain desert, where local extinctions are periodic. It may be predicted that such genotypes and behaviors will promote emigration of young fish and be accompanied by behavioral responses to evidence of relative permanence of habitat (e.g., aquatic plants) and population density.

DISCUSSION AND SUMMARY

Habitat size is the most important general factor in the control of survival, diversity, and life histories of desert fishes. Quantity of aquatic habit, however, though intuitively simple, is not easily defined or measured. A measure based on depth, though useful in certain systems (Sheldon 1968), is not reliable for many purposes because of the importance of substrate and productivity to most fishes (consider deep springs or the bathypelagic realm). Likewise, volume of water can be negatively correlated with volume of suitable habitat if the increased volume is moving through a confined channel, as in torrential rivers. A third unsatisfactory but simple measure, width of habitat, is used here as a first approximation of habitat size for desert stream fishes. Habitat width is regularly related to other measures of habitat size in streams; it fluctuates seasonally and allows comparison of seasonal variability.

Two arid regions are compared: the North American intermountain cold desert, where water originates on mountain islands and flows into lakes or temporary sinks, and the Paraguayan Chaco, where precipitation occurs during a hot, wet season and drains through swamps and low-gradient streams, leaving fishes in a severe desert climate for about three months each year. The contrast allows elucidation of ecological and selective effects of variation in habitat size on colonization, extinction, and species richness.

The Chaco is subjected to severe annual extinction and massive annual recolonization. A comparable North American system was studied by Harrel et al. (1967). The drainage direction is from unstable dry to stable wet, and the colonization source is rich with species adapted to the productivity of the annually flooded arid lands. Species number is low where streams are small (or low) and high in large aquatic habitats. An average habitat has about 15 species, but variations of 0 to 40 occur geographically and annually.

In the intermountain desert severe extinctions also occur because of fluctuations in habitat volume. Recolonization is so restricted by barriers that many small basins are now fishless. Few samples contain more than seven species. Annual fluctuations are less severe in the intermountain desert because topographic relief ameliorates habitat volume somewhat by groundwater regulation. Recolonization is measured in geological time because of the geological barriers between drainage basins. Most of the colonization has been restricted in and among these basins over the last several million years. There is evidence of some colonization from the north (Smith 1975) and from west coastal lowlands (Bell 1974) in the Pliocene.

The Chaco and intermountain samples provide extreme examples of short- and long-term instability. The species richness at any locality in either place is a product of both scales of history. The steepness of the curves in Figures 3 and 5 reflects short-term fluctuations and adaptations (e.g., migration) to them. The striking difference in maximum species richness reflects the negative effect of long-term fluctuations and barriers to colonization in the intermountain desert.

The geological and fossil record suggests that the interaction of intermountain pluvial habitats and isolation during the Pleistocene might have produced considerable evolutionary differentiation. Such differentiation exists and has been extensively documented (Hubbs and Miller 1948; Hubbs et al. 1974; Miller 1948, 1959; Miller 1965 and references therein; Smith 1966). Quantification of the amount of differentiation in relation to the duration of isolation will prove interesting. Indirect evidence presented here suggests insufficient long-term persistence of habitat for full speciation despite apparently ideal isolating conditions.

In summary, long- and short-term fluctuations in aquatic habitat size have resulted in the low and variable species richness that characterizes deserts. Long-term fluctuations like those in the Great Basin alternately increase and depress adaptations to desert conditions and divergence of isolated populations.

Habitat size directly effects body size in fishes by ecological effects on life history. Small habitats support small fishes, even among species that are capable of larger size. Therefore it is surprising that cyprinids and catostomids in the intermountain desert are usually larger than their counterparts in the Mississippi Basin. It might be suggested that the western forms are examples of ecological release; they fill the niches occupied by large species in more diverse eastern faunas. It might also be argued that large western forms exhibit residual adaptations to Pleistocene pluvial periods. These hypotheses may be rejected by comparison of data presented here. *Ptychocheilus* is smaller in restricted waters; it was not smaller in the presence of diverse large salmonid, centrarchid, and ictalurid predators in the Pliocene. *Gila* shows significant reduction of size at maturity in small waters of the Great Basin but individuals may average four times longer in large habitats. This suggests that potential for large size is not a Pleistocene relict but is selected by ecological consequences of local habitat size.

Body size is measured as maximum adult size in each sample. The results of this study are consistent with the hypothesis that individual size is partly heritable and strongly controlled by the effects of relative juvenile and adult mortality on late versus early reproductive investment. When seasonal fluctuations in habitat size result in heavy mortality; individuals that reproduce at small size leave more descendants than those that risk another round of mortality for another season of growth (Hirshfield and Tinkle 1975).

Reinvestment of the genotype in many small units increases the probability of survival, especially in small habitats in which large individuals might be more vulnerable to predation, oxygen depletion, or starvation. When adult mortality is low individuals leave more descendants by growing larger and producing more young over several seasons. Fish in the harsh, low-ration Colorado River (and productive, predator-rich lakes) probably show much lower adult mortality in relation to juvenile mortality. _Ptychocheilus_ and trout are the only large native predators in the intermountain desert. Therefore survival past a threshold in the neighborhood of 200 mm in standard length places an individual in a relatively predator-free environment. All but benthic fishes in this environment are larger than their relatives elsewhere.

Benthic fishes are not free to grow substantially larger in larger habitats. The reason is probably a consequence of the relationship between individual size and home range, territory, or foraging area (Sale 1980). This phenomenon could be related to restriction of ration as a result of lower mobility in benthic species but greater vulnerability of adults to predators is as likely an explanation in streams and lakes. Similar reasons probably account for the exclusively small sizes of surface- and substrate-restricted atheriniform fishes in deserts. The failure of these benthic, surface, and edge species to grow much larger in the presence of abundant food and space indicates that, unlike many desert fishes, they are genetically limited in their pattern of growth and early reproduction. The opportunity to breed more than once a year may increase selection toward early reproduction in cyprinodonts, poeciliids, small characins, and small cichlids by negative feedback of reproductive effort on survivorship. The exceptional size of some large annual cyprinodonts in the Chaco provides support for this hypothesis; the long growing season and predictable on-set of drought favors relatively large size and later reproduction.

The extremely small cyprinids in the intermountain region, especially Iotichthys and Eremichthys, may be the result of rarity of small competitors for microcrustaceans and aufwuchs. The existence of so few parallels to Notropis is probably a function of sparse terrestrial input to desert streams.

Despite the well-documented endemism in intermountain fishes, it is ecologically significant that the fauna is largely populated by representatives of the most widespread genera of freshwater fishes in North America and the Northern Hemisphere. Catostomus, Cottus, Rhinichthys, and Cyprinodon are the widest-ranging genera on the continent. Cottus, Salmo, Prosopium, and Catostomus, as well, have broad distributions in the remainder of the holarctic region. All have the capacity to succeed in the widest range of habitat sizes.

ACKNOWLEDGMENTS

Research in Paraguay was supported by the National Science Foundation and conducted by Jeff Taylor, Thom Grimshaw, Roger Myers, and Beverly Smith. I am grateful to the Museum of Paleontology, University of California, Berkeley, for the loan of fossils from Stewart Spring, Nevada, and to Robert R. Miller for reading the manuscript and for allowing me to study fossils collected by him. Charles Anderson and M. L. Smith also read the manuscript and made many helpful suggestions. Donna Robbins and R. L. Elder typed the manuscript and helped to prepare the figures.

REFERENCES

Axelrod, D. I. 1950. Evolution of desert vegetation in western North America. In: D. I. Axelrod (Ed.), Studies in Late Tertiary Paleobotany. Carnegie Institution of Washington Publication, Contributions to Paleontology Series 590, pp. 217-306.

Axelrod, D. I. 1956. Mio-Pliocene floras from west central Nevada. University of California Publications in Geological Sciences 33:1-316.

Axelrod, D. I. 1968. Tertiary floras and topographic history of the Snake River Basin, Idaho. Geological Society of America Bulletin 79:713-734.

Axelrod, D. I. 1979. Age and origin of Sonoran Desert vegetation. Occasional Papers of the California Academy of Sciences 132:1-74.

Barbour, C. D. and J. H. Brown. 1972. Fish species diversity in lakes. American Naturalist 108:473-489.

Bell, M. A. 1974. Reduction and loss of the pelvic girdle in *Gasterosteus* (Pisces): a case of parallel evolution. Los Angeles County Museum of Natural History, Contributions in Science 257:1-36.

Blackwelder, E. and E. W. Ellsworth. 1936. Pleistocene lakes of the Afton Basin, California. American Journal of Science 31:453-463.

Brakenridge, G. R. 1978. Evidence for a cold, dry full-glacial climate in the American southwest. Quaternary Research 9:22-40.

Brett, J. R., J. E. Shelbourn, and C. T. Shoop. 1969. Growth rate and body composition of fingerling sockeye salmon, *Oncorhynchus nerka*, in relation to temperature and ration size. Journal of the Fisheries Research Board of Canada 26:2363-2394.

Bright, R. C. 1967. Late-Pleistocene stratigraphy in Thatcher Basin, southeastern Idaho. Tebiwa 10:1-7.

Brown, J. H. and C. R. Feldmeth. 1971. Evolution in constant and fluctuating environments: thermal tolerances of desert pupfish (*Cyprinodon*). Evolution 25:390-398.

Bryson, R. A. and W. M. Wendland. 1967. Tentative climatic patterns for some late-glacial and post-glacial episodes in central North America. *In:* W. J. Mayer-Oakes (Ed.), Life, Land, and Water, University of Manitoba Press, Winnipeg, Canada, pp. 271-299.

Carter, G. S. and L. C. Beadle. 1930. The fauna of the swamps of the Paraguayan Chaco in relation to its environment. I. Physico-chemical nature of the environment. Journal of the Linnean Society of London (Zoology) 37:205-257.

Carter, G. S. and L. C. Beadle. 1931. The fauna of the swamps of the Paraguayan Chaco in relation to its environment. II. Respiratory adaptations in the fishes. Journal of the Linnean Society of London (Zoology) 37:327-368.

Chaney, R. W. 1944. The Troutdale flora (Oregon). Carnegie Institution of Washington Publication 553:323-351.

Cokendolpher, J. C. 1980. Hybridization experiments with the genus Cyprinodon (Teleostei:Cyprinodontidae). Copeia 1980:173-176.

Constantz, G. D. 1979. Life history patterns of a livebearing fish in contrasting environments. Oecologia 40:189-201.

Curry, R. R. 1966. Glaciation about 3,000,000 years ago in the Sierra Nevada. Science 154:770-771.

Dorf, E. 1930. Pliocene floras of California. Carnegie Institution of Washington Publication 412:1-108.

Dorf, E. 1959. Climatic changes of the past and present. Contributions from the University of Michigan Museum of Paleontology 13:181-210.

Emery, A. R. 1980. The basis of fish community structure: marine and freshwater comparisons. Environmental Biology of Fishes 3:33-47.

Emiliani, C. 1972. Quaternary paleotemperatures and the duration of the high-temperature intervals. Science 178:398-401.

Evans, J. W. and R. L. Noble. 1979. The longitudinal distribution of fishes in an east Texas stream. American Midland Naturalist 101:333-343.

Fischer, A. G. 1960. Latitudinal variations in organic diversity. Evolution 14:64-81.

Gerking, S. D. 1966. Length of the growing seasons of the bluegill sunfish in northern Indiana. Verhandlungen Internationale Vereinigung Limnologie 16:1056-1064.

Gorham, J. R. 1973. The Paraguayan Chaco and its rainfall. In: J. R. Gorham (Ed.), Paraguay: Ecological Essays. Academy of Arts and Sciences of the Americas, Miami, FL, pp. 39-60.

Gorman, O. T. and J. R. Karr. 1978. Habitat structure and stream fish communities. Ecology 59:507-515.

Harrel, R. C., B. J. Davis and T. C. Dorris. 1967. Stream order and species diversity of fishes in an intermittent Oklahoma stream. American Midland Naturalist 78:428-436.

Harrington, H. J. 1962. Paleogeographic development of South America. Bulletin of the American Association of Petroleum Geologists 46:1773-1814.

Hibbard, C. W. 1960. An interpretation of Pliocene and Pleistocene climates in North America. Michigan Academy of Science, Arts, and Letters, Report 1959-1960:5-30.

Hirshfield, M. F., C. R. Feldmeth and D. L. Soltz. 1980. Genetic differences in physiological tolerances of Amargosa pupfish (*Cyprinodon nevadensis*) populations. Science 207:999-1001.

Hirshfield, M. F. and D. W. Tinkle. 1975. Natural selection and the evolution of reproductive effort. Proceedings of the National Academy of Sciences USA 72:2227-2231.

Horwitz, R. J. 1978. Temporal variability patterns and the distributional patterns of stream fishes. Ecological Monographs 48:307-321.

Hubbell, S. P. 1979. Tree dispersion, abundance, and diversity in a tropical dry forest. Science 203: 1299-1309.

Hubbs, C. L. and R. R. Miller. 1948. The zoological evidence: correlation between fish distribution and hydrogeographic history in the desert basins of western United States. In: The Great Basin, with emphasis on glacial and postglacial times. Bulletin of the University of Utah 38(20), Biological Series 10:17-166.

Hubbs, C. L., R. R. Miller and L. C. Hubbs. 1974. Hydrogeographic history and relict fishes of the north-central Great Basin. Memoirs of the California Academy of Sciences 7:1-259.

Hunt, C. B. 1969. Geologic history of the Colorado River. United States Geological Survey Professional Papers 669C:59-130.

Kimmel, P. G. 1975. Fishes of the Miocene-Pliocene Deer Butte Formation, Southeast Oregon. University of Michigan Papers on Paleontology 14:69-87.

Kimmel, P. G. 1979. Stratigraphy and paleoenvironments of the Miocene Chalk Hills Formation and Pliocene Glenns Ferry Formation in the western Snake River Plain, Idaho. Unpublished Ph.D. dissertation, University of Michigan, Ann Arbor.

Langbein, W. B. et al. 1949. Annual runoff in the United States. United States Geological Survey Circular 52:1-14.

LaRivers, I. 1962. Fishes and Fisheries of Nevada. Nevada State Fish and Game Commission, Carson City, Nevada.

LaRivers, I. 1964. A new trout from the Barstovian (Miocene) of western Nevada. Occasional Papers of the Biological Society of Nevada 3:1-4.

LaRivers, I. 1966. Paleontological miscellanei. Occasional Papers of the Biological Society of Nevada 11:1-8.

Larson, E. E., M. Ozima and W. C. Bradley. 1975. Late Cenozoic basic volcanism in northwestern Colorado and its implications concerning tectonism and the origin of the Colorado River system. In: B. F. Curtis (Ed.), Cenozoic History of the Southern Rocky Mountains. Geological Society of America, Memoirs 144, pp. 155-178.

Leopold, L. B., M. G. Wolman, and J. P. Miller. 1964. Fluvial Processes in Geomorphology. Freeman, San Francisco.

Levins, R. and H. Heatwole. 1963. On the distribution of organisms on islands. Caribbean Journal of Science 3:173-177.

Longwell, C. R. 1928. Geology of the Muddy Mountains, Nevada, with a section through Virginia Range to Grand Wash Cliffs, Arizona. United States Geological Survey Bulletin 978:1-152.

Lowe-McConnell, R. H. 1964. The fishes of the Rupununi savanna district of British Guiana, South America. Part I. Ecological groupings of fish species and effects of the seasonal cycle on the fish. Journal of the Linnean Society (Zoology) 45:103-144.

Lucchitta, I. 1972. Early history of the Colorado River in the Basin and Range Province. Geological Society of America Bulletin 83:1933-1948.

Lugaski, T. 1979. Gila traini, a new Pliocene cyprinid fish from Jersey Valley, Nevada. Journal of Paleontology 53:1160-1164.

MacArthur, R. H. and E. O. Wilson. 1967. The Theory of Island Biogeography. Princeton University Press, Princeton, NJ.

Malde, H. E. and H. A. Powers. 1962. Upper Cenozoic stratigraphy of western Snake River Plain, Idaho. Geological Society of America Bulletin 73:1197-1220.

Mannion, L. E. 1962. Virgin Valley salt deposits, Clark County, Nevada. In: A. C. Bersticker (Ed.), Symposium on Salt. Northern Ohio Geological Society, Cleveland, Ohio, pp. 166-175.

Mares, M. A. 1976. Convergent evolution of desert rodents: multivariate analysis and zoogeographic implications. Paleobiology 2:39-63.

Miller, R. R. 1945. Four new species of fossil cyprinodont fishes from eastern California. Journal of the Washington Academy of Sciences 35: 315-321.

Miller, R. R. 1948. The cyprinodont fishes of the Death Valley system of eastern California and southwestern Nevada. Miscellaneous Publications of the University of Michigan Museum of Zoology 68:1-155.

Miller, R. R. 1959. Origin and affinities of the freshwater fishes of North America. In: C. L. Hubbs (Ed.), Zoogeography. American Association for the Advancement of Science Publications 51, pp. 187-222.

Miller, R. R. 1965. Quaternary freshwater fishes of North America. In: H. E. Wright, Jr. and D. E. Frey (Eds.), The Quaternary of the United States. Princeton University Press, NJ, pp. 569-581.

Miller, R. R. and C. L. Hubbs. 1960. The spiny-rayed cyprinid fishes (Plagopterini) of the Colorado River system. Miscellaneous Publications of the University of Michigan Museum of Zoology 115:1-39.

Miller, R. R. and G. R. Smith. 1981. Distribution and evolution of *Chasmistes* (Pisces:Catostomidae) in Western North America. Occasional Papers of the University of Michigan Museum of Zoology (In press).

Morrison, R. B. 1965. Quaternary geology of the Great Basin. *In*: H. E. Wright, Jr. and D. E. Frey (Eds.), The Quaternary of the United States. Princeton University Press, Princeton, NJ, pp. 265-285.

Murphy, G. 1968. Patterns in life history and the environment. American Naturalist 102:391-404.

Pianka, E. R. 1966. Latitudinal gradients in species diversity: a review of concepts. American Naturalist 100:33-46.

Renfro, J. and L. G. Hill. 1971. Osmotic acclimation in the Red River pupfish, *Cyprinodon rubrofluviatilis*. Comparative Biochemistry and Physiology 40A:711-714.

Sale, P. F. 1980. Coexistence of coral reef fishes-a lottery for living space. Environmental Biology of Fishes 3:85-102.

Sanchez, T. F. 1973. The climate of Paraguay. *In*: J. R. Gorham (Ed.), Paraguay: Ecological Essays. Academy of Arts and Sciences of the Americas, Miami, FL, pp. 33-38.

Sheldon, A. L. 1968. Species diversity and longitudinal succession in stream fishes. Ecology 49: 193-197.

Sigler, W. F. and R. R. Miller. 1963. Fishes of Utah. Utah State Department of Fish and Game, Salt Lake City, Utah.

Smiley, C. J. 1963. The Ellensburg flora of Washington. University of California Publications in Geological Sciences 35:159-276.

Smith, C. L. 1980. Coral reef fish communities: a compromise view. Environmental Biology of Fishes 3:109-128.

Smith, G. R. 1966. Distribution and evolution of the North American catostomid fishes of the subgenus Pantosteus, genus Catostomus. Miscellaneous Publications of the University of Michigan Museum of Zoology 129:1-132.

Smith, G. R. 1975. Fishes of the Pliocene Glenns Ferry Formation, southwest Idaho. University of Michigan Papers on Paleontology 14:1-68.

Smith, G. R. 1978. Biogeography of intermountain fishes. In: K. T. Harper and J. L. Reveal (Eds.), Intermountain Biogeography: A Symposium. Great Basin Naturalist Memoirs 2, pp. 17-42.

Smith, G. R. and R. R. Miller. Taxonomy of fishes from Miocene Clarkia lake beds, Idaho. In: C. J. Smiley and A. Leviton (Eds.), American Association for the Advancement of Science, Publications. In Press.

Smith, G. R., W. L. Stokes and K. F. Horn. 1968. Some late Pleistocene fishes of Lake Bonneville. Copeia 1968:807-816.

Snyder, C. T. and W. B. Langbein. 1962. The Pleistocene lake in Spring Valley, Nevada, and its climatic implications. Journal of Geophysical Research 67:2385-2394.

Solbrig, O. T. 1976. The origin and floristic affinities of the South American temperate desert and semi-desert regions. In: D. W. Goodall (Ed.), Evolution of Desert Biota. University of Texas Press, Austin, pp. 7-49.

Stevenson, M. M. 1981. Cytomorphology of several species of Cyprinodon. Copeia 1981 (In press).

Taylor, D. W. 1960. Distribution of the freshwater clam Pisidium ultramontanum: a zoogeographic inquiry. American Journal of Science 258A:325-334.

Taylor, D. W. and G. R. Smith. 1981. Pliocene molluscs and fishes from northeastern California and northwestern Nevada. University of Michigan Papers on Paleontology. In Press.

Turner, B. J. 1974. Genetic divergence of Death Valley pupfish species: biochemical vs. morphological evidence. Evolution 28:281-294.

Turner, B. J. and R. K. Liu. 1977. Extensive interspecific genetic compatibility in the New World killifish genus Cyprinodon. Copeia 1977:259-269.

US Department of Commerce. 1968. Climatic Atlas of the United States. United States Department of Commerce, Environmental Science Service Administration, Environmental Data Service.

Uyeno, T. and R. R. Miller. 1963. Summary of late Cenozoic freshwater fish records for North America. Occasional Papers of the University of Michigan Museum of Zoology 631:1-34.

Uyeno, T. and R. R. Miller. 1965. Middle Pliocene cyprinid fishes from the Bidahochi Formation, Arizona. Copeia 1965:28-41.

Uyeno, T. and R. R. Miller. 1973. Chromosomes and the evolution of the plagopterin fishes (Cyprinidae) of the Colorado River system. Copeia 1973:776-782.

Van Devender, T. R. and W. G. Spaulding. 1979. Development of vegetation and climate in the southwestern United States. Science 204:701-710.

Van Devender, T. R. and F. M. Wiseman. 1977. A preliminary chronology of bioenvironmental changes during the Paleoindian period in the monsoonal southwest. In: E. Johnson (Ed.), Paleoindian Lifeways. West Texas Museum Association, The Museum Journal 17, pp. 13-17.

Wheeler, H. E. and E. F. Cook. 1954. Structural and stratigraphic significance of the Snake River capture, Idaho-Oregon. Journal of Geology 62:525-536.

Williams, G. C. 1966. Adaptations and Natural Selection. Princeton University Press, Princeton, NJ.

Zoback, M. C. and G. A. Thompson. 1978. Basin and Range rifting in northern Nevada: clues from a mid-Miocene rift and its subsequent effects. Geology 6:111-116.

[illegible] United States Geological Survey [illegible]

[illegible] Michigan Museum of [illegible]

[illegible] Colorado [illegible]

Van Devender, T. R. and [illegible] of vegetation [illegible] in the southwestern United States [illegible]

Van Devender, T. R. and [illegible] chronology of [illegible] during the [illegible] period [illegible] southwest [illegible]

[illegible] significance [illegible] Journal of [illegible]

Williams, G. C. 1966. Adaptation and Natural Selection. Princeton University Press, Princeton, N.J.

[illegible] and [illegible] in northern Nevada [illegible] and [illegible]

6 The Role of Competition in the Replacement of Native Fishes by Introduced Species

ALLAN A. SCHOENHERR
Fullerton College
Fullerton, California

ABSTRACT

Throughout North American deserts there are many examples in which introduced species have replaced native species. Among the best known of these replacements is the near extirpation of the Gila topminnow (*Poeciliopsis occidentalis*) by mosquitofish (*Gambusia affinis*). Because of similar morphology and habitat preferences, it is widely believed that this is a classic case of competition. However, data from thorough life history studies of the two species during cohabitation indicate that competition for a common resource cannot be demonstrated. The mode of replacement appears to be predation of *Gambusia* females on young *Poeciliopsis*, accompanied by various behavioral interactions, such as aggression. In other examples of replacement and/or speciation by isolation, competition has also been implicated in *Gambusia* species groups from Texas. In the Salton Sea, California, Zill's cichlid (*Tilapia zillii*) and the sailfin molly (*Poecilia latipinna*) rapidly have replaced the native desert pupfish (*Cyprinodon macularius*) where they were once abundant. In this situation competition for a common resource cannot be proved and various behavioral interactions apparently are the cause. It is apparent that in social animals such as these fishes, the established dogma of replacement by competition is not always provable, and every situation of interaction between native and introduced species must be examined individually before processes of extirpation can be identified. If aggressive behavior, as an expression of territoriality, may be interpreted as competition for space, then careful observation of behavioral interactions must be documented. Aggressive activity of short duration probably serves to divide cohabiting fishes into separate niches. If no refuge is available in the process of niche partitioning, extirpation is the result. It is perhaps possible, though ironic, that the best evidence we have for competitive exclusion in nature may be that we rarely observe it in progress but rather infer its occurrence from documentable niche partitioning and character displacement.

INTRODUCTION

Deacon et al. (1979) included 251 fish taxa in North America that are endangered, threatened, or of special concern. Among the 251 taxa, 26 percent (65) may be considered desert species. Among the desert species more than two-thirds have been reduced in number while interacting with introduced forms. Most often these interactions have been coupled with habitat alterations but the influence of nonnative species cannot be denied.

Reduction in numbers of a native fish population by an introduced species may occur by hybridization, predation, and/or competition. Hybridization and predation usually are obvious but competition has been nearly impossible to prove. In the absence of obvious hybridization or predation it appears that most workers intuitively label the cause of replacement as competition.

In the 1930s introduced species began to form a large portion of the desert fauna as native species declined (Deacon and Minckley 1974; Miller 1961; Minckley and Deacon 1968). Among the desert species the following fishes seem particularly to have been influenced by introduced species. The spikedace (*Meda fulgida*) and the loach minnow (*Tiaroga cobitis*) appear to have been replaced by the red shiner (*Notropis lutrensis*) (Minckley and Deacon 1968). Decline of the Pahrump killifish (*Empetrichthys latos*) and extirpation of (*Empetrichthys merriami*) also are related to introduced forms, particularly goldfish (*Carassius auratus*), shortfin mollies (*Poecilia sphenops*), and mosquitofish (*Gambusia affinis*) (Deacon 1964; Pister 1974; Soltz 1979; Soltz and Naiman 1978). The Gila topminnow (*Poeciliopsis occidentalis*) declined as *Gambusia affinis* expanded its range (Miller 1961; Minckley 1973; Minckley and Deacon 1968; Schoenherr 1974). Last, the desert pupfish (*Cyprinodon macularius*) has been reduced in distribution in association with numerous introduced forms. In Arizona *Gambusia affinis* was implicated (Deacon and Minckley 1974), and in California the replacement has occurred in association with the sailfin molly (*Poecilia latipinna*) and Zill's cichlid (*Tilapia zillii*) (Black 1980; Moyle 1976; Schoenherr 1979).

THE MOSQUITOFISH AS A COMPETITOR

Gambusia affinis have been introduced worldwide as a "natural" control of mosquitos, despite the native larvivorous fish species that already inhabit many of the waters (Gerberich and Laird 1965). Many of these introductions may be unnecessary. In California, for example, Danielson (1968) demonstrated that Amargosa pupfish (*Cyprinodon nevadensis*) are superior to *Gambusia affinis* at capturing mosquito larvae, particularly among reeds.

The mosquitofish has been implicated in situations as a competitor probably more than any other fish species. Myers (1965) cited numerous examples of the impact of mosquitofish on other fish populations; for example about 40 years ago the California Department of Fish and Game stopped using mosquitofish as forage because they were eliminating newly hatched game fish (largemouth bass). In San Jose, California, *Gambusia affinis* was considered the primary factor in preventing growth of a goldfish population, ostensibly by predation on young. In regard to the effects on other fishes (most of which also eat mosquito larvae) the list is long and involves species from many parts of the world (Table 1). Furthermore it has been demonstrated that *Gambusia affinis* is such an effective predator on invertebrates that its presence is instrumental in altering trophic relationships in entire ecosystems and causes phytoplankton to increase in abundance as small aquatic invertebrates are removed (Hurlbert et al. 1972).

Hubbs and Springer (1957) considered isolation by drought with competition from *Gambusia affinis* as factors that promote speciation in clear, spring-fed habitats of species in what they defined as the *Gambusia nobilis* species group. They hypothesized that drought and/or invasion by natural means of *Gambusia affinis* into downstream habitats of western Texas and northern Mexico forced forms of the *nobilis* group into isolated headspring areas, where they became differentiated. In addition, Hubbs (1957) speculated that similar competition with *Gambusia affinis* in Menard County, Texas, was responsible for speciation of *Gambusia heterochir*, which he considered to be a member of the *G. nicaraguensis* group. He further proposed a competition hypothesis by noting different food habits in regions of sympatry (Hubbs 1971). *Gambusia gaigei* in Big Bend National Park, one of the *nobilis* group mentioned, has been eliminated several times by interaction with *Gambusia affinis* and has been reestablished from stocks held in the laboratory (Hubbs and Brodrick 1963).

TABLE 1. Fish Species Known to Have Been Reduced in Number by Interaction with *Gambusia affinis*

Species	Locality
Cyprinidae	
Carrasius auratus[a]	San Jose, California
Cyprinodontidae	
Aplocheilus panchax[a]	Bangkok, Thailand
Micropanchax schoelleri[a]	Lake Chad Basin, Egypt
Cyprinodon nevadensis calidae[b]	Tecopa Hot Springs, California
C. n. mionectes[b]	Ash Meadows, Nevada
C. n. pectoralis[b]	Ash Meadows, Nevada
C. radiosus[b]	Owens Valley, California
C. macularius[c]	Salton Sea, California
Empetrichthys merriami[i]	Ash Meadows, Nevada
Neostethidae	
Neostethus siamensis[b]	Chantabun, Thailand
Phenacostethus smithi[a,b]	Bangkok, Thailand
Gulaphallus mirabilis[a,b]	Laguna de Bay, Luzon
Poeciliidae	
Gambusia nobilis[d]	Pecos River, Texas
G. geiseri[d]	San Marcos River, Texas
G. senilis[d]	Rio Conchos, Mexico
G. hurtadoi[d]	Rio Conchos, Mexico
G. alvarezi[d]	Rio Conchos, Mexico
G. gaigei[d,e]	Big Bend National Park, Texas
G. heterochir[f,g]	San Saba River, Texas
G. amistadensis[h]	Rio Grande, Texas
Poeciliopsis o. occidentalis[b]	Gila River, Arizona

(Continued)

Table 1. (Continued)

Species	Locality
Gasterosteidae	
Gasterosteus aculeatus williamsoni[b]	Santa Clara River, California
Centrarchidae	
Micropterus salmoides[a]	Friant, California

[a]Myers (1965).
[b]Miller (1973).
[c]Evermann and Clark (1931).
[d]Hubbs and Springer (1957)
[e]Hubbs and Brodrick (1963).
[f]Hubbs (1957).
[g]Hubbs (1971).
[h]Peden (1973).
[i]Soltz (1979).

REPLACEMENT OF THE GILA TOPMINNOW BY MOSQUITOFISH

Gambusia affinis has been implicated as a competitor responsible for reduction of the Gila topminnow (*Poeciliopsis occidentalis*). In Arizona topminnows and a form of desert pupfish (now extinct) were present and widespread and doubtless quite effective in mosquito control. Elimination of these native larvivores is now well documented (Deacon 1968; Miller 1961; Miller 1973; Miller and Lowe 1964; Minckley 1969a,b, 1973; Minckley and Deacon 1968) but mechanics of the pattern of extirpation has been elusive. Usually competition has been implicated.

A trend toward the decline of the Gila topminnow began before the appearance of mosquitofish, beginning with activities of humans, such as diversion and drying of smaller creeks plus drainage of marshes and larger backwaters (Miller 1961). As late as the 1930s, however, long after initial changes in habitats, *Poeciliopsis* still was considered one of the most common fishes in the southern part of the Colorado River drainage (Hubbs and Miller 1941). Mosquitofish were introduced to California in 1922 and probably made their way to Arizona soon after (Miller and Lowe 1964, 1967). By 1926 *G. affinis* had been collected from the Gila River near its mouth and from the Salt River near Tempe (Miller 1961). From the 1930s on the occurrence of *Gambusia* increased as *Poeciliopsis* decreased.

Actual replacement of *Poeciliopsis occidentalis* by *Gambusia affinis* has been documented in four cases. First, Aravaipa Creek, Pinal County, Arizona, was fishless until about 1936 when, during a malaria outbreak, it was stocked by the Arizona State Health Department with what they thought were mosquitofish from the vicinity of Tucson (Miller 1961). Apparently they were *Poeciliopsis*, not *Gambusia*, because as late as April 12, 1957, Miller found only *Poeciliopsis* in Aravaipa Creek. Less than two years later on March 30, 1959, Gila topminnows were gone entirely and mosquitofish were abundant. Second, Minckley and Deacon (1968) reported a similar situation in ponds and canals fed by artesian springs near Safford, Arizona. In 1962 *Poeciliopsis* abounded. By 1963 specimens of *Gambusia affinis* including some sympatric with *Poeciliopsis*, were taken in the area. In 1966 an intensive study of the entire artesian system yielded only mosquitofish. Third, Minckley (1969a) recorded attempts to reestablish *Poeciliopsis* in its former range that were thwarted by later introductions of *Gambusia affinis*. Such instances near Tempe, Arizona, include introductions in 1965 into ponds in the Salt River bed, on the Pagago golf course, and in Papago Park. By 1966 *Gambusia affinis*, not *Poeciliopsis*, were found in the ponds. Fourth, in April 1969, a pond at the Arizona Game and Fish Department regional Headquarters in Deer Valley was stocked with Gila topminnows. A collection made on September 26, 1969, disclosed that *Gambusia affinis* had been introduced. The pond was swarming with *Poeciliopsis*, but one male and one female *Gambusia* were collected with 143 *Poeciliopsis*. By December 1970 Gila topminnows were all but eliminated, for the stock was represented only by a few large, old females. Figure 1 records this replacement.

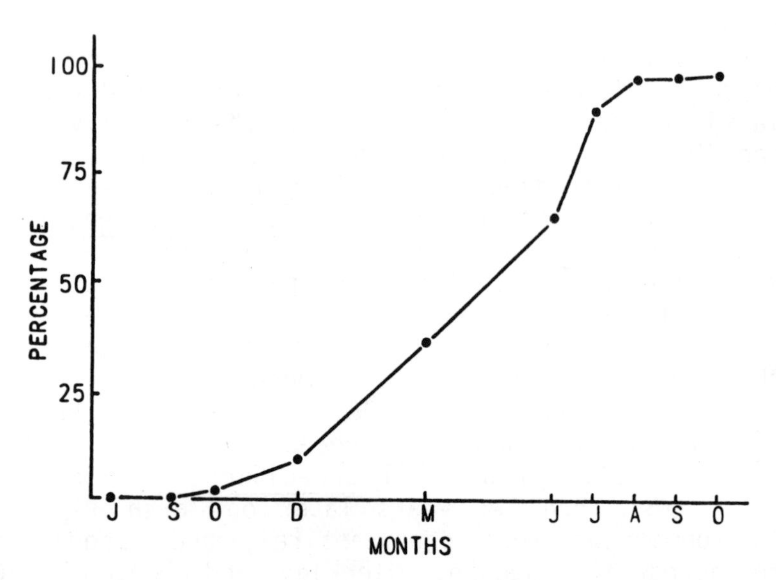

FIGURE 1. Percentage of Gambusia affinis in collections from an experimental pond in Deer Valley. Letters indicate months from July 1968 through October 1969.

Minckley et al. (1977), however, reported the co-occurrence of mosquitofish and Gila topminnows in southern Arizona in Sonoita Creek, and the upper reaches of the Santa Cruz River. In each case it was apparent that the source of the topminnows was an upstream population in a springhead. It appears that mosquitofish are selected against in clear spring-fed waters high in carbonates when those waters are occupied by Poeciliopsis. The situation is similar to that mentioned above for members of the genus Gambusia in Texas.

It is clear, however, that in waters other than springheads Gambusia has a selective advantage and rapidly replaces native species. In an attempt to isolate the actual mechanism of replacement, I conducted an intensive analysis of the life history of *Poeciliopsis occidentalis*, both in mono-species populations and during replacement by *Gambusia*. Analysis of fishes in natural and artificial ponds and in aquaria were conducted. I compared the life history of *Poeciliopsis occidentalis*, revealed by my research of *Gambusia affinis* and reported in the extensive literature. I also conducted studies in which comparative life histories during cohabitation were analyzed (Schoenherr 1974).

Gila topminnows and mosquitofish are similar in size, morphology, and habitat preferences, which might lead us to assume that the mechanism of replacement is competition. Predation on fry, documented by Myers (1965) for mosquitofish on other fish species, remained as an alternative.

Comparison of the relationships between the two species may be summarized as follows:

1. Both species prefer marshes, sloughs, and other quiet waters and have approximately the same notable tolerances for environmental extremes (Brett 1956; Heath 1964; Hubbs 1957, 1959; Hubbs and Springer 1957; Krumholz 1948; Jordan 1927; Minckley 1973; Minckley et al. 1977; Schoenherr 1974, 1977; Sicault 1939; Simpson and Gunter 1956; Smith 1960). *Gambusia affinis* can tolerate colder water than *Poeciliopsis* and avoids clear spring-fed waters high in carbonates, particularly when they are occupied by other species of poeciliids.

2. *Gambusia affinis* has a higher reproductive potential than *Poeciliopsis*. *Gambusia* brood sizes are larger (Figure 2). *Poeciliopsis* has superfetation, therefore a shorter interval between broods, but that does not compensate for the large number of fry produced by each *Gambusia* female. Brood sizes of *Poeciliopsis* dropped off noticeably during the height of interaction between the two species (Figure 3).

3. When the two species coexist mate selection of conspecific females (Table 2) and sperm storage prevent gamete waste. Although pregnancy of topminnow females appeared to occur normally, there is no evidence that the young were born.

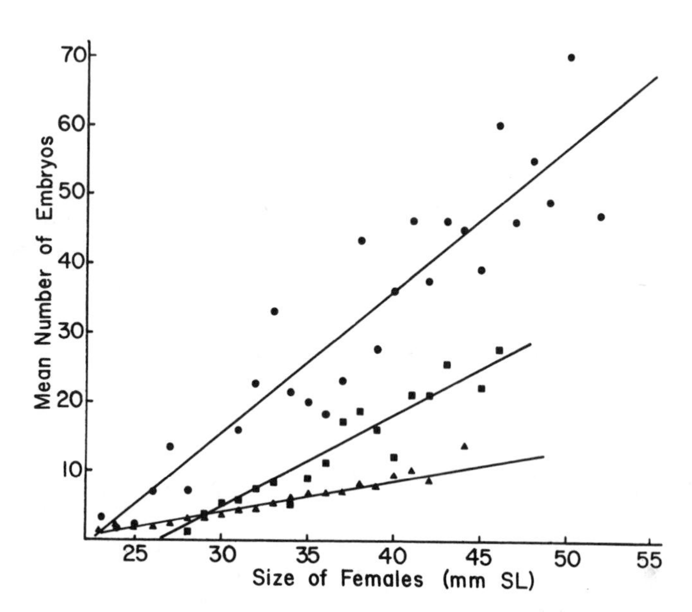

FIGURE 2. Comparison of mean numbers of embryos per size group of females of Poeciliopsis occidentalis and Gambusia affinis in Arizona. Closed circles indicate G. affinis from Deer Valley, squares indicate P. occidentalis from Deer Valley, and triangles indicate P. occidentalis from constant temperature Monkey Spring where G. affinis did not occur. Lines fitted by least squares regression (from Schoenherr, Ecology 58: 438, Copyright 1977 by the Ecological Society of America).

4. Gambusia females grew more rapidly and reached greater size than Poeciliopsis but during cohabitation surviving Poeciliopsis seemed to grow at normal rates (Figure 4). Note that recruitment of young Poeciliopsis did not occur after Gambusia affinis appeared in the pond.

5. In monospecies populations Poeciliopsis feeds mainly on detritus and aquatic arthropods, whichever is abundant (Schoenherr 1974). Gut length is 1.5 to 2.0 times its standard length, which implies that it is adapted for a diet high in roughage. Cohabitation seemed to have little influence on the food of

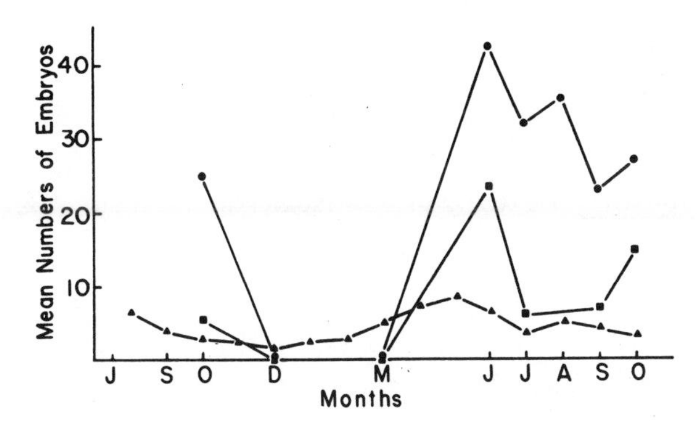

FIGURE 3. Seasonal variation in mean numbers of embryos from July 1968 through October 1969. Symbols as in Figure 2 indicate a comparison between *Gambusia affinis* and *Poeciliopsis occidentalis* in Deer Valley and Monkey Spring. Note the severe reduction in mean numbers of embryos for *Poeciliopsis* after June 1969 (from Schoenherr, Ecology 58: 438, Copyright 1977 by the Ecological Society of America).

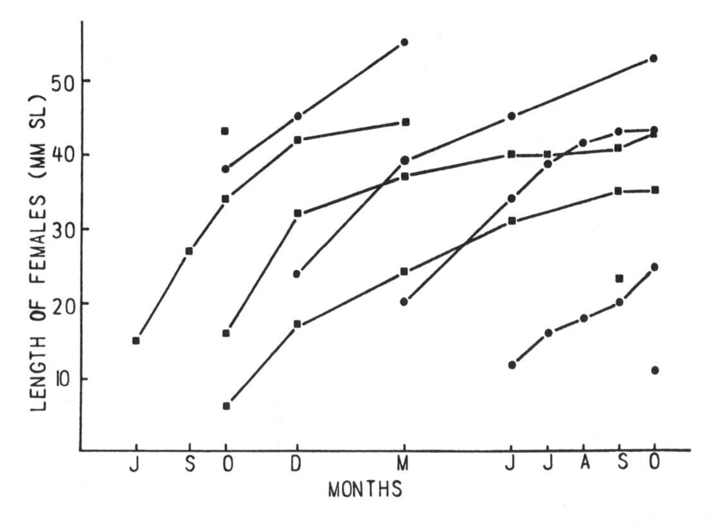

FIGURE 4. Comparative growth rates of *Poeciliopsis occidentalis* and *Gambusia affinis* in an experimental pond at Deer Valley. Closed circles indicate *Gambusia affinis* and squares indicate *Poeciliopsis occidentalis*. Letters indicate months from July 1968 through October 1969. Lines derived by connecting mode values from length-frequency data.

TABLE 2. Mate Preference of *Poeciliopsis occidentalis* Males and *Gambusia affinis* Males for Conspecific Versus Heterospecific Females

Total Distribution of Behavior Time (Minutes)				
Male	Trial	*Poeciliopsis* Female	Nondirected	*Gambusia* Female
Poeciliopsis 1	a	8.2	4.7	2.1
	b	5.9	8.3	0.8
	c	2.0	13.0	0.0
Poeciliopsis 2	a	7.6	6.7	0.7
	b	7.0	4.7	3.3
	c	0.0	15.0	0.0
Poeciliopsis 3	a	5.1	7.8	2.1
	b	3.9	9.2	1.9
	c	0.9	14.1	0.0
Poeciliopsis 3 (Same male as above)	a	2.3	12.7	0.0
	b	1.4	13.6	0.0
	c	0.0	15.0	0.0
Total Time		44.3	104.8	10.9
Total Time (%)		27.7	65.5	6.8
Directed Time (%)		80.3	-	19.7
Gambusia 1	a	4.2	4.3	6.5
	b	4.9	4.2	5.9
	c	7.1	3.7	4.2
Gambusia 2	a	2.5	0.0	12.5
	b	2.1	8.2	4.7
	c	6.2	0.7	8.1
Gambusia 3	a	3.7	1.2	10.1
	b	4.8	3.5	6.7
	c	6.2	9.6	0.0
Total Time		40.9	45.4	58.7
Total Time (%)		28.2	31.3	40.5
Directed Time (%)		41.1	-	58.9

Poeciliopsis. During cohabitation Gambusia fed primarily on arthropods and Poeciliopsis fed mainly on detritus (Table 3) but food was abundant and probably not limiting. Three forms of indirect evidence implied that food was not limiting: phytoplankton were abundant in the water; food was found in the stomachs of preserved specimens; and sex ratios continued to approximate 1:1. Gut length of Gambusia was only 69 percent of its standard length. In addition, Gambusia has open cephalic canals and strong conical teeth (Rosen and Mendelson 1960), which further implies a preferred carnivorous habit. Presence of young fish in gut contents of Gambusia implicated predation on fry as a component in the mode of replacement. Observations of Gambusia in aquaria confirm that they prey on the fry of Poeciliopsis as well as their own. Paul Winkler (personal communication) has observed, however, that when Gambusia affinis and Poeciliopsis are together in aquaria Poeciliopsis stop feeding and die within 60 days.

6. In aquaria a behavioral hierarchy was observed. Poeciliopsis males dominated Gambusia males but Gambusia females dominated over all others. In mixed populations in aquaria Poeciliopsis females seem to be chased constantly, particularly by Gambusia females.

7. Mortality of Poeciliopsis females occurred at a high rate during cohabitation with Gambusia, but a similar trend was not observed in Gambusia. Paul Winkler (personal communication) confirmed this observation and noted further that in an aquarium in which the two species were separated by a divider, mortality of Poeciliopsis did not occur until the divider was removed.

In the final analysis it appears that Gambusia affinis, aided by rapid growth and high reproductive potential, eliminates Poeciliopsis occidentalis during cohabitation by predation on fry and by reducing survivorship potential of adult females, likely by physiological stress due to aggression. Because food did not appear to be limiting, this interaction may be interpreted as competition for space.

TABLE 3. Comparison of Food Items in Stomachs of _Gambusia affinis_ and _Poeciliopsis occidentalis_ in Arizona

Locality	Filamentous Chlorophyta	Diatoms	Cladocera	Ephemeroptera	Tendipedidae (Chironomidae)	Heleidae (Ceratopogonidae)	Culicidae	Tipulidae	Terrestrial Arthropods	Fishes	Organic Detritus	Inorganic Debris	Empty Stomachs	Number of Stomachs
Percentage Frequency of Occurrence for Food Items in Stomachs														
G. affinis														
Deer Valley pond	30.8	tr	38.4	15.4	67.3	30.8	11.6	5.8	80.8	9.8	9.6	7.7	3.8	52
Page Spring raceway	33.3	6.7	50.0	-	3.3	13.3	16.7	-	70.0	10.0	33.3	6.7	6.7	30
P. occidentalis														
Deer Valley pond	76.4	tr	-	-	tr	-	tr	-	-	-	91.2	tr	18.7	64
Page Spring raceway	52.9	82.4	-	-	-	-	-	-	-	-	81.3	82.4	4.1	34
Percentage Total Relative Volume for Food Items in Stomachs														
G. affinis														
Deer Valley pond	4.2	tr	8.8	2.1	17.2	6.6	4.9	tr	47.1	6.6	1.5	tr		52
Page Spring raceway	3.3	tr	66.3	-	tr	-	5.4	-	12.3	5.4	6.3	tr		30
P. occidentalis														
Deer Valley pond	24.2	tr	-	-	tr	-	tr	-	-	-	73.7	tr		64
Page Spring raceway	19.3	tr	-	-	-	-	-	-	-	-	79.3	2.4		34

REPLACEMENT OF DESERT PUPFISH BY INTRODUCED SPECIES

The disappearance of the desert pupfish (*Cyprinodon macularius*) in Arizona has followed a pattern similar to that of *Poeciliopsis occidentalis* (Miller 1961; Minckley 1973). Implicated in its rapid decline have been the introduction of several nonnative fishes, notable among which is *Gambusia affinis* (Deacon and Minckley 1974; Everman and Clark 1931; Fisk 1972). Introduction and subsequent eradication of the golden shiner (*Notemigonus crysoleucas*) in the last stronghold of *C. macularius* in Quitobaquito Spring illustrates that even in the confines of a national monument protection from introductions is not always possible. In Arizona a related form, the Monkey Spring pupfish (*Cyprinodon* sp.), was extirpated by largemouth bass (*Micropterus salmoides*) introduced in 1969 (Minckley 1973; Schoenherr 1974).

In the Salton Sea and its tributaries, *Cyprinodon macularius* are being replaced rapidly by introduced fishes (Schoenherr 1979; Black 1980). Researchers studying the area formed a committee in 1977 that led to a listing package to consider *Cyprinodon macularius* for endangered species status. Quarterly surveys since March 1978 have revealed that desert pupfish constitute less than 5 percent of the fauna. The total number of introduced species in freshwater tributaries to the Salton Sea is now at least 22 species (Schoenherr 1979). In particular, the sailfin molly (*Poecilia latipinna*) and Zill's cichlid (*Tilapia zillii*) have replaced the pupfish in nearly all habitats except shallow, warm portions of irrigation ditches and San Felipe Creek, a tributary connected to the Salton Sea only during flooding. The mechanism of this replacement is not totally understood but replacement by behavioral means appears to be important. Margaret Matsui (personal communication) reported that in both instances it is a behavioral interaction that interferes with *Cyprinodon macularius*. Males of *Poecilia latipinna* court both male and female members of the *Cyprinodon* population and in so doing directly interfere with reproductive success. On the other hand, *Tilapia zillii* juveniles passively interfere with male

Cyprinodon because the latter constantly chase away the juvenile Tilapia; therefore spending considerable time patroling their territories while less time is spent courting, thus lowering reproductive success. Paul Loiselle (personal communication) reported that male Cyprinodon macularius are able to recognize their own spawn by its odor and that males will eat the fertilized eggs of others. It is possible, therefore, that while a dominant male is busy chasing juvenile Tilapia other male Cyprinodon, particularly subdominant males resembling females in color, are able to intrude and eat the eggs. Furthermore, adult Tilapia may prey on young Cyprinodon. In the Salton Sea area it appears that the only refuge for Cyprinodon, which is water less than 10 cm deep, is now being threatened by juvenile Tilapia. It remains to be seen, however, if Cyprinodon can tolerate higher temperatures than juvenile Tilapia, in which case shallow hot water may become the only refuge. In the King Street canal described below Cyprinodon was common in water 5 to 10 cm deep at 39°C (Schoenherr 1979).

In the King Street irrigation drain on the northwest corner of the Salton Sea Cyprinodon coexisted with four species of introduced fishes during March 1977 (Schoenherr 1979). Marked differences in flow, turbulence, and temperature had provided a diversity of habitats and each species was segregated into nearly pure species populations (Table 4). Seine hauls in each area yielded better than 99 percent of each dominant species. In this circumstance niche partitioning was evident, and it may be inferred from observations already noted that mechanisms of exclusion are related to a mixture of tolerances, preferences, and behavioral interactions and that segregation was rapid. In March 1978, after flooding and reconstruction, the canal had been severely altered to the degree that there were essentially only two habitats: flowing shallow water and a pond. Poecilia sphenops, Notropis lutrensis, and Cyprinodon were taken from the flowing water and Cyprinodon inhabited the pond with Gambusia affinis and Poecilia latipinna. With spatial diversity reduced, the species became sorted into less specialized niches. In July 1978 another visit to the area revealed adult

TABLE 4. Habitat Preferences for Fishes Inhabiting the King Street Canal, Riverside County, California, on March 3, 1977 (from Schoenherr, 1979, Bulletin of the Southern California Academy of Sciences 78: 46)

Species	Water Temperature (°C)	Diss. O_2 (mgl^{-1})	Depth (cm)	Flow	Physical Habitat
Cyprinodon macularius	39	6.8	5-10	Slow	Shallow wash
Gambusia affinis	32	7.0	25	Moderate	Stream margin
Poecilia sphenops	22	8.5	18	Rapid	Stream source
Poecilia latipinna	26	8.5	50	Slow	Undercut banks
Notropis lutrensis	26	8.7	8-15	Rapid	Riffles

Tilapia zillii had been introduced to the pond. At this time, along with juvenile *Tilapia*, mosquitofish, and sailfin mollies, *Cyprinodon* was taken only in shallow water at the margin of the pond. In March 1980 the area was visited again. Habitat diversity had returned and abundant vegetation had been reestablished. Species sorting was as it appeared in March 1977 except that the shallow wash habitat was occupied primarily by juvenile *Tilapia zillii*. *Cyprinodon macularius* was present but in apparently reduced numbers. Two hours of seining yielded only four adult specimens, taken from the margin of the wash among the roots of cattails. In the King Street Canal replacement by *Tilapia zillii* may now be complete.

DISCUSSION

This chapter may pose more questions than it answers. In particular, we must consider the definition of competition. Numerous definitions of competition have been proposed (Weatherly 1972). Most seem to agree with Larkin (1956), who suggested that competition is the demand, typically at the same time, of more than one organism for the same resource(s). When competition becomes labeled as the cause of species replacement we must ask about the common resource. In none of the cases mentioned above can it be demonstrated that a common resource (such as food) was being used by one species more efficiently than another, which is not to say that common resource utilization never has been demonstrated in fishes. In that respect certain studies on salmonids (Andrusak and Northcote 1971; Nilsson 1960, 1963; Schutz and Northcote 1972), and on centrarchids (Werner and Hall 1976, 1977) have demonstrated that competition for a common food source was the apparent mechanism causing coexisting species to shift to different food sources, a form of niche partitioning.

It is noteworthy, however, that in spite of numerous examples of one fish species causing sharp reductions in abundance of another, actual competition rarely has been demonstrated (Larkin 1956; Ivlev 1961).

Weatherly (1972), in defining competition primarily as the demand for common foods, implies that it probably never truly occurs in fishes. It is often the case that part of the difficulty lies in how one perceives competition. Are behavioral interactions forms of competition? Is aggression competition for space?

Many biologists think of competition as comprehensive interactions that take place not only in relation to resources but also as encompassing all commonly occupied space-time of an ecosystem. Within this frame of reference, when competition occurs between species, theoretically species better suited to the endeavor ultimately will extirpate less effective forms. This has been called the principle of competitive exclusion, or Gause's axiom (Gause 1934). Since that time numerous experiments have been conducted in the laboratory (Park 1948, 1954) and in the field (Connell 1961; Harger 1972) to show that extirpation need not occur and under certain circumstances "coexistence" is possible, despite apparent competition. In many of these studies it became apparent that the most important criterion for coexistence of similar species was environmental diversity (Ayala 1972; Miller 1965), although some experiments illustrated that coexistence occurred, at least for a short time in instances of superabundance of food (Fryer 1959; Ross 1957; Weatherly 1963).

Hutchinson (1957), in perceiving the ecologic niche as an *n*-dimensional hypervolume, generalized a concept useful in competition analysis. McNaughton and Wolf (1970) pointed out that most modern niche theory derives from efforts to relate the competitive exclusion principle to Hutchinson's *n*-dimensional niche. For example, in situations displaying environmental diversity, rather than extirpation, coexistence is brought about by niche partitioning and/or character displacement. These two phenomena are highly interrelated. In complex ecosystems many niches are available, and when similar species occur together it is possible for them to "coexist" by use of different portions of the total niche. Under selection pressures reproduction is favored among those forms best suited to the partitioned portions of the niche; hence character displacement logically follows niche partitioning.

These phenomena have been demonstrated in many desert fishes and the phenomenon of competitive exclusion is believed to be the cause. Among the examples is that of Miller (1965), who discussed the occurrence of four closely related species of whitefish (*Prosopium*) in Bear Lake, Utah. These species feed and breed in different portions of the lake and are separated by size and breeding season. Niche partitioning is manifested by different habitat preferences and different breeding seasons, whereas character displacement is exemplified by morphological and behavioral differences. Among poeciliids character displacement has been demonstrated in the *Gambusia nobilis* species group in which different species have evolved in association with drought and exclusion by *Gambusia affinis* (Hubbs and Springer 1957). Niche partitioning has been demonstrated by division of the breeding season for sympatric populations of *Poecilia latipinna* and the all-female form *Poecilia formosa* which competes for *latipinna* males (Hubbs 1964). *Poecilia latipinna* males tend to mate first with *latipinna* females and later with *formosa* females. Harrington and Harrington (1961) demonstrated that several poeciliid species coexisted in a salt marsh by feeding on food items that included mosquito larvae and pupae in different stages of development. May we infer that competition in these cases is responsible for the observed avoidance phenomena?

Pristine desert habitats are occupied typically by few species and when they occur single-species populations exploit all available space (Deacon and Minckley 1974). When several species occur together they often segregate into distinct niches, as in Aravaipa Creek, Arizona, where seven native species are ecologically allopatric (Barber and Minckley 1966), or in the Cuatro Cienegas area of Mexico, where 10 to 14 species live in large springheads (Minckley 1969c). Niche separation among introduced species in the King Street canal near the Salton Sea, California, illustrated that the degree of separation was related directly to habitat diversity.

When few species exist in an area (e.g., most desert habitats) introduced species seem to establish themselves easily, by causing native species to occupy a fragment of their former range or extirpating them entirely. In the Salton Sea area introduced species have been segregated into rather specific habitats in the irrigation drains (Schoenherr 1979) and the desert pupfish (Cyprinodon macularius) has become greatly reduced in numbers. Desert pupfish now occupy only a portion of their former habitat and appear to have been forced into shallow warm water by behavioral interactions with sailfin mollies (Poecilia latipinna) and Zill's cichlids (Tilapia zillii). It remains to be seen whether this habitat will be a refuge from the Tilapia. In Arizona, interactions with mosquitofish (Gambusia affinis) have forced the native Gila topminnow (Poeciliopsis occidentalis) from their preferred habitat in sloughs and backwaters into clear, carbonaterich springheads.

As near to true coexistence as two species can become is probably represented by those forms that are separated by behavioral features. Wynne-Edwards (1965) has emphasized that social behavior also is important in character displacement. In poeciliids, among which internal fertilization is necessary, this displacement is represented by different patterns of mating behavior and mate discrimination. These roles also have been studied as isolating factors in Poecilia (Liley 1966), Xiphophorus (Clark et al. 1954; Schlosberg et al. 1949) Lebistes (Haskins and Haskins 1949), and Gambusia (Arnold 1966; Hubbs and Delco 1960, 1962; Peden 1970; Warburton et al. 1957). Sympatric populations of closely related forms seem to retain genetic identity by different kinds of mating behavior and effective mate discrimination. Is the breakdown of mate discrimination a form of competition? May we consider the result of breakdown as competition for mates?

Among the behavioral interactions that cause niche partitioning in nature is aggression. The loser is expelled from a territory and resource utilization is maximized by the winner. Lorenz (1963) and Eibl-Eibesfeldt (1961) suggest that aggressive activity dispurses animals in the environment and prevents overuse of resources. If food is abundant, is aggression competition for space? If food were abundant, why would live-bearing poeciliid fishes fight for space?

As exemplified by studies of thrushes in the eastern United States birds, breeding territories are established each spring by fighting within the first week after their return from southern wintering grounds (Morse 1971). Therefore energy is not wasted in fighting after the first week. Diamond (1978) noted that it is highly unlikely that fighting would ever be observed in a tropical situation because birds are on their territories year-round and live longer than birds in temperate zones. Compare them with desert fishes. Although most desert fishes are short-lived, many species do maintain year-round territories. Thus it may be difficult to demonstrate actual competition for a resource, whereas aggressive interactions are readily observable by experimentation in aquaria. In extirpation, in which a common resource is not demonstratable, it is highly likely that the competitive interaction is behavioral. Ironically, our best evidence of that relationship is that because of short-term aggression the two species seem not to interact.

According to that logic, perhaps we are justified, in the absence of a demonstratable shared resource such as food, to use niche partitioning as indirect though valid evidence that competition occurs. Most experimentation in which competition for a shared resource has led to exclusion has been on invertebrates such as flour beetles (*Tribolium*) which lack complex behavioral interactions (Park 1948, 1954). It is in social animals like fishes and birds that behavioral phenomena (e.g., mate discrimination and aggression) rapidly lead to niche partitioning, and the actual period of direct competition is so reduced in time that it is rarely demonstratable. Gatz (1979), in constructing a mathematical model to explain the relationship of fish morphology to community organization, has developed several hypotheses to explain his observations of niche partitioning and character displacement. He concluded that competition cannot be established as the cause of the observed separations but we must assume that it is the most realistic choice. Whatever the case, in the absence of direct proof, it may be interpreted as rare in aquatic habitats, or simply it may be a manifestation of a rapid, difficult-to-prove phenomenon.

Larkin (1956) contended that freshwater habitats are geologically short-lived, highly diversified, and subject to rapid ecological succession, all of which tend to promote generalization rather than specialization in fishes. Pristine desert waters are ecologic islands characterized by a paucity of species that often are ecologically segregated. As frequently found among island species, they are adapted superbly for coping with the physical environment but unable to contend with intruders. If generalized species with wide ranges of tolerances such as carp, goldfish, mosquitofish, and sailfin mollies are introduced, they find it relatively easy to become established, especially by aggressive behavior. The species that avoid extirpation are those that escape to a niche too extreme for the generalist to tolerate. In desert habitats these refugia may be carbonate or carbon dioxide-rich headsprings, or they may be areas too shallow, too hot, too alkaline, oxygen-poor, or any of these conditions in combination.

The fact remains that in spite of overwhelming evidence that it is harmful to native species, introductions of new fish species continue. If the scientific community is to play a significant role in the preservation of native species and the prevention of introductions of unnecessary nonnative species, we must continue to expand our studies of life histories including behavioral and ecological interactions, and use the results to discourage this practice. A case in point: why must we continue to introduce *Gambusia affinis* to eat mosquito larvae when there are native species to do the job?

REFERENCES

Andrusak, H. and T. G. Northcote. 1971. Segregation between adult cutthroat trout (*Salmo clarki*) and dolly varden (*Salvelinus malma*) in small coastal British Columbia lakes. Journal of the Fisheries Research Board of Canada 28:1259-1268.

Arnold, E. T. 1966. Comparative studies of mating behavior in gambusiin fishes (Cyprinodontiformes: Poecilidae). M. S. Thesis, Arizona State University, Tempe.

Ayala, F. J. 1972. Competition between species. American Scientist 60:348-357.

Barber, W. E. and W. L. Minckley. 1966. Fishes of Aravaipa Creek, Graham and Pinal Counties, Arizona. Southwestern Naturalist 11:313-324.

Barnickol, P. E. 1941. Food habits of *Gambusia affinis* from Reelfoot Lake, Tennessee, with special reference on malaria control. Journal of the Tennessee Academy of Science 16:5-13.

Black, G. F. 1980. Status of the desert pupfish, *Cyprinodon macularius* (Baird and Girard), in California. State of California, Department of Fish and Game, Inland Fisheries Endangered Species Program, Special Publication 80-1:1-42.

Brett, J. R. 1956. Some principles in the thermal requirements of fishes. Quarterly Review of Biology 31:75-81.

Clark, E., L. R. Aronson and M. Gordon. 1954. Mating behavior patterns in sympatric species of xiphophorin fishes, their inheritance and significance in sexual isolation. Bulletin of the American Museum of Natural History 103:135-226.

Connell, J. H. 1961. The influence of interspecific competition and other factors on the distribution of the barnacle *Chthamalus stellatus*. Ecology 42: 710-723.

Danielson, T. L. 1968. Differential predation on *Culex pipiens* and *Anopheles albimanus* mosquito larvae by two species of fish (*Gambusia affinis* and *Cyprinodon nevadensis*) and the effects of simulated reeds on predation. Ph.D. Dissertation. University of California, Riverside.

Deacon, J. E. 1968. Endangered non-game fishes of the West: causes, prospects, and importance. *In*: Proceedings of the 48th Annual Conference of the Western Association of Game Fish Commission, pp. 534-549.

Deacon, J. E., C. Hubbs and B. J. Zahuranec. 1964. Some effects of introduced fishes on the native fish fauna of southern Nevada. Copeia 1964:384-388.

Deacon, J. E., G. Kobetich, J. D. Williams and S. Contreras. 1979. Fishes of North America endangered, threatened, or of special concern: 1979. Fisheries 4:29-44.

Deacon, J. E. and W. L. Minckley. 1974. Desert Fishes. In: G. W. Brown, Jr. (Ed.), Desert Biology, Vol. II. Academic, New York, pp. 385-488.

Diamond, J. M. 1978. Niche shifts and the rediscovery of interspecific competition. American Scientist 66:322-331.

Eibl-Eibsfeldt, I. 1961. The fighting behavior of animals. Scientific American. 205:112-123.

Evermann, B. W. and H. W. Clark. 1931. A distributional list of the species of freshwater fishes known to occur in California. California Department of Fish and Game, Fish Bulletin 35:1-67.

Fisk, L. O. 1972. Status of certain depleted inland fishes. California Department of Fish and Game Inland Fisheries Administration Report 72-1:1-13.

Fry, F. E. J. 1960. The oxygen requirements of fish. In: C. M. Tarzwell (Ed.), Biological Problems in Water Pollution. Public Health Service Technical Report W60-3, pp. 106-109.

Fryer, G. 1959. The trophic interrelationships and ecology of some littoral communities of Lake Nyassa with special reference to the fishes, and a discussion of the evolution of a group of rock-frequenting cichlidae. Proceedings of the Zoological Society of London 132:153-281.

Gatz, J. A. Jr. 1979. Community organization in fishes as indicated by morphological features. Ecology 60:711-717.

Gause, G. F. 1934. The struggle for existence. Williams and Wilkins, Baltimore.

Gerberich, J. B. and M. Laird. 1965. An annotated bibliography of papers relating to the control of mosquitos by the use of fish. World Health Organization WHO/EBL/66.71:1-107.

Harger, J. R. E. 1972. Competitive coexistence among intertidal invertebrates. American Scientist 60:600-607.

Harrington, R. W. Jr. and E. D. Harrington. 1961. Food selection among fishes invading a high subtropical salt marsh: from onset of flooding through the progress of a mosquito brood. Ecology 42: 646-666.

Haskins, C. P. and E. F. Haskins. 1949. The role of sexual selection as an isolating mechanism in three species of poeciliid fishes. Evolution 3: 160-169.

Heath, W. G. 1964. Maximum temperature tolerance as a function of constant temperature acclimation in the Gila topminnow, _Poeciliopsis occidentalis_. Ph.D. Dissertation, University of Arizona, Tucson.

Hubbs, C. L. and R. R. Miller. 1941. Studies of the fishes of the order Cyprinodontes. XVII. Genera and species of the Colorado River system. Occasional Papers of the Museum of Zoology of the University of Michigan 433:1-9.

Hubbs, C. 1957. _Gambusia heterochir_ a new poeciliid fish from Texas, with an account of its hybridization with _Gambusia affinis_. Tulane Studies in Zoology 5:3-16.

Hubbs, C. 1959. High incidence of vertebral deformities in two natural populations of fishes inhabiting warm springs. Ecology 40:154-155.

Hubbs, C. 1964. Interactions between a bisexual fish species and its gynogenetic sexual parasite. Bulletin of the Texas Memorial Museum 8:1-72.

Hubbs, C. 1971. Competition and isolation mechanisms in the _Gambusia affinis_ X _Gambusia heterochir_ hybrid swarm. Bulletin of the Texas Memorial Museum 19:1-47.

Hubbs, C. and H. J. Brodrick. 1963. Current abundance of _Gambusia gaigei_, an endangered fish species. Southwestern Naturalist 8:46-48.

Hubbs, C. and J. L. Deacon. 1964. Additional introductions of tropical fishes into southern Nevada. Southwestern Naturalist 9:249-252.

Hubbs, C. and E. A. Delco, Jr. 1960. Mate preference in males of four species of gambusiin fishes. Evolution 2:145-152.

Hubbs, C. and E. A. Delco, Jr. 1962. Courtship preference of *Gambusia affinis* associated with the sympatry of the parental populations. Copeia 1962:396-400.

Hubbs, C. and V. G. Springer. 1957. A revision of the *Gambusia nobilis* species group, with descriptions of three new species, and notes on their variation, ecology, and evolution. Texas Journal of Science 9:279-327.

Hurlbert, S. H., J. Zedler and D. Fairbanks. 1972. Ecosystem alteration by mosquitofish (*Gambusia affinis*) predation. Science 175:639-641.

Hutchinson, G. E. 1957. Concluding remarks. Cold Spring Harbor Symposium on Quantitative Biology 22:415-427.

Ivlev, V. S. 1961. Experimental Ecology of the Feeding of Fishes. Yale University Press, New Haven, Conn.

Jordan, D. S. 1927. The mosquitofish (*Gambusia*) and its relation to malaria. Smithsonian Report 1926: 361-368.

Krumholz, L. A. 1948. Reproduction in the western mosquitofish, *Gambusia affinis affinis* (Baird and Girard) and its use in mosquito control. Ecological Monographs 18:1-43.

Larkin, P. A. 1956. Interspecific competition and population control in freshwater fish. Journal of the Fisheries Research Board of Canada 13:327-342.

Liley, N. R. 1966. Ethological isolating mechanisms in four sympatric species of poeciliid fishes. Behavioral Supplement 13:1-197.

Lorenz, K. 1963. On Aggression. Bantam Books, New York.

McNaughton, S. J. and L. L. Wolf. 1970. Dominance and the niche in ecological systems. Science 167: 131-139.

Miller, R. R. 1961. Man and the changing fish fauna of the American Southwest. Papers of the Michigan Academy of Science, Arts, and Letters 46:365-404.

Miller, R. R. 1965. Quaternary freshwater fishes of North America. In: H. E. Wright, Jr. and D. G. Grey (Eds.), The quaternary of the united States: A review volume for the 7th Congress of the International Association for Quaternary Research. Princeton University Press, Princeton, NJ, pp. 569-581.

Miller, R. R. 1973. Red data book, Volume 4 -- Pisces, freshwater fishes. International Union for Conservation of Nature and Natural Resources, Survival Service Commission. Morges, Switzerland.

Miller, R. R. and C. H. Lowe. 1964. Annotated checklist of the fishes of Arizona. In: C. H. Lowe (Ed.), Vertebrates of Arizona. University of Arizona Press, Tucson, pp. 133-151.

Miller, R. R. and C. H. Lowe 1967. Fishes of Arizona. In: C. H. Lowe (Ed.), Vertebrates of Arizona. University of Arizona Press, Tucson, pp. 133-151.

Minckley, W. L. 1969a. Attempted re-establishment of the gila topminnow within its former range. Copeia 1969:193-194.

Minckley, W. L. 1969b. Native Arizona Fishes, Part I, Livebearers. Wildlife Views 16:6-8.

Minckley, W. L. 1969c. Environments of the Bolson of Cuatro Cienegas, Coahuila, Mexico, with special reference to the aquatic biota. Texas Western Press, University of Texas at El Paso, Science Series 2:1-65.

Minckley, W. L. 1973. Fishes of Arizona. Arizona Game and Fish Department, Phoenix.

Minckley, W. L. and J. E. Deacon. 1968. Southwestern fishes and the enigma of "Endangered Species." Science 159:1424-1432.

Minckley, W. L., J. N. Rinne and J. E. Johnson. 1977. Status of the Gila topminnow and its co-occurrence with mosquitofish. United States Deparment of Agriculture, Forest Service Research Paper RM-198: 1-8.

Moore, W. S., R. R. Miller and R. J. Schultz. 1970. Distribution, adaptation and probable origin of an all-female form of *Poeciliopsis* (Pisces: Poeciliidae) in northwestern Mexico. Ecology 24:789-795.

Morse, D. H. 1971. Effects of the arrival of a new species upon habitat utilization by two forest thrushes in Maine. Wilson Bulletin 83:57-65.

Moyle, P. B. 1976. Inland Fishes of California. University of California Press, Berkeley.

Myers, G. S. 1965. *Gambusia* the fish destroyer. Tropical Fish Hobbyist 13:31-32, 53-54.

Nilsson, N. 1960. Seasonal fluctuations in the food segregation of trout, char and whitefish in 14 North-Swedish lakes. Report of the Institute for Freshwater Research, Drottingholm 41:185-205.

Nilsson, N. 1963. Interaction between trout and char in Scandinavia. Transactions of the American Fisheries Society 92:276-285.

Park, T. 1948. Experimental studies of interspecies competition. 1. Competition between populations of the flour beetles *Tribolium confusum* (Duval) and *Tribolium castaneum* (Herbst). Ecological Monographs 18:265-308.

Park, T. 1954. Experimental studies of interspecies competition. 2. Temperature, humidity, and competition of two species of *Tribolium*. Physiological Zoology 27:177-238.

Peden, A. E. 1970. Courtship behavior of *Gambusia* (Poeciliidae) with emphasis on isolating mechanisms. Ph.D. Dissertation. University of Texas, Austin.

Peden, A. E. 1973. Virtual extinction of *Gambusia amistadensis* n. sp., a poeciliid fish from Texas. Copeia 1973:210-221.

Pister, E. P. 1974. Desert fishes and their habitats. Transactions of the American Fisheries Society 103:531-540.

Rice, L. A. 1941. *Gambusia affinis* in relation to food habits from Reelfoot Lake, with special emphasis on malaria control. Journal of the Tennessee Academy of Science 16:77-87.

Rosen, E. E. and J. R. Mendelson. 1960. The sensory canals of the head in poeciliid fishes (Cyprinodontiformes) with reference to dentitional types. Copeia 1960:203-210.

Ross, H. 1957. Principles of natural co-existence indicated by leafhopper populations. Evolution 11:113-129.

Schlosberg, J., M. C. Duncan and B. H. Daitch. 1949. Mating behavior in two live-bearing fish, Xiphophorus helleri and Platypoecilus maculatus. Physiological Zoology 22:148-161.

Schoenherr, A. A. 1974. Life history of the topminnow Poeciliopsis occidentalis (Baird and Girard) in Arizona and an analysis of its interaction with the mosquitofish Gambusia affinis (Baird and Girard). Ph.D. Dissertation. Arizona State University, Tempe.

Schoenherr, A. A. 1977. Density dependant and density independent regulation of reproduction in the Gila topminnow, Poeciliopsis occidentalis (Baird and Girard). Ecology 58:438-444.

Schoenherr, A. A. 1979. Niche separation within a population of freshwater fishes in an irrigation drain near the Salton Sea, California. Bulletin of the Southern California Academy of Sciences 78:46-55.

Schutz, D. C. and T. G. Northcote. 1972. An experimental study of feeding behavior and interaction of coastal cutthroat trout (Salmo clarki clarki) and dolly varden (Salvelinus malma). Journal of the Fisheries Research Board of Canada 29:1531-1536.

Sicault, G. 1934. Note sur l'adaptation du Gambusia holbrooki aux eaux salees. Bulletin Societie Pathologique Exotique 27:485-488.

Simpson, D. G. and G. Gunter. 1956. Notes on habitats, systematic characters, and life histories of Texas salt water Cyprinodontes. Tulane Studies in Zoology 4:113-134.

Smith, D. L. 1960. The ability of the topminnow, Gambusia affinis (Baird and Girard) to reproduce and over-winter in an outdoor pond at Winnipeg, Manitoba, Canada. Mosquito News 20:55-56.

Soltz, D. L. 1979. Our disappearing desert fishes. The Nature Conservancy News 29(6):8-12.

Soltz, D. L. and R. J. Naiman. 1978. The Natural History of Native Fishes in the Death Valley System. Natural History Museum of Los Angeles County, Science Series 30:1-76.

Theobald, P. V. K. 1959. The relation of thyroid function to upper lethal temperature in *Gambusia affinis*. Catholic University of America, *Biological Studies* 50:1-37.

Warburton, B., C. Hubbs and D. W. Hagen. 1957. Reproductive behavior of *Gambusia heterochir*. Copeia 1957:299-300.

Weatherly, A. H. 1963. Notions of niche and competition among animals with special reference to freshwater fish. Nature 197:14-17.

Weatherly, A. H. 1972. Growth and Ecology of Fish Populations. Academic, London.

Werner, E. E. and D. J. Hall. 1976. Niche shifts in sunfishes: experimental evidence and significance. Science 191:404-406.

Werner, E. E. and D. J. Hall. 1977. Competition and habitat shift in two sunfishes (Centrarchidae). Ecology 58:869-876.

Wynne-Edwards, V. C. 1965. Self-regulating systems in populations of animals. Science 147:1543-1548.

7 Variable Breeding Systems in Pupfishes (Genus *Cyprinodon*): Adaptations to Changing Environments

ASTRID KODRIC-BROWN
University of Arizona, Tucson

ABSTRACT

Pupfish of the genus *Cyprinodon* exhibit four basic types of breeding system: territoriality, dominance hierarchy, group spawning, and consort pair. In a territorial breeding system and dominance hierarchy most breeding males are highly aggressive and female choice is an important selective force in determining male reproductive success; male nuptial coloration is well developed, only a small proportion of males are able to breed, and the variance in their reproductive success is high. Group spawning and consort pair are nonterritorial breeding systems in which the males show low levels of aggression, low variance in reproductive success, and poorly developed nuptial coloration.

Each type of breeding system is associated with certain kinds of habitat. Territoriality, the most common type of breeding system in *Cyprinodon*, occurs in habitats of large physical dimensions with high primary productivity, limited breeding substrates, and high population densities. In productive habitats of small physical dimensions, which support small but dense populations, the breeding system is a dominance hierarchy. The breeding habitat is partitioned unequally among the few adult males according to body size and social status. Environmental parameters, such as a strong current or limited breeding substrate which increases the cost of territoriality, favor group spawning. Consort pair breeding is characteristic of populations in habitats of low primary productivity, low population density, and abundant breeding habitat.

Natural perturbations and experimental manipulations indicate that these breeding systems represent facultative responses rather than genetically fixed adaptations because individual males switch breeding strategies as environmental conditions change population density. As population density decreases the breeding system changes from territoriality to consort pair. Reduction of the physical dimensions of a habitat and a corresponding decrease in population size changes a breeding system from territoriality to a dominance hierarchy. Individuals that exhibit group spawning in flowing water become territorial or form a dominance hierarchy in calm water.

Alternative male breeding strategies in variable environments can be conceptualized in terms of a cost-benefit model in which a male adopts an optimal strategy based on an assessment of his potential reproductive success for a certain reproductive effort. These costs and benefits depend on the number of potentially competing males and the availability of oviposition substrates as well as other factors. The model predicts that when two alternative breeding systems confer similar reproductive success males should adopt a mixed breeding strategy and switch back and forth between them. Males that exhibited both territoriality and consort pair spawning achieved identical reproductive success with each strategy. Mixed male breeding strategies within populations may reflect adaptations to short-term changes in environmental conditions or to heterogeneous habitats.

INTRODUCTION

The breeding system of a group of organisms reflects the constraints of evolutionary history and the effects of ecological and social environments. The selective pressures operating on the breeding system can be inferred by studying the effects of social and ecological parameters on the reproductive success of both sexes. In organisms in which sexes invest unequally in their offspring (Trivers 1972) the nature of the breeding system will reflect the action of sexual selection operating most strongly on the sex that invests proportionally less in its offspring.

Polygynous breeding systems, in which the parental investment of males is virtually confined to the production of sperm, are well suited to studies of the effects of ecological and social constraints on the reproductive performance of males because such breeding systems can be manipulated and male reproductive success can be used to assess the intensity of different selective pressures responsible for the evolution and maintenance of the breeding system. Lek breeding systems are characterized by conspicuous aggregates of males that establish discrete territories used primarily for courtship and mating (Buechner and Roth 1974; Campanella and Wolf 1974; Emlen and Oring 1977; Hogan-Warburg 1966; Lill 1974; Loiselle and Barlow 1978). Typically functional sex ratios are highly skewed in favor of females because many males are excluded from breeding altogether. Access to females is determined by competitive interactions among males; consequently the variance in male reproductive success is potentially great. Males should be under strong selection pressure to evolve behavioral adaptations, such as alternative breeding strategies, to compensate for environmental or social conditions that affect their reproductive success. Alternative breeding strategies have been reported in many polygynous breeding systems (Dominey 1980; Drewry 1962; Emlen 1976; Gross 1979; Orians and Collier 1963; Parker 1974; Verner and Wilson 1966; Warner and Hoffman 1980; Wells 1977; Wiley 1974).

Pupfish of the genus Cyprinodon provide an excellent system of investigating the selective pressures important in the evolution of polygynous breeding systems and the underlying behavioral adaptations that maximize individual reproductive success in diverse ecological and social environments.

Populations occur in a variety of habitats that range from small stenothermal springs to large eurythermal streams, lakes, and estuaries and vary greatly in size and density. Pupfish show diverse breeding systems that range from territoriality to dominance hierarchies and group spawnings to consort pairs.

This chapter documents variation in the breeding systems among Cyprinodon populations; it also shows that many of the differences are facultative responses to local environments and discusses the adaptative significance of these patterns in terms of a cost-benefit model. The study is based on my published (Kodric-Brown, 1975, 1977, 1978) and unpublished observations and on other investigations of the ecology and behavior of pupfish (Barlow 1958, 1961; Cowles 1934; Cox 1966; Echelle 1973; Itzkowitz 1974, 1977, 1978; Itzkowitz and Minckley 1969; James 1969; Liu 1969; Loiselle personal communication; Minckley and Deacon 1973; Soltz 1974).

THE BIOLOGY AND BREEDING SYSTEM OF CYPRINODON

Pupfish are small fishes that belong to the family Cyprinodontidae. They inhabit diverse coastal habitats along the Atlantic coast of North America and south through the West Indies to northern South America, inland habitats such as tributaries of the Colorado and Rio Grande Rivers, and isolated streams and springs in the Mojave, Sonoran, and Chihuahuan deserts. Pupfish occur in freshwater and hypersaline habitats where temperatures may be remarkably constant or fluctuate on a daily and seasonal basis (Brown and Feldmeth 1971; Soltz 1974). Pronounced differences are apparent in the absolute size and density of local populations. In most populations highest densities and the peak of the breeding season occur in the summer, thus coinciding with warm temperatures and high food availability (James 1969; Kodric-Brown 1974, 1977; Naiman 1975, 1976; Soltz 1974). The breeding season for many populations extends throughout the spring and summer months and in thermally stable habitats it may be almost continuous throughout the year.

Adults of most *Cyprinodon* species range in size from 20 to 50 mm in standard length. Sexes are dimorphic and males are deeper bodied and have a flatter head profile than females. Both sexes are olive-brown marked by a series of dark lateral stripes; breeding males of many species often develop a brilliant blue nuptial coloration during the breeding season.

In *Cyprinodon* parental investment by both sexes is limited to the production of gametes. Both sexes are promiscuous and breed sequentially with several partners during a single day as well as throughout the breeding season. Females typically lay a few large demersal eggs each day but may not lay on consecutive days. The eggs, deposited singly, adhere to the substrate by means of a sticky coating. Fertilization is external. Males defending breeding territories probably provide some protection for the eggs from potential predators. Reproductive success of females is constrained by the metabolic requirements of egg production and is limited by the number of eggs that can be produced. Males can maximize their reproductive success by fertilizing as many eggs as possible. Therefore males show a greater variance in their reproductive success than females.

TYPES OF BREEDING SYSTEM

Breeding systems of *Cyprinodon* are characterized by extreme variability within and among populations. They differ in the intensity and spatial organization of male-male and male-female interactions, in the intensity and duration of courtship, and in the reproductive investment of males. I classify these breeding systems into four basic types: (1) territoriality, (2) dominance hierarchy, (3) group spawning, and (4) consort pair. In the first two types breeding males spend much time and energy courting females and interacting aggressively with other males and there is great variation in their reproductive success. In the last two types males allocate relatively small amounts of time and energy to reproduction and variation in their reproductive success is low.

Male reproductive success can be estimated by counting the number of females spawned with or, whenever possible, the actual number of eggs fertilized. In organisms that provide no parental care investments in time and energy on breeding activities can be defined as reproductive effort (Fisher 1930; Parker 1974; Tinkle 1969; Williams 1966). Agonistic interactions between males, courtship activities, physiological changes, including the development and maintenance of nuptial coloration, and the more obvious expenditures such as the production of gametes are considered to be part of the male's reproductive effort. A description of the four types of breeding systems and representative habitats in which they occur follows.

Territorial Breeding Systems

Territoriality. Territorial breeding systems have been described for *C. rubrofluviatilis* (Echelle 1973), *C. atrorus* (Itzkowitz and Minckley 1969), *C. variegatus* (Itzkowitz 1977, 1978; Raney et al. 1953), *C. macularius* (Barlow 1958, 1961; Cox 1966), *C. nevadensis* (Kodric-Brown 1975; Soltz 1974), and *C. pecosensis* (Kodric-Brown 1977, 1978). During the breeding season males establish territories that function in courtship and mating. Territories are established over substrates with some topographic complexity that are favored as oviposition sites by females (Kodric-Brown 1977). Females visit these sites and spawn with the resident male. Territorial males engage in both overt and covert types of agonistic interaction with other individuals to maintain exclusive use of these breeding sites. Nuptial coloration is well developed in males that are successful in maintaining territories. The intensity of male courtship on territories is inversely correlated with the number of gravid females in the population. At the beginning and end of the breeding season, when few females are gravid, male courtship is both intense and prolonged, but it becomes much shorter at the peak of the breeding season when gravid females are numerous (Echelle 1973). Territory size is inversely correlated with population density and territories are defended as long as temperatures remain in the 15 to 35°C range (Echelle 1973). Territories enable a resident to court and spawn with minimal disturbance by other males and to provide some protection for the eggs from predation by conspecifics when population density is high. Because substrates suitable for oviposition are limited, in many habitats a large proportion of the mature male population is excluded from breeding and males successful in defending oviposition sites most preferred by females have the highest reproductive success (Kodric-Brown 1977). Male reproductive success depends on the physical features of the breeding habitats, such as macro- and microtopography, the location of territories on the breeding ground, and the personal characteristics of the individuals, such as the intensity of their nuptial coloration (Kodric-Brown 1977, 1978, and unpublished data).

Territoriality is the most common breeding system in _Cyprinodon_. Populations with this type of breeding system are found in lakes, ponds, and large springs in which high primary productivity and other environmental conditions (e.g., absence of predators) result in high population densities. Mirror Lake in Bottomless Lakes State Park, New Mexico, and Big Spring in Ash Meadows, Nevada, represent habitats in which a territorial breeding system occurs (Table 1).

A comparison of the territorial breeding system of _C. pecosensis_ at Bottomless Lakes, New Mexico, with that of _C. nevadensis mionectes_ at Big Spring, Nevada, is instructive because it illustrates the extremes in environmental conditions under which territorial breeding occurs. _C. pecosensis_ was in a thermally fluctuating environment, where temperatures are highest in the summer and lowest in winter. The breeding season was short and restricted to the summer months and coincided

TABLE 1. Some Characteristics of the Habitat, Demography, and Breeding Systems of Five populations of Pupfish (*Cyprinodon*)

Breeding System	Locality	Species	Physical Parameters of Habitat: Temperature (°C)	Size (m^2)	Substrate Suitable for Territories(%)
Territoriality	Mirror Lake, Bottomless Lakes, New Mexico	*C. pecosensis*	Variable (10.0-32.0)	1256	5.3
Territoriality and occasionally consort pair	Big Spring, Ash Meadows, Nevada	*C. nevadensis mionectes*	Constant (27.0-29.0)	314	10.0
Dominance hierarchy	Mexican Spring, Ash Meadows, Nevada	*C. nevadensis pectoralis*	Variable (12.0-30.0)	3.25	55.2
Group spawning and some territoriality	Amargosa River, Tecopa, California	*C. nevadensis amargosae*	Variable (4.0-40.0)	>>1000	1.6[a]
Consort pair	Devil's Hole, Ash Meadows, Nevada	*C. diabolis*	Constant (33.4-34.0)	60	30.0

Breeding System	Demographic Parameters			References
	Population Size	Population Density (m^2)	Breeding Season	
Territoriality	35,000-55,000 (summer)	150-225	May-August	Kodric-Brown (1975; 1977; 1978)
Territoriality and occasionally consort pair	950-30,000	4-150	January-December	Soltz (1974); Kodric-Brown (unpublished data)
Dominance hierarchy	12-50	3-15	January-September	Soltz (1974); Kodric-Brown (unpublished data)
Group spawning and some territoriality	>>50,000 (summer)	60-300[a]	March-September	Soltz (1974)
Consort pair	100-700	2-12	March-October	James (1969); Minckley and Deacon (1973; 1975); Soltz (1974); Kodric-Brown (unpublished data)

[a] In a 30 m^2 section of the river.

with high population densities (Table 1). Competition among males for limited oviposition substrate was intense, territories were generally small, and only a small proportion of males was reproductively active (Kodric-Brown 1977). On the other hand, *C. n. mionectes* inhabited a warm, constant-temperature spring and had a long breeding season that extended virtually throughout the year, although most productivity was high (Soltz 1974). With a long breeding season and overlapping generations the cohort of adults was smaller in relation to the availability of suitable breeding substrate, even at the peak of the breeding season, and competition among males for breeding substrate presumably was less intense than among *C. pecosensis*. Less intense competition among males for access to breeding substrate resulted in larger territories, even over substrates with a high degree of topographic complexity, in portions of the breeding habitat that were unoccupied, and in a less compact spatial arrangement of territories. Mean male reproductive success was also lower than for *C. pecosensis* (Table 2).

The variance in size of territorial males was also greater and small males defended territories in the shallow portions of the breeding habitat. In the *C. pecosensis* population small males did not defend territories; instead they became satellites of larger territorial males.

Seasonal changes in the size of territories in response to adult densities were similar to those for *C. pecosensis*. Territories were smallest, most rigorously defended, and continuously occupied during the summer months but they increased in size and were no longer contiguous during the fall and spring. At the height of the breeding season the population was territorial but in the spring and fall, consort pair spawning, an alternative breeding strategy, was observed (Kodric-Brown 1975).

Dominance Hierarchy. In small bodies of water in which total available habitat as well as population size is small but population density is high and breeding habitat is limiting unequal partitioning of breeding habitat among the few adult males results in a

TABLE 2. Differences in Male Reproductive Success in Representative Populations of *Cyprinodon* with Different Breeding Systems

Breeding System	Example	Number of Females Mated per Male per Hour				Total Number of Hours of Observations	References
		x	SD	n			
Territoriality	*C. pecosensis*	3.7	2.8	38		10	Kodric-Brown (1975, 1977)
Territoriality and occasionally consort pair	*C. nevadensis mionectes*	3.0	3.2	15	(territoriality only)	10	Soltz (1974)
		0.78	-	6	(territoriality[a])	10	Kodric-Brown (1975)
		0.72	-	6	(consort pair)		
Dominance hierarchy	*C. nevadensis pectoralis*	0.9	-	4		5	Soltz (1974)
Group spawning and some territoriality	*C. nevadensis amargosae*	Not available		-	(group)		
		0.7	-	-	(territoriality)	3	Soltz (1974)
Consort pair	*C. diabolis*	0.6	-	6		2	Kodric-Brown (unpublished data)

[a]Spawnings by males with a mixed strategy of consort pair and territoriality.

dominance hierarchy. Breeding habitat is partitioned among males according to social status, based on the outcome of aggressive interactions (Itzkowitz 1977; Soltz 1974). Because competitive outcome between contestants is primarily determined by size (Borowsky and Kallman 1976), the largest, brightest, and most aggressive male typically defends most of the oviposition substrate. Smaller, subordinate individuals with less well developed nuptial coloration, usually hold smaller areas of less suitable oviposition substrate. Frequently the dominance hierarchy consists of a single domiant male and one or more subordinate males that defend peripheral or spatially isolated areas. Subordinate males maximize their reproductive success by spawning surreptitiously with females or sneaking up to join spawning pairs that contain dominant males in an apparent attempt to fertilize some of the eggs. The dominant male is extremely aggressive toward conspecifics, including even gravid females. Despite his ability to defend most of the superior oviposition substrate, the dominant male may chase away potential females and consequently his reproductive success may sometimes be even lower than that of subordinate males (Soltz 1974). Mexican Spring, Ash Meadows, Nevada, was a representative habitat that supported a stable dominance hierarchy (Table 1). The breeding system consisted of a large dominant male and three to four smaller, subordinate males. The dominant male controlled approximately half the substrate suitable for oviposition and territorial defense.

A dominance hierarchy thus may be regarded as a distortion of a territorial breeding system with an increase in variance of size of territory and aggressiveness. Because the size of the population is often too small to contain males of matched aggressive abilities, the most aggressive male can control most of the breeding substrate; thus he increases his reproductive success. Itzkowitz (1978) showed that when _C. variegatus_ males were introduced into aquaria with _a_ homogeneous substrate that consisted of sand, male reproductive success was positively correlated with size of territory. A similar mechanism may operate in naturally occurring dominance hierarchies in which male quality is determined by the size of territory defended (but see Soltz 1974).

Nonterritorial Breeding Systems

Breeding systems in which males do not defend territories have been described for several species of *Cyprinodon* (James 1969; Liu 1969; Soltz 1974). In these breeding systems intermale aggression is reduced or entirely absent and males do not defend breeding substrates; they have poorly developed breeding coloration and the mean and variance in their reproductive success is often low. Two types of nonterritorial breeding systems have been described: group spawning and consort pair.

Group Spawning. In this type of breeding system all courtship and spawning activities take place in groups. These groups are primarily feeding aggregations that vary in size and may consist of as many as several hundred individuals of both sexes. Courtship activity is extremely brief and usually consists of a male steering a receptive female to a spawning site at the periphery of the group; however, spawning may also occur near the center of the group (Liu 1969; Soltz 1974). Occasionally spawnings have been observed between a single female and several males (Liu 1969; Raney et al. 1953). Soltz (1974) described the spawning activity of males in aggregates of *C. n. amargosae* and recorded mean spawning rates of 1.5 females fertilized per male per hour. The breeding system of this population is primarily group spawning but 18 percent of the spawnings take place on territories in shallow pools that presumably are suboptimal breeding sites because of high water temperatures. Territorial males spawned with 0.9 females per male per hour. Although no data are available on the variance in male reproductive success, virtual absence of male aggression and the high frequency of male-initiated spawnings suggest that female choice is of relatively minor importance in determining male reproductive success.

A breeding system based on group spawnings occurs typically in habitats in which population densities are high but conditions are not suitable for territorial establishment (Barlow 1958; Kodric-Brown unpublished data; Liu 1969; Soltz 1974). Portions of the Amargosa River, for example, the area near Tecopa, California, are representative of habitats that support group spawning (Table 1); a strong current and limited breeding substrate appear to be important in favoring group spawning in a system that presumably would otherwise be territorial.

Consort Pair. In this breeding system a gravid female is closely followed by one or more males. The male usually maintains a distance of less than one body length behind a female and follows her for as long as one hour. The pair periodically descends to the substrate and spawns. The male consort often blocks the approach of intruding males by decreasing his distance from the female or by interposing his body between the female and the approaching male. There is almost no overt aggression between males. Courtship behavior may last several minutes and will consist of a male sidling up to or nudging a female. Reproductive success of consort males is low but more males probably participate in spawning than in territorial breeding systems. During three hours of observations of six _C. diabolis_ consort pairs the mean male reproductive success was 0.6 spawning per male per hour and the maximum rate of spawnings recorded was 1.5 per male per hour (Kodric-Brown unpublished data). The only pupfish species known to have an exclusively consort-pair breeding system is _C. diabolis_ in Devils Hole, Ash Meadows, Nevada. The spring is located in a deep chasm and receives only about two hours of direct sunlight daily during the summer months. Primary productivity is low at all times of the year (James 1969). The consort-pair breeding system usually occurs in populations at low densities and has been observed in _C. n. mionectes_ at Big Spring, Ash Meadows, Nevada (Kodric-Brown 1975; Soltz 1974) in _C. variegatus_ (Kodric-Brown unpublished data; Raney et al. 1953), and _C. macularius_ in aquaria (Barlow 1961).

ADAPTIVE SIGNIFICANCE OF ALTERNATIVE BREEDING SYSTEMS

What are the selective pressures responsible for maintaining variable breeding systems in *Cyprinodon*? This variability may be understood in terms of optimization of male reproductive success under spatially and temporally unpredictable environments. In *Cyprinodon* parental investment of both sexes is virtually limited to the production of gametes. Since female investment is greater than male investment, because eggs are metabolically more expensive to produce than sperm, eggs should be the limiting resource and males should compete to fertilize them. Selection should operate on females to protect their more costly investment by choosing oviposition sites in which there is low egg mortality from predation or adverse physical factors like silting, anoxia, and extremes in temperature (Kinne and Kinne 1962; Kodric-Brown 1977; Shrode 1975; Shrode and Gerking 1977). Females should also choose to mate with males of high quality, as indicated by their bright nuptial coloration and appropriate body size under certain ecological conditions (Kodric-Brown 1977 and unpublished data; Loiselle personal communication).

Male reproductive success is extremely variable, limited primarily by the availability of and access to females. Because males are under strong selective pressure to adopt behaviors that maximize the number of eggs they fertilize, the reproductive success of males should vary more both within and among populations than that of females. Because environmental and social conditions in the habitats often change and affect male access to gravid females, natural selection should operate particularly strongly on males and the variability in the breeding system of *Cyprinodon* should reflect male adaptations to different environmental and social conditions.

If males are able to assess their potential reproductive success and modify their behavior accordingly, they should exhibit the optimal breeding strategy for a particular habitat or set of environmental conditions. If these conditions change, then male breeding behavior should also change. If no single reproductive strategy is optimal over the entire range of ecological and demographic conditions experienced by a single population, individual males should be selected to switch facultatively between alternative breeding strategies.

A Model of Male Breeding Strategies

Male strategies in variable environments are perhaps best conceptualized in terms of an optimization model in which a male adopts a particular breeding strategy based on an assessment of his potential reproductive success for a certain reproductive effort (Figure 1).

Average male reproductive success is known for the different types of breeding systems (Table 2) but the reproductive effort can only be estimated. Figure 1 indicates expected changes in the *Cyprinodon* breeding system as a function of population density and availability of oviposition substrate. The two parameters are often but not always interdependent because competition for limited oviposition substrate becomes more intense with increasing population density.

Because one male can fertilize the eggs of many females, variance in reproductive success of individual males should be an increasing function of availability of potential mates, which in turn should increase with absolute population size and population density. At low population densities both females and potential male competitors are scarce; access to suitable spawning sites is comparatively easy and a male should devote most of his reproductive effort to locating and courting females. The intensity and duration of male courtship is the most conspicuous feature of the consort pair breeding system. Males may spend as long as one hour courting a single female (Kodric-Brown personal observation). Because male reproductive success at low population densities is low (Table 2), as is the variance, reproductive effort should be correspondingly low, below the threshold level for territoriality to be an adaptive strategy (Figure 1).

Female investment in eggs increases with body size and large females carry a greater number of ripe eggs than smaller ones (Iles 1974); therefore, whenever

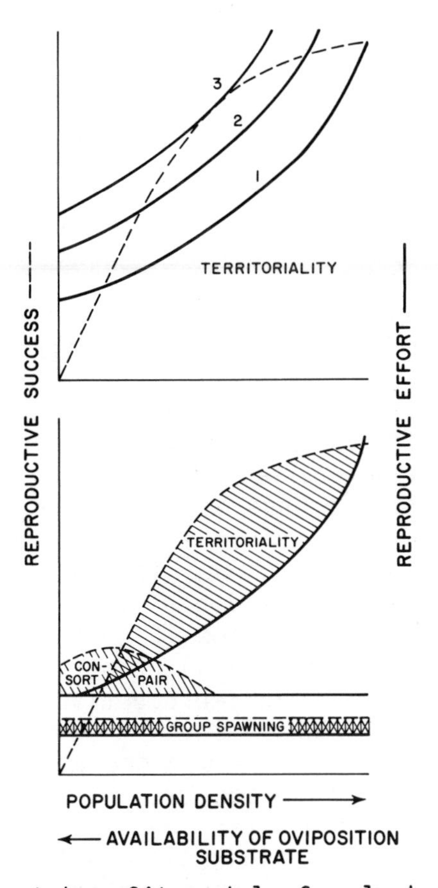

FIGURE 1. A cost-benefit model of male breeding strategies. Reproductive success is indicated by a broken line, reproductive effort by a solid line. Arrows indicate the direction of increase in population density and availability of oviposition substrate. In the lower graph the area between the intercepts of curves for reproductive effort and reproductive success indicates the range of population densities and oviposition substrate over which a particular breeding strategy would be favored. Partial overlap between territoriality and consort pair breeding strategy is indicated by double cross hatching. The upper graph shows the effects of progressively increasing (from 1 to 3) the cost (reproductive effort) of territoriality. As the area between the cost and benefit curves decreases, males should switch to a less costly breeding strategy. Note that reproductive success is correlated with reproductive effort, with group spawning having the lowest, consort pair intermediate, and territoriality the highest costs and benefits.

environmental conditions favor a consort pair breeding system, males should preferentially invest more time and energy in courtship of large females. Unfortunately, information on male choice, which would be expected to be highly developed in populations with a consort pair breeding system, is not available.

With increases in population density the number of potential mates increases but so does competition among males for access to females and spawning sites. Males should invest their reproductive effort in behavioral strategies that maximize access to receptive females and minimize interference with courtship and matings. When populations are large and dense in relation to suitable spawning substrate, it becomes advantageous for males to defend a portion of the breeding ground to the exclusion of other males (Figure 1). In a territorial breeding system a small proportion of males does all the breeding and mean reproductive success for territorial males is high (Table 2). However, the variance in reproductive success among males is also high because not all territories are equally attractive to females as oviposition sites and not all males are able to acquire territories (Kodric-Brown 1977).

In a territorial breeding system a male's initial reproductive effort is considerable because he must secure and defend a territory before he can reproduce. The duration of occupancy of a territory may indicate the magnitude of reproductive effort because occupancy of a territory by an individual male is inversely proportional to the number of competitors. The greater the cost of defending a territory, the higher the turnover rate of males. Although data on duration of occupancy of territories are not available for tagged males, observations of territorial males with natural markings indicate that at relatively low population densities males may occupy territories for several months (Cox 1966; Kodric-Brown 1975), but at high densities change of resident males on territories occurs at intervals that average about a week (Kodric-Brown 1975). At high densities males also pay an additional cost by the loss of opportunity to feed outside the territory, for even a temporary vacancy may result in a takeover by another male. Temporary abandonment and reoccupation of territories occurs frequently at low population densities and even at high population densities on suboptimal breeding sites (Kodric-Brown 1975).

Because of high reproductive effort, a territorial male should not be highly selective in his choice of spawning partners but should attempt to mate with every gravid female that visits his territory. Observations of male courtship and spawning behavior confirm this expectation. In observing hundreds of spawnings by territorial males I have never seen one reject a female.

The availability of suitable oviposition substrate should also directly affect male breeding strategy. In habitats in which oviposition substrate is relatively abundant the consort pair breeding system should persist longer and at higher population densities than in habitats in which oviposition substrate is more limited. Whenever a territorial breeding system is favored, both mean and variance in male reproductive success should increase as oviposition substrate becomes more restricted.

In habitats in which oviposition substrate is extremely scarce or population densities are exceptionally high competition for occupancy among males could become so intense that the cost of defending an exclusive spawning area would be prohibitive; therefore the breeding system of such a population should be group spawning. Unfortunately I know of few data with which to test this prediction. Aquarists have found, however, that the aggressiveness of territorial male *Cyprinodon* is detrimental to maintaining breeding stocks in aquaria. In these circumstances territoriality can be reduced or eliminated by keeping the fish at sufficiently high densities.

Figure 1 predicts changes in male breeding strategies with an increase in male reproductive effort. If factors such as a strong current, predation, or excessive competition increase the costs of a particular male strategy (e.g., territoriality) to the point at which they exceed the benefits (measured in terms of reproductive success), males should switch to an alternative breeding strategy that requires less reproductive effort. The model predicts that males should switch from territoriality to consort pair and finally to a group spawning strategy as costs increase in relation to benefits.

ADAPTATIONS OF THE BREEDING SYSTEMS TO CHANGING ENVIRONMENTS

A comparison of the habitats and demography of several species of Cyprinodon for which sufficient data on the breeding system are available shows that they differ most conspicuously in physical dimensions, availability of suitable oviposition substrates, primary productivity of the habitats, and size and density of the pupfish populations (Table 1). These patterns are interrelated and suggest that physical and demographic parameters interact to determine patterns of male reproductive success and, as a consequence, the nature of the breeding system. I offer the following generalizatons and interpretations of the observed patterns:

1. Territoriality is the primary breeding system in habitats of large physical dimensions in which high primary productivity and few predators result in large populations. When densities are high and oviposition substrates limited, males can maximize their reproductive success by defending space on the spawning ground. Exceptions to this pattern should occur when a factor like a strong current or perhaps extremely limited oviposition substrates makes territorial defense energetically too costly; then the breeding system should be group spawning.

2. A dominance hierarchy is the predominant breeding system under conditions that would favor territoriality, except that the absolute population size is too small to contain males of matched aggressive abilities. Whenever the physical dimensions are small total population size is correspondingly small, even though population density is relatively high. The most aggressive male will be able to control most of the limited spawning area.

3. Consort pair breeding systems occur when small populations inhabit habitats of large physical dimensions. This is likely to be when primary productivity is low or when competitors and/or predators are abundant. When population density is low, oviposition substrates are relatively widespread and males maximize reproductive success by locating and following gravid females.

If these interpretations of the patterns are correct and particular breeding systems represent facultative responses to local environments, it should be possible to predict changes in the breeding systems in response to natural environmental fluctuations or experimental manipulations.

Although the interpretation of patterns of response by the breeding systems to changing environments was derived *a posteriori*, it is supported and strengthened by observations of responses to natural and experimental manipulations.

Changes in Population Density. If the density of adults decreases, the breeding system should change from territoriality to consort pair. The density of *C. n. mionectes* in Big Spring flucutated seasonally in apparent response to photoperiod-induced changes in primary productivity (Soltz 1974). In summer, at the peak of the breeding season, population density was high (150 adults$\cdot m^{-2}$), oviposition substrates were limited, and the breeding system was male territoriality. In the spring and fall population density was low (10 adults$\cdot m^{-2}$), oviposition substrates were abundant, and the breeding system changed to consort pair (Table 3).

An experimental manipulation of this population produced identical results. Fifty adult *C. n. mionectes* of each sex were introduced from Big Spring into nearby Jackrabbit Spring on January 18, 1970 (Table 3). Jackrabbit Spring is a eurythermal spring that measures approximately 30 m^2. Before 1969 it supported a large, dense population of *C. n. mionectes* which was territorial during the peak of the breeding season (Miller 1948, personal observation). At the time of

TABLE 3. Effects of Environmental Manipulations on the *Cyprinodon* Breeding System

Species	Source of Population	Environmental Change	Breeding System		References
			Original	New	
C. nevadensis mionectes	Big Spring Ash Meadows, Nevada	Decrease in population density (from 150 to 10 individuals$\cdot m^{-2}$)	Territoriality	Consort pair	Kodric-Brown (1975 and unpublished data); Soltz (1974)
C. nevadensis mionectes	Big Spring Ash Meadows, Nevada	Decrease in population density; introduction of 100 adults into Jackrabbit Spring (30 m^2) in January 1970	Territoriality	Consort pair March 1970 Territoriality July 1970	Kodric-Brown (unpublished data)
C. nevadensis mionectes	Big Spring Ash Meadows, Nevada	Decrease in size of habitat; introduction of 30 adults into wading pool (12.5 m^2)	Territoriality	Dominance hierarchy	Kodric-Brown (unpublished data)
C. nevadensis amargosae	Amargosa River Tecopa, California	Decrease in size of habitat; introduction of 30 adults into wading pool (12.5 m^2)	Group spawning and some territoriality	Dominance hierarchy	Kodric-Brown (unpublished data); Soltz (1974)

the introduction the spring had recently recovered from being pumped dry; its pupfish population had been exterminated. In March, two months after the introduction, population density was low (0.3 adults· m^{-2}) and pupfish were breeding in consort pairs. By July the population was estimated at seven adults per square meter, suitable oviposition substrate was limiting, and all reproduction took place on territories.

Changes in the Size of the Habitat. If the physical dimensions of a productive *Cyprinodon* habitat are reduced, the breeding system should change from territoriality to dominance hierarchy. Experimental studies with *C. variegatus* (Itzkowitz 1977), *C. atrorus* (Itzkowitz and Minckley 1969), and *C. n. mionectes* (Kodric-Brown unpublished data) have shown that if a small number of individuals from populations with a territorial breeding system are confined in environments of small physical dimensions such as aquaria or small artificial ponds a dominance hierarchy is established. One experiment is described briefly here. I introduced 15 individuals of each sex, matched in size, of *C. n. mionectes* from Big Spring, Ash Meadows, Nevada, into wading pools (Table 3). The substrates of these pools consisted of sand, rooted aquatic vegetation, and eight rocks positioned equidistantly. There were three replicates of the experiment. Each experiment lasted two weeks. The experiments were also repeated with *C. n. amargosae* from the Amargosa River (Table 3). Males established territories with rocks as focal points, but females aggregated in one area of the pool. For all three replicates for each population only one male defended a permanent territory, which included three or four rocks, for the duration of the experiment. This dominant, aggressive male had well-developed nuptial coloration and performed most of the spawnings (72 percent of 120 spawnings in *C. n. mionectes* and 80 percent of 160 spawnings in *C. n. amargosae*). The other males were pale, spawned infrequently, and defended the remaining rocks for only short periods of time, during which they frequently displaced one another. The dominant male often interrupted the spawnings of the subordinate males. These results demonstrate that by reducing the physical dimensions of the habitat and correspondingly reducing absolute population size it is possible to induce a dominance hierarchy in populations that characteristically exhibit other breeding systems. It is interesting that similar results were obtained with _C. n. mionectes_, which is typically territorial, and with _C. n. amargosae_, which is primarily a group spawner.

ALTERNATIVE MALE BREEDING STRATEGIES

The variability of the _Cyprinodon_ breeding systems suggests that males are extremely opportunistic and capable of rapid, facultative changes in their breeding behavior in response to short-term changes in environmental conditions or heterogeneous habitats. Thus they appear to act as short-term maximizers of reproductive success. In habitats that fluctuate to favor two breeding systems in succession, during the transition periods both breeding strategies should confer a similar reproductive success and require a similar reproductive effort. At such times males should become opportunistic and switch back and forth between two breeding strategies or each strategy should be adopted by some individuals in the population. The model (Figure 1) predicts a mixed breeding system of consort pair-territoriality at low to intermediate population densities.

At relatively low densities the population of _C. n. mionectes_ in Big Spring had a mixed breeding system of territoriality and consort pair. Males abandoned their territories and pursued passing females with which they formed consort pairs. Individual males switched back and forth between the two behavioral strategies, presumably because of the low cost of territorial defense and the availability of unoccupied breeding substrate. Such behavioral versatility also suggests that the adoption of either strategy confers a similar reproductive success. Data for six males engaged in a mixed reproductive strategy indicate that average reproductive success for consort pair breeding was essentially the same as for territoriality (Table

2). Presumably this mixed behavioral strategy increased individual overall reproductive success. Unfortunately, no data are available for males who adopted an exclusively territorial or consort pair breeding strategy.

The mixed breeding strategy of *C. n. mionectes* males is interesting because it requires rapid change from a resource-based behavioral repertoire with emphasis on topographic cues to behavior directed exclusively toward females. Why does the Big Spring population exhibit a mixed breeding strategy? Among the populations studied it has the longest breeding season (Table 1). With a prolonged breeding season and overlapping generations the cohort of adults is smaller at the peak of the breeding season, even during the summer months. Breeding substrate is also less limiting and intermale competition for breeding sites is less intense than in populations with an exclusively territorial breeding system (e.g., *C. pecosensis* at Mirror Lake). These factors should favor a distribution of reproductive effort over longer periods and a corresponding reduction in the mean and variance in male reproductive success.

Daily and seasonal changes in male breeding behavior, such as size of territories during the breeding season, reflect the resident's compromises in response to varying competitive pressures and the availability of breeding habitat. Short-term, diel changes in male breeding behavior in response to fluctuating water temperatures have been documented for several species of pupfish (Barlow 1958; Echelle 1973). Typically, males begin to defend their territories as water temperatures rise above 15°C but abandon them as temperatures approach 35°C to move to deeper water, where they may switch to consort pair or group spawning, which are less costly breeding strategies. Because metabolic rates of fish rise with an increase in water temperature, temporary abandonment of a territory may actually optimize male reproductive success because it reduces cost of reproduction when metabolic costs are high.

If several types of breeding habitat are available, males should show breeding behaviors that maximize their reproductive success within each habitat type; for example, the males of *C. n. amargosae* in the

Amargosa River show a mixed breeding system of group spawning and territoriality that separates out according to habitat (Soltz 1974). Whether there is a genetically based, frequency-dependent behavioral dimorphism or whether it is facultative and individual males can switch back and forth from territoriality to group spawning has not been determined experimentally, although a mechanism of facultative behavioral flexibility is more likely.

In habitats in which seasonal environmental conditions necessitate a short breeding season adult males that have not yet attained a size at which they can successfully compete for territories on the breeding ground become satellites of territorial males or delay their reproduction until the following breeding season (i.e., if they can survive the winter). A combination of these breeding strategies may occur. The high reproductive success of satellite males of _C. pecosensis_ and their relative abundance on the prime breeding grounds (limestone embankments) suggests that the delayed reproductive strategy is the less viable one, especially when the reproductive success of large males that appear to have overwintered is quite low (Kodric-Brown 1977). Therefore under conditions of high population density and low breeding substrate availability size-dependent breeding strategies are favored. Intense competition results in a narrow range of sizes of males that are able to defend territories successfully. Small males, which might under less intense competition be territorial, adopt a satellite strategy that still enables them to reproduce at the peak of the breeding season rather than at the end or in the following year. Reproductive success of satellites shows patterns similar to those of territorial males.

Presumably the satellite strategy is adopted only in habitats in which all available breeding substrates are defended and larger competitively superior males prevent the small ones from establishing territories. Thus satellite males should occur in populations with extremely short breeding seasons in which breeding substrate is limited (ex. _C. pecosensis_ at Bottomless Lakes). Small males, which may have been satellites, were described for _C. macularius_ in the Salton Sea (Barlow 1961).

CONCLUSIONS

The concept of evolutionarily stable strategies (ESS), developed by Maynard Smith and Parker (1976) and applied to breeding strategies by Dawkins (1979), is useful to explain some of the evolutionary mechanisms operating on the *Cyprinodon* breeding system. Two evolutionary mechanisms may account for the observed variation in male reproductive startegies:

1. *Stable polymorphisms*. Observed variation in male breeding behavior may be due to a genetic polymorphism. The underlying assumption is that to be maintained the fitness of one behavioral morph must equal the fitness of the other (Gadgil 1972).
2. *Mixed breeding strategies*. Variation in male reproductive behavior often appears to represent a facultative response of individuals to their immediate ecological or social environment. Age and/or size-dependent reproductive behavior, such as that shown by satellite males of *Cyprinodon pecosensis* (Kodric-Brown 1977), "sneak" males of *Cyprinodon macularius* (Matsui personal communication), and males of *Cyprinodon nevadensis mionectes*, which switch back and forth between two alternative behavioral repertoires such as territoriality and consort pair breeding, may be examples of conditional male strategies. In this case the immediate fitness of the different behavioral types need not be the same but the behaviors adopted must serve to maximize individual reproductive success over a lifetime if they are to be adaptive. Although such behaviors may be genetically determined, it seems most likely that they are facultative mixed breeding strategies. Definitive breeding and rearing experiments need to be done.

ACKNOWLEDGMENTS

I thank R. Warner, R. Thornhill, T. C. Gibson, and D. Gori for valuable discussions and J. H. Brown, M. Itzkowitz, and A. Echelle for critical comments on the manuscript. The study was supported by the National Science Foundation Grant DEB 76-09499.

REFERENCES

Barlow, G. W. 1958. Daily movements of desert pupfish, Cyprinodon macularius, in shore pools of the Salton Sea, California. Ecology 39:580-587.

Barlow, G. W. 1961. Social behavior of the desert pupfish, Cyprinodon macularius in the field and in the aquarium. American Midland Naturalist 65:339-359.

Borowsky, R. L. and K. K. Kallman. 1976. Patterns of matings in natural populations of Xiphophorus (Pisces: Poeciliidae). I: X. maculatus from Belize and Mexico. Evolution 30:693-706.

Brown, J. H. and C. R. Feldmeth. 1971. Evolution in constant and fluctuating environments: thermal tolerances of desert pupfish (Cyprinodon). Evolution 25:390-398.

Buechner, H. K. and H. D. Roth. 1974. The lek system in Uganda Kob antelope. American Zoologist 14: 145-162.

Campanella, P. J. and L. Wolf. 1974. Temporal leks as a mating system in a temperate zone dragonfly (Odonta:Anisoptera). I. Plothemis lydia (Drury). Behaviour 51:4-87.

Cowles, R. B. 1934. Notes on the ecology and breeding habits of the desert minnow, Cyprinodon macularius Baird and Girard. Copeia 1934:40-42.

Cox, T. J. 1966. A behavioral and ecological study of the desert pupfish Cyprinodon macularius in Quitobaquito Springs, Organ Pipe Cactus National Monument, Arizona. Ph.D. Thesis, University of Arizona, Tucson.

Dawkins, R. 1979. Good strategy or evolutionarily stable strategy? In: G. W. Barlow and J. Silverberg (Eds.), Sociobiology: Beyond Nature/Nurture? Westview Press, Boulder, Colorado.

Dominey, W. J. 1980. Female mimicry in male bluegill sunfish (Lepomis macrochirus): A genetic polymorphism? Nature 284:546-548.

Drewry, G. E. 1962. Some observations of courtship behavior and sound production in five species of Fundulus. M.S. Thesis, University of Texas, Austin.

Echelle, A. A. 1973. Behavior of the pupfish, *Cyprinodon rubrofluviatilis*. Copeia 1973:68-76.

Emlen, S. T. 1976. Lek organization and mating strategies in the bullfrog. Behavioral Ecology and Sociobiology 1:283-313.

Emlen, S. T. and L. W. Oring. 1977. Ecology, sexual selection, and the evolution of mating systems. Science 197:215-223.

Fisher, R. A. 1930. The genetical theory of natural selection. University Press, London.

Gross, M. R. 1979. Cuckoldry in sunfishes (*Lepomis*: Centrarchidae). Canadian Journal of Zoology 57: 1507-1509.

Hogan-Warburg, A. J. 1966. Social behavior of the Ruff *Philomachus pugnax* (L.) Ardea 54:109-229.

Iles, T. D. 1974. The tactics and strategy of growth in fishes. *In*: F. R. Harden Jones (Ed.), Sea Fisheries Research. Wiley, New York, pp. 331-345.

Itzkowitz, M. 1974. The effects of other fish on the reproductive behavior of the male *Cyprinodon variegatus* (Pisces: Cyprinodontidae). Behaviour 48:1-22.

Itzkowitz, M. 1977. Interrelationships of dominance and territorial behaviour in the pupfish, *Cyprinodon variegatus*. Behavioral Processes 2:283-391.

Itzkowitz, M. 1978. Female mate choice in the pupfish, *Cyprinodon variegatus*. Behavioral Processes 3:1-8

Itzkowitz, M. and W. L. Minckley. 1969. Qualitative behavior of a pupfish (*Cyprinodon atrorus*) in differing environments. The Great Basin Naturalist 29:169-180.

James, C. J. 1969. Aspects of the ecology of the Devil's Hole pupfish, *Cyprinodon diabolis*, Wales. M.S. Thesis, University of Nevada, Las Vegas.

Kinne, O. and E. M. Kinne. 1962. Rates of development in embryos of a cyprinodont fish exposed to different temperature-salinity-oxygen combinations. Canadian Journal of Zoology 40:231-253.

Kodric-Brown, A. 1975. Breeding territories in two freshwater fishes of the genus *Cyprinodon* (Pisces, Cypriodontidae) in the southwestern United States. Ph.D. Thesis, University of Southern California, Los Angeles.

Kodric-Brown, A. 1977. Reproductive success and the evolution of breeding territories in pupfish (*Cyprinodon*). Evolution 31:750-766.

Kodric-Brown, A. 1978. Establishment and defense of breeding territories in a pupfish (Cyprinodontidae:*Cyprinodon*). Animal Behaviour 26:818-834.

Lill, A. 1974. Social organization and space utilization in the lek-forming white-bearded manakin, *M. manacus trinitatis* Hartert. Zeitschrift fur Tierpsychologie 36:513-530.

Liu, R. K. 1969. The comparative behavior of allopatric species (Teleostei-Cyprinodontidae:*Cyprinodon*). Ph.D. Thesis, University of California, Los Angeles.

Loiselle, P. V. and G. W. Barlow. 1978. Do fishes lek like birds? *In*: E. Reese and F. Lighter (Eds.), Contrasts in Behavior. Wiley, New York, pp. 31-75.

Maynard Smith, J. and G. A. Parker. 1976. The logic of asymetric contest. Animal Behaviour 24:159-175.

Miller, R. R. 1948. The Cyprinodont fishes of the Death Valley system of eastern California and southwestern Nevada. Miscellaneous Publications, Museum of Zoology, University of Michigan 68:1-55.

Minckley, C. O. and J. E. Deacon. 1973. Observations on the reproductive cycle of *Cyprinodon diabolis*. Copeia 1973:610-613.

Naiman, R. J. 1975. Food habits of the Amargosa pupfish in a thermal stream. Transactions of the American Fisheries Society 104:536-538.

Naiman, R. J. 1976. Productivity of a herbivorous pupfish population (*Cyprinodon nevadensis*) in a warm desert stream. Journal of Fish Biology 9: 125-137.

Orians, G. H. and G. Collier. 1963. Competition and blackbird social systems. Evolution 17:449-459.

Parker, G. A. 1974. The reproductive behavior and the nature of sexual selection in *Scatophaga stercoraria* L. IX. Spatial distribution of fertilization rates and evolution of male search strategy within the reproductive area. Evolution 28:93-108.

Raney, E. C., R. H. Backus, R. W. Crawford and C. R. Robins. 1953. Reproductive behavior in Cyprinodon variegatus Lacepede, in Florida. Zoologica 38:97-105.

Shrode, J. B. 1975. Developmental temperature tolerance of a Death Valley pupfish (Cyprinodon nevadensis). Physiological Zoology 48:378-389.

Shrode, J. B. and S. D. Gerking. 1977. Effects of constant and fluctuating temperatures on reproductive performance of a desert pupfish, Cyprinodon n. nevadensis. Physiological Zoology 50:1-10.

Soltz, D. 1974. Varition in life history and social organization of some populations of Nevada pupfish, Cyprinodon nevadensis, Ph.D. Thesis, University of California, Los Angeles.

Tinkle, D. W. 1969. The concept of reproductive effort and its relation to the evolution of life histories in lizards. American Naturalist 103: 501-516.

Trivers, R. L. 1972. Parental investment and sexual selection. In: B. Campbell (Ed.), Sexual Selection and the Descent of Man, 1871-1971. Aldine, Chicago, pp. 136-179.

Verner, J. and M. Wilson. 1966. The influence of habitats on mating systems of North American passerine birds. Ecology 47:143-147.

Warner, R. R. and S. G. Hoffman. 1980. Population density and the economics of territorial defense in a coral reef fish. Ecology 772-780.

Wells, K. D. 1977. The social behavior of anuran amphibians. Animal Behaviour 25:666-693.

Wiley, R. H. 1974. Evolution of social organization and life history patterns among grouse (Aves: Tetraonidae). Quaterly Review of Biology 49:201-227.

Williams, G. C. 1966. Adaptation and Natural Selection. Princeton University Press, Princeton, NJ.

8 Life History Patterns of Desert Fishes

GEORGE D. CONSTANTZ
Academy of Natural Sciences, Philadelphia

ABSTRACT

Because desert springs and streams seem little influenced by events outside the arid zone, fishes native to these habitats were chosen as most likely to exhibit life history adaptations to desert conditions. Desert springs are characterized by small volume, constant physicochemical parameters, a depauperate fauna, and gradual and predictable seasonal changes. In contrast, desert stream habitats are permanent pools that fluctuate dramatically and unpredictably in volume. The larger habitat volume and irregular inputs characteristic of streams contribute to great spatial and temporal heterogeneity.

Relative to stream fishes, spring forms exhibit the following tendencies: chubby body shape; a long, even year-round, breeding season; small body; short time to first breeding; low fecundity; and large investment per offspring. Territoriality, brightly colored bodies, and alternate mating strategies seem more prevalent among males of spring forms.

After rain some species of stream species disperse from permanent pools to temporary waters. I predict that fish should not disperse when an increase in fecundity does not compensate for the increased probability of death by desiccation. Further, there may be sedentary and dispersing individuals within populations.

In terms of life history traits, fishes of North American deserts do not appear to be qualitatively different from species of surrounding mesic biomes.

INTRODUCTION

Desert fishes display a fascinating diversity of life histories. Some species thrive under catastrophic flash flooding and habitat desiccation, undergo explosive dispersal, produce many small offspring, and are ecologically ubiquitous; other forms are restricted to a single small spring, are never subjected to catastrophes, and produce few large offspring. What environmental factors have determined the life history patterns of desert fishes?

My specific approach to answering this question places three limits on the scope of this review. First, primary interest is on ultimate, evolutionary causes of genetically restricted characteristics; ontogenetically plastic traits that respond to proximate, contemporaneous stimuli are considered only as they shed light on adaptations. Second, I focus on fishes that are native to desert springs and streams because this combination of fauna and ecosystems appears to be little influenced by events outside the arid zone. For example, major rivers often arise in montane forests, and most desert lakes are man-made habitats with transplanted fishes; their inclusion would confound the search for adaptations to desert waters. Third, I have restricted this chapter geographically to the Great

Basin, Mohave, and Sonoran deserts and taxonomically to 16 species that are well known. In essence, my goal is to compare the traits of fishes indigenous to the springs and streams of the deserts of North America; I hope this process will lead to a general understanding of the life histories of desert fishes.

CHARACTERISTICS OF DESERT STREAMS AND SPRINGS

Noy-Meir (1973) characterized a desert as a water-controlled ecosystem with infrequent, discrete, and largely unpredictable water inputs. As the following description suggests, these factors appear to be important influences of fishes in desert arroyos (used synonymously with streams and washes).

I have drawn the following generalized description of a desert stream from personal observations and literature sources (Cole 1968; Constantz 1976; Fisher and Minckley 1978; Harrell 1978; Minckley 1969a; Minshall 1978; Soltz and Naiman 1978; Sutton and Sutton 1966). As a wash progresses downstream from its origins as a spring or surface run-off, it may receive tributaries of surface or spring origin, flow underground through channels, gravel interstices, or sands, or be forced to the surface by impermeable strata (e.g., clay, igneous extrusion). For most of its length the stream bed may be dry, but if isolated pools and short flowing reaches occur, they support permanent biological communities. The stream may terminate obscurely undergound, flow onto a playa where it will evaporate, or empty into an ocean.

During most of the year the pools of desert streams are relatively closed ecosystems characterized by no surface in- or outflow, stratification of temperature and oxygen, modest autochthony, and internal recycling of nutrients. Seasonal variation within pools may be pronounced (Figure 1). Each pool along a streambed may be unique, possibly resulting in differential phenotypic selection among populations (Greenfield and Deckert 1973). Within generally predictable rainy seasons, the timing and magnitude of discharge may be random. Two levels of discharge seem important to fishes

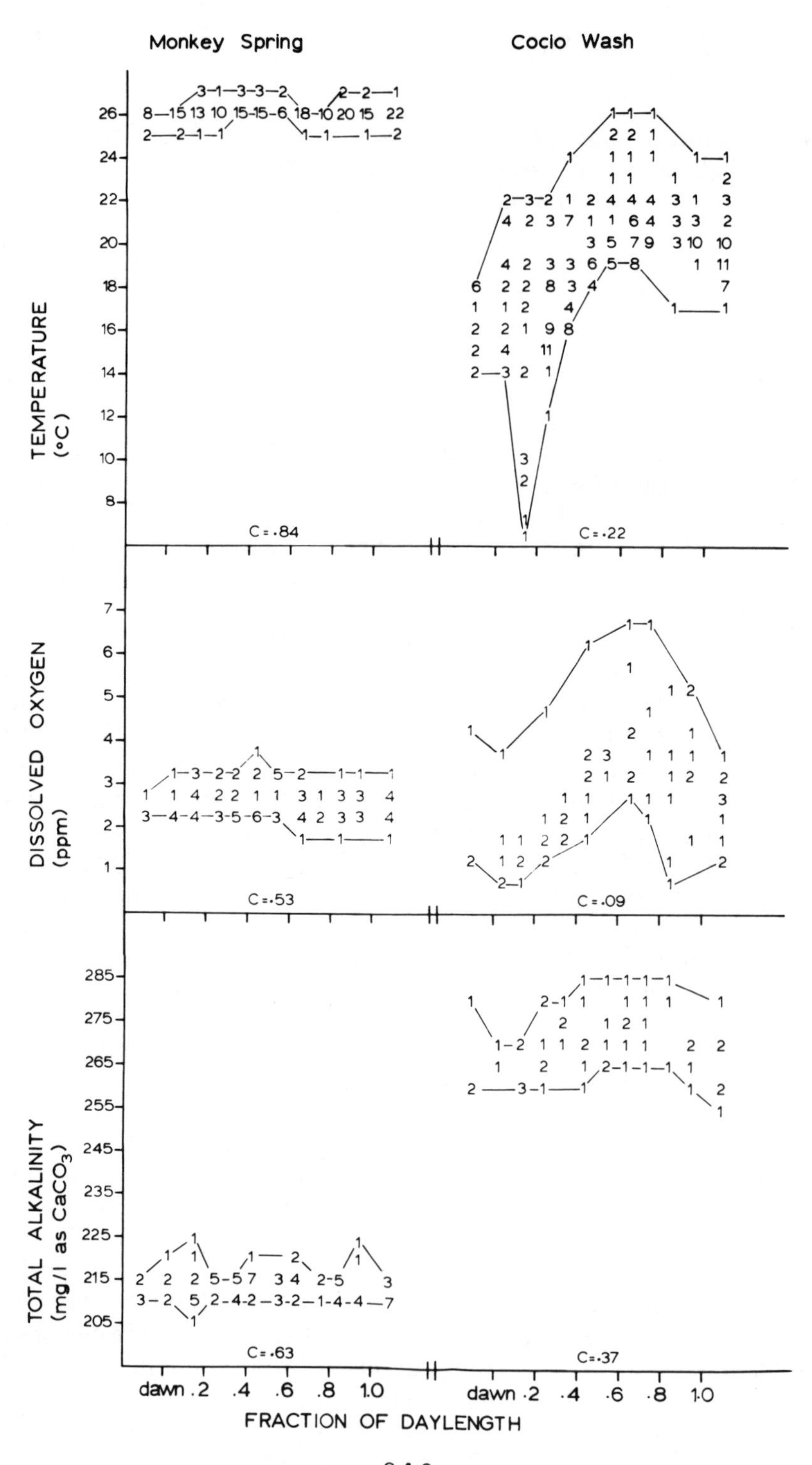
Monkey Spring
Cocio Wash
TEMPERATURE (°C)
DISSOLVED OXYGEN (ppm)
TOTAL ALKALINITY (mg/l as $CaCO_3$)
C = .84
C = .22
C = .53
C = .09
C = .63
C = .37
dawn .2 .4 .6 .8 1.0
FRACTION OF DAYLENGTH

in pools of desert washes. First, low flow may import plant material and dissolved nutrients which recharge the detrital cycle in the pool. Second, flash floods rearrange pool basins, scour algae and invertebrates, and displace fishes downstream.

Torrential spates have obvious consequences for fishes. Discharge ($m^3 \cdot s^{-1}$) may increase suddenly by five orders of magnitude. As flow subsides fish may disperse into vast temporary reaches that seem to offer lower fish density and more food. Then, with the coming of the dry season the original dispersers and their offspring may undergo a roulette competition (*sensu* Sale 1977) for refuge in the few permanent pools. Fundamental consequences of these phenomena are large interannual variation in reproductive success and population density (e.g., Barber and Minckley 1966; Lehtinen and Echelle 1979).

When compared with arroyos, desert springs and their immediate outflows are more closed, more autochthonous ecosystems, that fluctuate less in the face of local weather changes (Brown 1971; Constantz 1976; Cox 1966; Deacon 1968; Deacon and Deacon 1979; Deacon and Minckley 1974; Schoenherr 1974; Soltz 1974; Sumner and Sargent 1940). Least variable are outflows that have no surface drainage and are constantly warm because of deep origin (Figure 1). In such systems, where subterranean inputs are physicochemically constant, seasonal solar cycles are the primary agents of change in the ecosystem. Such temporal changes are gradual, noncatastrophic, and predictable, at least from the human observer's point of view. Because spring habitats are

---

FIGURE 1. Constancy of three environmental variables in desert springs and streams. Monkey Spring, a constant-condition rheocrene, and Cocio Wash, a permanent pool in a desert arroyo, both in the Sonoran Desert, Arizona, illustrate the difference in environmental variation between these two types of habitats. Each datum represents the number of times a value was observed at that particular hour throughout an annual cycle. Lines connect annual minima and maxima. C = constancy, calculated as defined by Colwell (1974).

usually small in volume (< 20 m across, < 3 m deep), the physicochemical environment within them may be spatially as well as temporally homogeneous.

These habitat characteristics of springs are correlated with an interesting biological consequence: few resident species and, therefore possibly low levels of interspecific interaction. Furthermore, the isolation of many springs contributes to their high degree of endemism (e.g., Hubbs and Peden 1969; Miller 1976; Stevenson and Buchanan 1973).

As defined above, desert streams and springs represent the ends of a continuum in degree of response to local meteorological events. Intermediate points are represented by variable springs of shallow origin, streams with intermittent subterranean input, and spring flow as it proceeds downstream. Cole (this volume) and Naiman (this volume) provide further details on fish habitats in deserts.

LIFE HISTORIES OF THE FISHES OF NORTH AMERICAN DESERTS

I have assembled information on specific environments, life history traits, and behavior of 16 fishes native to springs and streams of North American deserts (Table 6; see p.). This subset of the ichthyofauna was chosen because its natural history is best described in the literature. Accordingly, this sample may be biased in several ways. First, the geographic representation is Great Basin-8 species, Mohave-3, and Sonoran-11 (the total is greater than 16 because several species occur in more than one desert). In terms of taxonomic distribution, five cyprinodontids, three poeciliids, six cyprinids, and two catostomids appear in Table 6. Eight species occur in streams, four inhabit springs, and four taxa are apparently able to complete their life cycles in both habitats. Thus, I have categorized species into three habitat groups. In addition to the 16 focal species, data on other desert fishes are included when available. Procedurally, information on the 16 focal species is referenced in Table 6, whereas sources on additional fishes are cited in text.

As the informational center of this chapter Table 6 deserves description. In column 1 the 16 focal species are listed alphabetically within habitat categories. The suite of life history traits which requires explanation is included under columns 6 to 10. Two additional life history traits treated in text are not included in the table because so few data were available. Other columns provide information potentially useful for interpreting trends among the life history data. Columns 2, 3, 5, and 13 list various population characteristics. Body form (column 4) is integrally related to behavioral options such as social interactions (column 11) and dispersal ability (column 12).

The remainder of this and the next major section compare life history and other traits among species of the three habitat groups.

External Morphology

Stream fishes are predominantly fusiform (definitions follow Lagler et al. 1962), sometimes attenuated, in contrast to the more truncated shapes of spring species (Table 6, column 4). Generalist species may be fusiform, truncated, or of intermediate shape. Spring fishes have dorsal and anal fins in a position more posterior than that of stream species. Generalists fins are midway along the longitudinal axis. Fin roundness is also clearly habitat-specific: stream species have pointed fins, whereas spring taxa have rounded fins. It is not surprising that generalists exhibit both and intermediate types. Species with forked caudal fins are found primarily in streams; round-tailed forms inhabit springs. Again generalists are drawn from both groups. Some spring dwellers lack pelvic fins. In terms of mouth orientation spring fishes tend to have superior or terminal mouths and streams contain taxa with mouths at all orientations. Last, some spring endemics have partly degenerate lateral lines. These differences are illustrated in Figure 2.

The results of the above interspecific inference are consistent with the experience of others. The relict dace (*Relictus solitarius*) exemplifies the morphology of fishes native to desert springs (Hubbs et

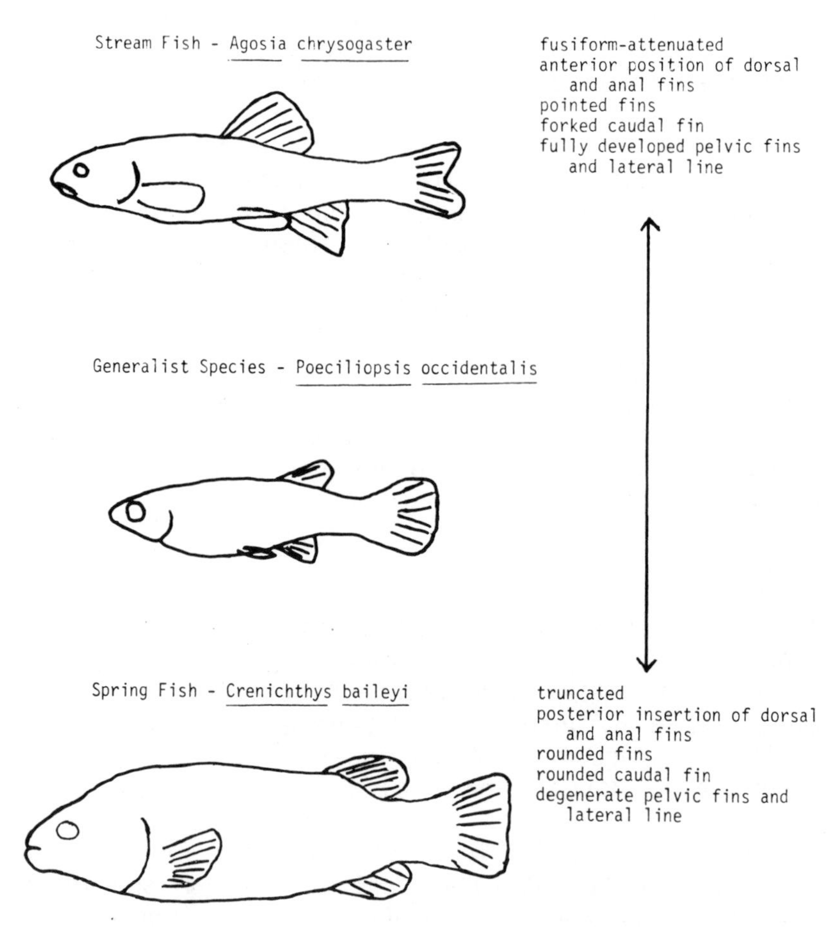

FIGURE 2. External morphology of desert fishes.

al. 1974): chubby body, rounded fins, and reduced sensory organs. Further, the external morphology of the speckled dace (*Rhinichthys osculus*) allows prediction of habitat type: spring dwellers are blunt-nosed, whereas riverine forms are more attenuated (Minckley 1973; Hubbs et al. 1974).

Food Habits

Food quality influences the evolution of life histories through its effects on growth rate, fat cycling, and degree of exposure to predators during feeding. Of the 16 focal species information on food habits was available for 12 (Table 6, column 5). Among five stream fishes one is carnivorous and the remainder is omnivorous; two spring and five generalist species are omnivorous.

By reviewing five state fish manuals (Koster 1957; LaRivers 1962; Minckley 1973; Moyle 1976; Sigler and Miller 1963) enough information was obtained to categorize a total of 34 desert fishes (Table 1). The number of species in each feeding group, by descending rank, is 25 omnivores, 7 carnivores, and 2 herbivores. The data (Table 1) indicate that all spring species are omnivorous, suggesting that specialized feeders cannot persist in desert springs. In contrast streams seem to support a few exclusively carnivorous and herbivorous fishes. It is possible that the more complex and productive stream ecosystems allow diet specialization, whereas the simple, short food chains of springs enforce omnivory.

Breeding Season

As suggested in subsequent sections, length of breeding season influences several life-history parameters such as size and age at sexual maturity, growth rate, and reproductive output. Adequate estimates are available for 13 focal species (Table 6, column 6).

Six stream fishes have restricted, although variable, breeding seasons. The longfin dace (*Agosia chrysogaster*) and woundfin (*Plagopterus argentissimus*) spawn over a longer period (January-November and April-August, respectively) than the roundtail chub (*Gila robusta*) and the spikedace (*Meda fulgida*) (June-July and May-June, respectively). In the riverine squawfish (*Ptychocheilus lucius*) the spawning season is quite restricted (June-August; Vanicek and Kramer 1969). In the Red River of northern Texas, *Cyprinodon rubrofluviatilis* displays a protracted breeding season (February-November; Echelle 1973); this apparent exception may be due to unpredictable periods of optimal breeding within a prolonged season of warm water (Echelle et al. 1973).

TABLE 1. Relationship Between Habitat Type and the Number of Desert Fish Species in Each Feeding Category. Two Taxa Illustrate Each Category

Habitat	Food Habits		
	Carnivory	Omnivory	Herbivory
Stream	5 *Tiaroga cobitis* *Ptychocheilus lucius*	14 *Agosia chrysogaster* *Xyrauchen texanus*	2 *Dionda episcopa* *Catostomus latipinnis*
Generalist	2 *Iotichthys phlegethontis* *Cyprinodon radiosus*	7 *Poeciliopsis occidentalis* *Rhinichthys osculus*	0
Spring	0	4 *Moapa coriacea* *Cyprinodon diabolis*	0

Spring fishes exhibit comparatively protracted, even year-round breeding. The White River killifish (*Crenichthys baileyi*), Devils Hole pupfish (*Cyprinodon diabolis*), and spring populations of Amargosa pupfish (*Cyprinodon nevadensis*) and Gila topminnow (*Poeciliopsis occidentalis*) apparently spawn throughout the year. Stream populations of the latter two generalists have comparatively restricted breeding seasons. Year-round reproduction, or at least significantly protracted breeding seasons, occurs in spring-dwelling fishes of other arid zones [*Etheostoma lepidum* (Hubbs et al. 1968); *Cyprinodon bovinus* (Kennedy 1977); *Etheostoma fonticola* (Schenck and Whiteside 1977); *Gambusia geiseri* (Davis 1978)]. The relict dace breeds from June through September in springs and may therefore be an interesting exception.

This comparison suggests that fishes in constant-condition desert springs breed throughout a longer portion of the year, than species of desert streams. Quantitatively, the comparison is as follows: (a) stream, $\bar{x}$ = 4.3 mo, range = 2-8 mo, n = 9; and (b) spring, $\bar{x}$ = 9.8 mos, range = 4-12 mo, n = 6. Viewed at this level, a longer spawning season does not indicate continuous spawning by individuals; instead it could be the collective result of asynchronous individuals.

Body Size at Sexual Maturity

Data on adult body size of desert fishes are rich enough to evaluate at three levels: interspecific and intraspecific differences between habitat types, and intrapopulation differences among seasons.

Estimates are available for all but one of the 16 focal species (Table 6, column 7). In addition, I have supplemented these data with information from four state manuals (LaRivers 1962; Minckley 1973; Moyle 1976; Sigler and Miller 1963). The collective results (Table 2) suggest that stream species mature at an average size larger than spring forms and that trends among focal species are consistent with the collective direction. Half the stream species are small-bodied (28 to 71 mm TL). Therefore both small and large species occur in streams but springs are occupied only by

TABLE 2. Distribution of Body Lengths of Fishes in Desert Streams and Springs[a]

	$\bar{x}$	Range	n
Streams			
Focal species	172.5 mm TL	28-800 mm TL	8
Supplemental species	203.1 mm TL	40-495 mm TL	7
Combined	186.8 mm TL	28-800 mm TL	15
Springs			
Focal species	31.3 mm SL	26-66 mm SL	3
Supplemental species	64.2 mm SL	51-76 mm SL	5
Combined	54.9 mm SL	26-76 mm SL	8

[a]TL = total length; SL = standard length; SL ≃ 0.85 TL.

small fishes (26 to 76 mm TL). It seems reasonable that the larger volume and greater habitat diversity of streams maintain a greater variety of body sizes than the smaller springs.

At the intraspecific level different selective regimes may favor different body lengths among allopatric populations. M. F. Hirshfield and D. L. Soltz (this volume) have developed some provocative data on size at sexual maturity of the Amargosa pupfish. By standardizing the environment of test individuals, minimizing maternal effects, and standardizing the body size of mothers they were able to attribute observed differences in body length at first breeding to genetic dissimilarities between stocks derived from a warm spring and a desert stream. The averages of their statistically significant differences from replicated tests were 24.0 and 25.9 mm SL for spring and stream females, respectively. This suggests that in contrasting environments selection has favored different alleles which influence growth rates.

The story of size at maturity is further complicated by the fact that minimum size at first breeding may vary seasonally. In the Gila topminnow I observed that such changes were a dominant feature of population dynamics (Constantz 1979). Males may also mature at different sizes throughout the year as an apparent function of the availability of fertilizable females (*Gambusia affinis*, Hubbs 1971). However, because these data come only from the Poeciliidae, these observations may have little to do with desert habitats.

In conclusion, fishes of desert springs and streams differ in body size at first breeding because of a variety of factors, depending on the level of analysis. Large-bodied species occur only in streams apparently because they require habitats more voluminous than those offered by springs. Individuals of allopatric populations may differ in size at maturity because each environment may favor different alleles that influence minimum size at breeding. Furthermore, natural selection may favor developmental plasticity in size of maturity because its optimum varies seasonally.

Age at Sexual Maturity

Generation times of desert fishes are poorly understood; among the 16 focal species estimates were available for four stream fishes, four generalists, and only one spring species (Table 6, column 8). Three forms that inhabit streams (*Agosia*, *Meda*, *Plagopterus*) spawn at one year of age, whereas *Gila*, the fourth, does not reproduce until five years old. A stream population of the generalist *Rhinichthys* bred at one year. In contrast, fishes in warm springs mature variably at a few months or at about one year. In warm springs the desert pupfish (*Cyprinodon macularius*), the Amargosa pupfish, and Gila topminnow mature at ages of a few (1 to 5) months or at slightly less than one year; maturation at intermediate ages appears to be unusual. The apparent exception of the spring-dwelling relict dace maturing at one year is based on superficial study. Thus field observations consistent at both inter- and intraspecific levels suggest a fundamental difference between spring and stream fishes of arid zones: fishes of desert streams mature later and with less plasticity, compared with spring dwellers that mature variably in a few months or at about one year of age.

It seems counterintuitive that fish in constant-condition springs should exhibit more variable generation times than stream forms. I suggest that from the point of view of an individual fish "constant-condition" springs may be less predictable than fluctuating desert arroyos. Because environmental cycles such as solar input and dependent primary production have a wavelength longer than the organisms's life history, a young spring fish cannot predict the phase (e.g., high or low food) into which it will be born. Conversely, the synchrony between longevity and annual cycle seen in stream fishes indicates that a desert wash may actually be more constant and predictable than a warm spring. As always, the scale of view is critical to relevant interpretation.

Fecundity

Among the 16 focal species estimates of the number of eggs carried per female were available for five stream, two spring, and four generalist species (Table 6, column 9). Five stream species carry an average of 2088 eggs/female in a range of 11 to 10,000, and two spring forms averaged 9 in a range of 4 to 19. Comparing arithmetic means, however, is misleading because in this case one species with huge egg numbers skews the distribution.

A more relevant comparison results when the variation among stream fishes is analyzed. The limited data suggest three levels of fecundity among fishes of desert streams: (a) fecundities of 1 to 10, _Poeciliopsis lucida_ and _P. monacha_, (b) 10 to 100, _Agosia chrysogaster_ and _Meda fulgida_ and stream populations of the generalists _Cyprinodon macularius_ and _Rhinichthys osculus_, and (c) 1000 to 10,000, _Gila robusta_.

Estimates of mean offspring number are also available for two generalist species, each with two populations in contrasting environments. The Amargosa pupfish carries an average of 10 eggs in Big Spring and 40 in the Amargosa River. Gila topminnows display a similar trend: 5 embryos in a warm spring and 14 in a wash.

Thus both inter- and intraspecific comparisons suggest that fishes of desert streams produce more young than the smaller bodied spring dwellers.

My interpretation of this pattern draws from both levels of comparison. As suggested before, desert streams seem more complex than springs and may offer a greater diversity of niches, which in turn support a wider variety of body sizes. Because reproductive mass is a function of body size, the same would apply. Further, in populations subjected to catastrophic mortality the amount of food per individual may be high. In such situations the gonadal mass could be limited by coelomic space, not food, in which case fecundity would be the result of gonadal mass divided by egg size. In contrast, in stable springs gonadal mass may be limited by food availability. This difference in the limits of gonad volume may be fundamental between spring and stream fishes. See Hubbs et al. (1968) and Constantz (1979) for fuller treatments.

Investment per Offspring

Estimates of offspring size are available for five stream species, two spring taxa, and three species that occur in both habitat types (Table 6, column 10; Table 3). Egg diameter of six stream fishes averaged 1.5 mm in a range of 0.9 to 2.1 and two spring populations averaged 1.7 mm in a range of 1.5 to 1.9. The limited data on this and other measures of size of young (Table 3) suggest that the mean of investment per offspring may not differ by habitat type.

Note that species with both small and large eggs are found in streams but that only large-egged taxa occur in springs. In another arid zone stream *Notropis boops* produces a relatively small (0.9 mm diameter) egg (Lehtinen and Echelle 1979).

Data are available on two generalist species in contrasting habitats (Table 3). Gila topminnow females produce eggs of 2.2 mg in a warm spring and 1.6 mg in a wash. Recent experiments by Hirshfield and Soltz (personal communication) suggest that significant differences in investment per offspring between spring and stream populations of the Amargosa pupfish are due to genetic influences. Egg diameters of fish derived from Big Spring and the Amargosa River were 1.39 and 1.34 mm, respectively.

TABLE 3. Offspring Size of Desert Fishes. This is a summary of Table 6, Column 13[a]

Taxa	Egg Diameter (mm)	Egg Volume (mm^3)	Egg Weight (mg)	Fry Length (mm)	Fry Weight (mg)
		Stream Species			
Agosia chrysogaster	1.6	2.1	3.0 wet	4.0 SL	
Gila robusta	0.8	0.3			
Meda fulgida	1.7	2.4			
Poeciliopsis lucida	1.4	1.4	0.5 dry	6.8 TL	0.7 dry
Poeciliopsis monacha	2.1	4.5	2.1 dry	8.5 TL	1.3 dry
		Generalist Species			
Cyprinodon macularius	~2			3.7–5.3 TL	
Cyprinodon nevadensis					
Spring habitat	1.39				
Stream habitat	1.34				
Poeciliopsis occidentalis	1.9	3.6			~3 dry
Spring habitat			2.2 dry	~8 SL	
Stream habitat			1.6 dry	~7 SL	

Spring Species			
Crenichthys baileyi	1.5–1.9	1.8–3.6	4.3–5.3 TL
Cyprinodon diabolis			6.5 TL

[a]Dry = dry weight, wet = wet weight, SL = standard length, TL = total length.

Diversity in egg sizes of stream fishes is consistent with the complexity of their environments. Two examples of this relationship follow. Poeciliopsis monacha, which inhabits torrential headwater arroyos, yolks eggs that are large in relation to other stream species, possibly an adaptation for producing strong-swimming offspring and as compensation for scarce food for juveniles. Etheostoma spectabile produces relatively large eggs in apparent adaptation to short growing seasons (Hubbs 1958).

In springs, where food may be scarce, the fundamental environmental influence of investment per offspring may be the amount of energy required to produce competitive offspring. However, as with stream fishes, subtle factors may be important; for example, in warm springs an upper limit on egg size may be imposed by a surface area-volume constraint on embryonic respiration (Hubbs 1958).

Data on offspring size is fraught with problems such as the lack of a standard stage for measurement and of an understanding of variation within populations. For example, seasonal variation in egg size in a single population could lead to biased estimates of a population. The Gila topminnow produces eggs that vary 22 percent by weight over an annual cycle (Constantz 1976); this is independent of female size and stage of egg development. Fishes in other arid zones also vary egg size seasonally (Etheostoma spectabile and E. lepidum produce larger eggs in cooler months; Hubbs et al. 1968).

The preceding observations suggest that investment per offspring may reflect a compromise among several selective factors: amount of food for fry, current speed, kind and density of predators, and temperature and oxygen concentration of the habitat. An even more subtle constraint may be a tradeoff with egg number when gonadal mass is limited by coelomic space. Even though the patterns of investment per offspring are richer than that predicted by r- and K-selection (MacArthur and Wilson 1967, Pianka 1970), the expectation that females will produce larger offspring in habitats of greater constancy and predictability is generally supported.

Breeding Frequency

Two life history traits, breeding frequency and reproductive effort, were not included in Table 6 because few estimates were available for desert fishes.

Any estimate of total reproductive output over a life cycle must incorporate estimates of instantaneous fecundity and rate of egg production. Estimates of breeding frequency of desert fishes in nature are rare. Spikedace produce one clutch per year as yearlings and two per year thereafter. Otherwise, laboratory estimates of interbrood intervals for three *Poeciliopsis* species range from 10 to 12 days. This is insufficient data for generalization.

Reproductive Effort

Reproductive output may be estimated by egg number or may be viewed at the energetic level. Reproductive effort is the proportion of the total energy budget allocated to reproduction (Williams 1966a). Its theoretical importance stems from its presumed tradeoff with subsequent adult survival and reproduction (Williams 1966b); for example, individuals that allocate more energy to reproduction might have fewer resources for somatic maintenance and thereby incur lower survivorship to the next breeding season. The concept of reproductive effort has assumed a central place in theory on the evolution of life histories (e.g., Stearns 1976; Tinkle 1969). Unfortunately measurements of reproductive effort are lacking for all but one desert fish (Constantz 1979).

In a desert wash Gila topminnows allocated 4.8 percent of their total energy budget to reproduction, whereas in a warm spring they invested 2.1 and 3.5 percent, depending on life history type. It is interesting that the direction of these differences is consistent with the prediction of *r*- and K-selection (MacArthur and Wilson 1967) that in environments imposing density-independent mortality food would be plentiful, maintenance requirements easily satisfied, and consequently reproductive output would be high.

These and other data (Deacon and Wilson 1967;

Hubbs et al. 1967; Sumner and Lanham 1942) suggest that reproduction in warm springs may be limited by high metabolic costs (e.g., oxygen uptake and digestion) which yield only a tiny remainder of energy for gonads. In contrast, the cooler, relatively oxygen-rich streams, with occasional density-independent mortality, may allow lower metabolism and provide more food *per capita*. It would be fascinating to have estimates of reproductive effort for relict dace and woundfin to evaluate this hypothesis.

Social Behavior

There appears to be significant coevolution between life history traits and social structure (Wilson 1975: Chapter 4); for example, competition in a social hierarchy may favor males that mature at a large body size and, consequently, at an older age. Of interest are the following variables: spawning system, territoriality, sexual dichromatism, and alternate mating strategies. Although only limited data (Table 6, column 11) are available for five stream fishes, four spring forms, and five generalists, they are provocative.

All five stream taxa have external fertilization and spawn in aggregate, a condition in which two or more males attempt to spawn simultaneously with each female. *Fundulus kansae* of semiarid streams (Koster 1948) and *Xyrauchen texanus* of the Colorado River (Sigler and Miller 1963) also fit this pattern. Aggregative spawning has not been reported for any spring dwellers. This difference appears to be correlated with the degree of territoriality and sexual dichromatism: male spring fishes are often intensely territorial and brightly colored. It would seem that monopolizing a female is more difficult in larger flowing environments than in smaller standing waters. Thus selection in springs might favor males that defend areas in which females could be courted with little interference (*sensu* Arnold 1976), in contrast to the scramble competition that occurs in streams.

When large males aggressively monopolize females, selection may favor some small males that forego agonistic competition and instead constantly pursue females. A hierarchical system, plus long breeding seasons in which males may mature at a wide range of body sizes, may allow selection of alternative mating strategies (*sensu* Alcock 1979). In spring populations of generalist species (Gila topminnow, Amargosa pupfish) males exhibit such a polymorphism: large males defend territories in which they court females and small males avoid fighting, remain in satellite positions, and pursue females throughout the habitat (Constantz 1975; Soltz 1974). Yet satellite males are not obvious in stream populations of these generalists. It is possible that this dimorphism does not occur in fishes of streams because the short breeding season produces a new generation of males that matures in one year, as a relatively unified cohort, and within a small range of body size.

Dispersal Ability

I distinguish between two types of movement by fish in desert streams: displacement and dispersal. Displacement occurs when individuals are unable to maintain position in heavy flow. When displacement occurs regularly populations are presumably maintained when upstream reproduction exceeds downstream displacement (e.g., John 1964). I do not interpret this form of movement to be adaptative because there appears to be no chance of displaced individuals or their descendants returning to permanent water.

Dispersal is the movement of individuals under their own power from permanent pools to temporary waters. Although mass mortalities may occur in drying pools, temporary waters probably offer some benefit such as more food or fewer predators. A subsequent section develops an hypothesis on the adaptive significance of dispersal.

The extent to which spring species would disperse, given removal of their geologic barriers, is unknown but their chubby shapes and rounded fins suggest poor dispersal potential. Two stream forms for which information is available, longfin dace and stream populations of speckled dace, disperse over vast reaches of temporary water. Their fusiform body shape and wide geographical range are consistent with this behavior. *Lepidomeda vittata* is an analogous montane example of a stream fish that has spread from tiny refugia to repopulate entire drainages (Minckley and Carufel 1967).

Three generalists (desert pupfish, Amargosa pupfish, and Gila topminnow) have modest dispersal abilities: after rains they swim into local (~0.5 km) temporary waters.

Population Characteristics

Limited information on two population parameters, sex ratio and population density, was available for desert fishes.

Estimates of sex ratio are available for four focal and three additional arid zone fishes (Table 4). Two views of these data are interesting. First, in terms of habitat type, three of four stream forms had males in equal or greater proportion than females, whereas the warm spring relict dace populations were predominantly females. Second, populations in warm waters contained fewer males (male/female ratios of 0.28, 0.43, 0.66) than populations in cooler water (0.38, 2.56, 0.94, 0.47, 1.0). These limited data suggest that populations in cooler water may have greater proportions of males than those in warmer areas. The low proportion of males in the stream-dwelling spikedace devalues this generalization.

At the intraspecific level differences similar to the preceding are seen in the Gila topminnow and mosquitofish (*Gambusia affinis*) (Table 4). There are greater proportions of males in a stream population of Gila topminnows (Constantz 1976) and in the cool reaches of a thermal gradient inhabited by mosquitofish (Winkler 1973) than in warmer habitats. This suggests that water temperature may affect the sex ratio of these fishes. Winkler suggests that females are physiologically more resistant to heat; I conjecture that male copulatory behavior may be accelerated by warmer temperatures, increasing maintenance costs, and thereby accelerating aging.

TABLE 4. Sex Ratios of Desert Fishes

Taxa	Sex Ratio (M/F)	Reference
Stream Populations		
Agosia chrysogaster (desert rheocrene)	0.94	Lewis (1978)
Meda fulgida (desert stream)		Barber et al. (1970)
Age group 0	1.0	
Age group II	0.47	
Ptychocheilus lucius (river)	>1.0	Vanicek and Kramer (1969)
Etheostoma fonticola (warm headwaters)	1.39[b]	Schenck and Whiteside (1977)
Spring Populations		
Relictus solitarius (warm springs)		Hubbs et al. (1974)
Immatures	0.78	
Adults	0.66	
Generalist Species		
Gambusia affinis (artesian well outflow)		Winkler (1973)
Warmer reach	0.43	
Cooler reach	2.56	
Poeciliopsis occidentalis		Constantz (unpublished data)
Desert stream	0.38 (range = .31 - .52)	
Warm spring	0.28 (range = .05 - .41)	

[a]Not significantly different from 1.0.
[b]Significantly different from 1.0.

Five of the six stream forms (longfin dace, desert pupfish, desert sucker, *Poeciliopsis monacha*, and speckled dace) for which information is available exhibit large variations of reproductive success and survival over several years (Table 6, column 13). At the population level these individual responses result in large interannual changes in local density, even extinction, yet some stream species seem unaffected. For example, *Poeciliopsis monacha* exhibits large interannual variation in local abundance and periodic local extinctions, whereas the roundtail chub displays no apparent relationship between discharge pattern and year-class strength. This comparison is not parallel, but it is relevant because the former relationship probably contributes to the latter in desert streams. The speckled dace may exhibit an intermediate response, and even though its populations are reduced by catastrophes local extinctions seem rare.

In contrast, population densities seem to change less in warm springs. In the Devils Hole and Amargosa pupfishes, population numbers and structures are comparatively stable. The Devils Hole pupfish cycles between 200 individuals in October and 700 in August, primarily as a function of solar input and dependent algal production. In small warm springs the Amargosa pupfish maintains relatively stable numbers and size distribution throughout the year, compared with the variation seen in conspecific populations in streams.

This limited qualitative information suggests that the size and structure of fish populations may be more stable in desert springs than streams. Large interannual fluctuations in population characteristics have also been documented for desert lizards (Turner et al. 1970; Zweifel and Lowe 1966) and birds (Keast 1959; Ohmart 1973) in apparent response to the vagaries of rainfall.

GENERALIZED LIFE HISTORIES OF DESERT FISHES

Under the first two subheadings of this section I integrate the characteristics induced in preceding sections into generalized life histories of fishes in desert streams and springs. The last section reviews the life histories of two generalist species.

Stream Fishes

Permanent pockets of water along dry desert streambeds are generally closed, internally recycling ecosystems which are infrequently subjected to unpredictable water inputs. I recognize two levels of flow: (1) modest discharge that imports nutrients; (2) flash floods that reshape pools and displace organisms. When compared with springs, arroyos are typically larger and more variable in volume, cooler, more oxygen-rich, more variable in physicochemical characteristics, and biogeographically less isolated. Fluctuations in habitat volume may be enormous and somewhat unpredictable. The relatively complex structure of streams seems to support more species and therefore more kinds of biotic interactions than springs. Irregular allochthony, a richer fauna, and larger habitat volume contribute to an ecosystem that is more heterogeneous spatially and temporally than springs. Seasonal changes in desert streams are effected by variations of solar input, air temperature, and precipitation. Occasional catastrophic mortality of fish caused by their downstream displacement and desiccation in isolated pools may result in high food per capita and in reproductive output limited by body space. Population densities may be at levels below carrying capacity during much of the year. Last, comparatively high species diversity could contribute to biotic interactions such as predation and interspecific competition, which may be the less true for desert springs.

Fishes of desert streams are fusiform and have pointed fins, both of which enable persistence in turbulent waters and promote wide dispersal. Even though many of the fishes in desert streams are omnivorous, the ecosystem complexity and abundant food seem to support a few relatively specialized herbivores and carnivores. The relatively short period of warm water, controlled by local weather, restricts breeding to an average of four months. Streams include large-bodied species not observed in springs primarily because of their larger habitat volume. A large proportion of stream fishes are of the same body size as spring fishes, probably because of small-volume niches in streams, such as those of the littoral zone. Age at maturity follows the pattern described for body size.

Because the complex habitat of streams harbors a fairly large diversity of body sizes, gonadal masses are equally variable. Small-egged fishes occur only in streams. Thus when a large-bodied stream fish produces small eggs huge fecundity may result. Consequently, among desert fishes large numbers of eggs are produced only by some stream species. In addition, many stream taxa carry eggs of comparable number to that of spring fishes. This illustrates the generalization that the greater complexity of streams is correlated with a richer diversity of life history traits than found in springs.

A higher proportion of stream fishes exhibits aggregative spawning, possibly because sequestering a female is difficult in flowing, open habitat. Concomitantly, male fish in streams show little territoriality and are less brightly colored than spring species. A relatively shorter breeding season may restrict the period in which males mature, thereby minimizing the variance in body size. Habitat openness and similar size may contribute to a low selective advantage for alternative mating strategies by males.

After rain individuals disperse throughout temporary waters to feed and reproduce. Fusiform bodies and omnivorous diets enhance their dispersal. All three adaptations contribute to large geographical ranges.

Stream taxa may consist of a higher proportion of males than spring species because males may live longer in the cooler water of streams. Stream populations display major fluctuations in density in response to changes in food levels, habitat volume, and catastrophes.

Spring Fishes

Desert springs are typically small in volume; issue water that is warm, often oxygen-poor, and relatively constant in its physicochemical characteristics;

and are often geographically isolated. These environmental features have significant biogeographic consequences: few species and high rates of endemism. The closed, autochthonous habitat, combined with a depauperate fauna, results in a relatively simple ecosystem. Absence of large-bodied fishes and low species diversity appears to result in low levels of biotic interactions such as predation and interspecific competition. Because surface inflow is often insignificant and the subterranean source displays only minor changes in discharge volume, variations in habitat volume are trivial. Their small volume, ecological simplicity and constant water input result in an environment that is spatially and temporally homogeneous. Seasonal changes appear to be effected primarily by changes in the solar cycle, an exceedingly gradual and predictable progression. The stable environment and lack of catastrophic mortality allow individuals to maximize reproduction by tracking food availability. Thus food supply may limit the size of spring populations.

Fishes native to spring habitats seem to have been selected, at inter- and intraspecific levels for a specific suite of life history characteristics. Chubby body shape, posterior dorsal and anal fins, and rounded fins seem adaptive to maintaining a midwater position in standing or slowly flowing water. A lack of overt behavioral interaction with predators or competitors may have allowed the lateral line of some forms to degenerate. Omnivory seems to be enforced by the limited availability of a few dominant food species of algae and invertebrates. Breeding occurs over a season that averages 10 months and frequently may continue year-round; this is primarily in response to long periods of warm water and adequate food for young. Springs are inhabited exclusively by small-bodied fishes because of the small habitat volume and possibly because of low food availability. These small-bodied taxa mature within one year.

Spring fishes have few offspring because such small-bodied adults produce a small, possibly food-limited gonadal mass, which is then divided among fairly large eggs. Investment per offspring is high, apparently because of the priority of stocking young in a food-limiting environment. An egg's maximum size may be constrained by its respiratory requirements. Small-egged fishes do not occur in springs.

Large dominant males of spring forms defend their territories aggressively and display bright colors. Sexual monopolization of females is apparently facilitated in standing shallow water; this favors territoriality among spring fishes. Alternative mating strategies by males seem to be a consequence of variable size at maturity, a result of long breeding seasons, and of the hierarchical nature of male interactions.

At the population level spring forms had fewer males than stream populations, possibly because warm water accelerates aging of males. Spring populations of desert fishes seem to undergo smaller fluctuations in density, possibly because influences of birth and death rates, such as levels of food and predation, may be fairly constant throughout the year.

Generalist Species

If the differences outlined above between spring and stream species have an ecological basis, similar differences, but of lesser magnitude, should occur among conspecific populations in similarly contrasting habitats. Two generalist species have been studied with the express purpose of evaluating such life history differences (Constantz 1979; Soltz 1974).

The Amargosa pupfish and Gila topminnow occur in warm constant-condition springs, desert arroyos, marshes, ephemeral pools, and quiet shallows of rivers. Their body forms are intermediate along the attenuated-truncated continuum, which is consistent with their modest abilities to disperse. Both species are omnivorous. In spring habitats female *C. nevadensis* and *P. occidentalis* mature sexually at 19 and 20 mm SL, respectively. In flowing waters topminnow females matured at the same size as they did in springs (20 mm SL), whereas riverine pupfish matured at 24 mm SL, 4 mm larger than conspecifics in springs. Both have generation times that are more variable in springs than in streams; and both fishes reproduce year-round in warm springs but restrict their breeding in streams to the period between spring and fall. Both produce more and smaller offspring in washes than in springs. Although a measure of reproductive effort was not available for *C. nevadensis*, female *P. occidentalis* in the wash showed higher effort than spring females.

In terms of social behavior both species displayed greater territoriality in the more stable shallow areas of springs and littoral areas of rivers. Alternate mating strategies by males of both species occurred in springs. Their powers of dispersal seem comparable: both invade local ephemeral areas but have not been reported to repopulate entire drainages.

One inconsistency appeared in the preceding comparison: Amargosa pupfish females matured sexually at larger body sizes in a stream than in a spring, whereas Gila topminnows revealed no variation between populations, a difference for which I am unable to offer an explanation. Other than this single dissimilarity, there is a striking concordance in the life histories between spring and between stream populations of *Cyprinodon nevadensis* and *Poeciliopsis occidentalis*.

This section has revealed consistency at inter- and intraspecific levels. I consider this consistency to be strong evidence that the suites of life history traits previously identified for fishes of springs and of streams in deserts are real and general.

Causes of the Trends in Life History Patterns

What accounts for the general differences in the life histories of fishes in desert streams and springs? Three potential causes may explain the trends: (1) water temperature, (2) homology, and (3) analogy.

All eight focal stream species inhabit waters of cooler and more variable temperatures than spring waters and all four focal spring forms occur in constantly warm water. Data were not available on the life histories of fishes in cool fluctuating springs or in constantly warm streams, possibly because such habitats do not exist in North American deserts. This lack of cross-habitat and cross-temperature comparisons confounds my attempt to understand the contribution of these two factors to the general patterns. Life history features that could be potentially affected include maintenance costs and thereby reproductive effort, length of breeding season, and time to first breeding.

Similarity among inhabitants of one habitat type may be due to evolutionary homology. The important fish families of North American deserts and those treated in this review are Cyprinidae, Cyprinodontidae, Catostomidae, and Poeciliidae. Assuming that the fish community of this geographical area in the pluvial period was comprised of these four families (see G. R. Smith, this volume), the process of regional desiccation may have resulted in certain types of fish (families) that persisted in each habitat type. Cyprinids and catostomids persisted in streams, whereas cyprinodontids and poeciliids may have been preadapted to spring habitats. To the extent that family characteristics resulted in such interspecific selection between habitat types my inferences of habitat-specific life histories may simply be artifacts of family characteristics. Among the focal species three-fourths of the stream fishes are cyprinids and catostomids, whereas a similar proportion of the spring species are cyprinodontids. This proportional difference is consistent with a process of interspecific selection, which indicates that some of the general inferences are probably due to homology among species within habitat types.

Once deserts arose in North America, similarities among species within habitats may have evolved as a result of similar selection pressures. Two sets of hypotheses are available to evaluate the possibility that convergence has resulted in analogies: (1) *r*- and K-selection (MacArthur and Wilson 1966); (2) bet-hedging (Stearns 1976). Present data speak little to the latter because its predictions are based on measures of the constancy and predictability of age specific survival, parameters that are poorly understood for desert fishes.

Generally, the inferred differences between spring and stream fishes are consistent with predictions of the theory of *r*- and K-selection. Relative to desert streams, spring environments that presumably offered constant and predictable density-dependent mortality were inhabited by species and populations characterized by fewer and larger offspring. Even the early-maturing cohort in spring populations seems to be consistent with expectations if we accept the possibility that the period of optimal breeding is unpredictable to a young fish. I state this overall generalization with reservation because measures of relative food availability, the parameter on which these predictions are based, were not available.

In conclusion, the general differences between the life histories of fishes in desert streams and springs may be due to one or all three of the following reasons. First, the warmer temperatures of springs probably result in little energy available for growth and reproduction; this could be totally unrelated to environmental constancy, food availability, or other factors considered in this chapter. Second, persistence of related taxa within each habitat type has surely contributed to the similarity of life histories among species within habitats. Third, those same conditions that selected among species to create homologies may also have been selecting for differences among conspecific populations to produce some degree of analogy within habitats. Although I believe that these three factors account for most of the observed trends, it would be conjecture to rank their relative importance.

SPECIAL TOPICS

An Hypothesis on Dispersal in Temporary Waters

Following rains and enlargement of habitat volume, some desert fishes quickly occupy inundated areas; for example, *Clarias gariepinus* migrates onto flooded plains for spawning (Bruton 1979), longfin dace repopulate many kilometers of ephemeral washes (Minckley and Barber 1971), and Gila topminnows swim from permanent pools to local, temporary reaches (Constantz 1976). Welcomme (1969) documents other cases. If massive deaths often occur in drying pools (Barlow 1958a; Constantz, personal observation; Deacon and Minckley 1974; Miller 1976), why should fish leave the security of permanent pools? In other words, what are the evolutionary costs and benefits of occupying temporary waters?

I have made an initial attempt (Constantz 1976) to quantify the cost-benefit balance for dispersing and sedentary subsets of the Gila topminnow population at Cocio Wash, a small stream in the Sonoran Desert. Specifically, I evaluated the hypothesis that risk of desiccation to dispersing individuals is compensated by increases in offspring production; in contrast, sedentary individuals should be playing a low risk-low profit game.

At Cocio Wash topminnows persisted over an extended dry season in one permanent pool. With flow during the rainy season some individuals swam upstream and occupied temporary pools. Maximum longevity of such pools without additional flow was 60 days. The odds of a female and her offspring gaining refuge in the permanent pool before the dry season sets in is the probability of flow 60 days after occupancy of temporary pools. In August this probability was 0.55; conversely, the probability of no flow, and therefore of death by desiccation, is 0.45. At this time females in the temporary pool carried 52 percent more embryos and reproductive mass than females in the permanent pool (Table 5). This single observation is consistent with the hypothesis that the benefits of dispersal (more food, which allowed a 52 percent increase in fecundity) equal or exceed its costs (0.45 probability of death by desiccation). Dempster (1968) described an analogous example in which dispersing insects had higher reproductive success than sedentary individuals. Conversely, fish should not disperse when the fecundity differential does not compensate for the odds of death by desiccation.

I observed that only part of the Gila topminnow population at Cocio Wash invaded ephemeral waters. If such behavioral variability were widespread among fishes of desert streams, it would present an interesting question. What determines sedentary versus dispersing individuals? Although admittedly conjecture, both genetic and physiological mechanisms could be involved. It is possible that a genetic polymorphism for sedentary and dispersing genotypes may be maintained by

TABLE 5. Comparison of *Poeciliopsis occidentalis* in Permanent and Ephemeral Pools, Cocio Wash, AZ[a]

Dependent Variable	Regression Statistics[b]					Fecundity (calculated values)	
	r	*p*	*a*	*b*	*n*	30 mm	35 mm
Reproductive weight (mg dry weight)							
Home pool	0.23	0.24	-9.10	6.69	12	6.1	17.0
Temporary pool	0.60	0.01	-7.42	5.76	14	12.3	29.8
Fecundity (number of eggs and embryos)							
Home pool	0.32	0.16	-9.93	7.10	12	3.6	10.8
Temporary pool	0.63	0.01	-8.65	6.44	14	7.3	19.7

[a]Regression analyses relate the independent variable of body length ($\log_{10}$ mm SL) to two dependent life history variables ($\log_{10}$) for female *Poeciliopsis occidentalis* larger than or equal to the minimum reproductive size on the day of sampling (18 August 1974).

[b]*r* = correlation coefficient; *p* = probability that *r* is not significantly different from zero; *a* = Y-intercept; *b* = slope; *n* = sample size. Fecundity was calculated with the corresponding regression equations.

a cyclical combination of frequency- and density-dependent selection. Thus the genetic-control hypothesis (Chitty 1960; Krebs et al. 1973) proposed for rodent population cycles could apply to the boom-bust populations of fishes in desert streams. Overlaying this genetic base could be individual response to proximate cues, that is, the individual's immediate physiological condition (_sensu_ Wilbur and Collins 1973); for example, it is possible that thin individuals with relatively low reproductive output may have more to gain by dispersing than fat, highly fecund females.

Life History Adaptations of Fishes in Other Deserts

There appear to be no striking differences between the life histories of fishes in North American deserts and fishes in surrounding biomes. Unique adaptations are evident in other arid zones, however; two such traits improve resistance to desiccation: (1) diapause of eggs (Myers 1952; Wourms 1972) and (2) estivation by adults (Smith 1956, 1961). Much less specialized is the drought-coping behavior of the longfin dace in the Sonoran Desert: when daytime flow is low, adults remain under moist algae, emerging only at night to forage in a few millimeters of water (Minckley and Barber 1971).

I suggest that the fishes of North American deserts are not unique among their continental ichthyofauna for two reasons. First, taxa that colonized North American deserts did not possess precursors for such traits. Second, new adaptations have not evolved in situ because North American deserts are probably only 12,000 years old (see M. L. Smith, this volume), seemingly an insufficient time for evolving traits uniquely different from those of the original invaders.

ACKNOWLEDGMENTS

During the preparation of this review I was supported by research funds from the Division of Limnology and Ecology, Academy of Natural Sciences. R. Vannote,

Director of the Stroud Water Research Center, provided office space for writing. B. Weir, Academy librarian, cornered evasive references. For unpublished observations I thank J. N. Cross, J. E. Deacon, M. F. Hirshfield, J. N. Rinne, R. E. Thibault, and C. D. Vanicek. The following people offered helpful criticisms of the manuscript: C. E. Goulden, M. F. Hirshfield, R. J. Horwitz, R. J. Naiman, D. Reznick, and two anonymous referees. N. Ailes helped to prepare the final draft.

REFERENCES

Alcock, J. 1979. Animal behavior: an evolutionary approach. 2nd ed. Sinauer Associates, Sunderland, MA

Arnold, S. J. 1976. Sexual behavior, sexual interference and sexual defense in the salamanders *Ambystoma maculatum*, *Ambystoma tigrinum*, and *Plethodon jordani*. Zeitschrift fur Tierpsychologie 42:247-300.

Barber, W. E. and W. L. Minckley. 1966. Fishes of Aravaipa Creek, Graham and Pinal counties, Arizona. Southwestern Naturalist 11:313-324.

Barber, W. E., D. C. Williams and W. L. Minckley. 1970. Biology of the Gila spikedace, *Meda fulgida*, in Arizona. Copeia 1970:9-18.

Barlow, G. W. 1958a. High salinity mortality of desert pupfish, *Cyprinodon macularius*. Copeia 1958:231-232.

Barlow, G. W. 1958b. Daily movements of desert pupfish, *Cyprinodon macularius*, in shore pools of the Salton Sea, California. Ecology 39:580-587.

Barlow, G. W. 1961. Social behavior of the desert pupfish, *Cyprinodon macularius*, in the field and in the aquarium. American Midland Naturalist 65: 339-359.

Brown, J. H. 1971. The desert pupfish. Scientific American 225:104-110.

Brown, J. H. and C. R. Feldmeth. 1971. Evolution in constant nd fluctuating environments: thermal tolerances of desert pupfish (*Cyprinodon*). Evolution 25:390-398.

Bruton, M. N. 1979. The breeding biology and early development of Clarias gariepinus (Pisces:Clariidae) in Lake Sibaya, South Africa, with a review of breeding in species of the subgenus Clarias (Clarias). Transactions of the Zoological Society of London 35:1-45.

Chitty, D. 1960. The natural selection of self-regulatory behavior in animal populations. Proceedings of the Ecological Society of Australia 2:51-78.

Cole, G. A. 1968. Desert limnology. In: G. W. Brown, Jr. (Ed.), Desert Biology I. Academic Press, New York. pp. 423-485.

Colwell, R. K. 1974. Predictability, constancy, and contingency of periodic phenomena. Ecology 55: 1148-1153.

Constantz, G. D. 1974. Reproductive effort in Poeciliopsis occidentalis. Southwestern Naturalist 19:47-52.

Constantz, G. D. 1975. Behavioral ecology of mating in the male Gila topminnow, Poeciliopsis occidentalis (Cyprinodontiformes:Poeciliidae). Ecology 56:966-973.

Constantz, G. D. 1976. Life history strategy of the Gila topminnow, Poeciliopsis occidentalis: a field evaluation of theory on the evolution of life histories. Ph.D. Dissertation, Arizona State University, Tempe.

Constantz, G. D. 1979. Life history patterns of a livebearing fish in contrasting environments. Oecologia 40:189-201.

Constantz, G. D. 1980. Energetics of viviparity in the Gila topminnow. Copeia 1980:876-878.

Courtois, L. A. and S. Hino. 1979. Egg deposition of the desert pupfish, Cyprinodon macularius, in relation to several physical parameters. California Fish and Game 65:100-105.

Cowles, R. B. 1934. Notes on the ecology and breeding habits of the desert minnow, Cyprinodon macularius Baird and Girard. Copeia 1934:40-42.

Cox, J. C. 1966. A behavioral and ecological study of the desert pupfish (Cyprinodon macularius) in Quitobaquito Springs, Organ Pipe National Monument, Arizona. Ph.D. Dissertation, University of Arizona, Tucson.

Crear, D. and I. Haydock. 1971. Laboratory rearing of the desert pupfish, Cyprinodon macularius. Fishery Bulletin 69:151-156.

Cross, J. N. 1978a. Status and ecology of the Virgin River roundtail chub, Gila robusta seminuda (Osteichthyes:Cyprinidae). Southwestern Naturalist 23:519-528.

Cross, J. N. 1978b. Contributions to the biology of the woundfin, Plagopterus argentissimus (Pisces: Cyprinidae), an endangered species. Great Basin Naturalist 38:463-468.

Davis, J. R. 1978. Reproductive seasons in Gambusia affinis and Gambusia geiseri (Osteichthyes: Poeciliidae). Texas Journal of Science 30:97-99.

Deacon, J. E. 1968. Ecological studies of aquatic habitats in Death Valley National Monument, with special reference to Saratoga Springs. Final Report of Research to National Park Service under Contract No. 14-10-0434-1989.

Deacon, J. E. and W. G. Bradley. 1972. Ecological distribution of fishes of Moapa (Muddy) River in Clark County, Nevada. Transactions of the American Fisheries Society 101:408-419.

Deacon, J. E. and M. S. Deacon. 1979. Research on endangered fishes in the national parks with special emphasis on the Devils Hole pupfish. Proceedings of the 1st Conference on Scientific Research in National Parks, Vol. 1. U.S. Department of Interior, National Park Service Transactions and Proceedings Series No. 5:9-19.

Deacon, J. E. and W. L. Minckley. 1974. Desert fishes. In: G. W. Brown, Jr. (Ed.), Desert Biology II. Academic, New York pp. 385-488.

Deacon, J. E. and B. L. Wilson. 1967. Daily activity cycles of Crenichthys baileyi, a fish endemic to Nevada. Southwestern Naturalist 12:31-44.

Dempster, J. P. 1968. Intra-specific competition and dispersal: as exemplified by a psyllid and its anthocorid predator. *In*: T. R. E. Southwood (Ed.), Insect abundance. Symposium of the Royal Entomological Society of London No. 4, pp. 8-17.

Echelle, A. A. 1973. Behavior of the pupfish, *Cyprinodon rubrofluviatilis*. Copeia 1973:68-76.

Echelle, A. A., S. Wilson and L. G. Hill. 1973. The effects of four temperature-daylength combinations on ovogenesis in the Red River pupfish, *Cyprinodon rubrofluviatilis* (Cyprinodontidae). Southwestern Naturalist 18:229-239.

Espinosa, F. A., Jr. 1968. Spawning periodicity and fecundity of *Crenichthys baileyi*, a fish endemic to Nevada. M.S. Thesis, University of Nevada, Las Vegas.

Fisher, S. G. and W. L. Minckley. 1978. Chemical characteristics of a desert stream in flash flood. Journal of Arid Environments 1:25-33.

Gerking, S. D., R. Lee and J. B. Shrode. 1979. Effects of generation-long temperature acclimation on reproductive performance of the desert pupfish, *Cyprinodon n. nevadensis*. Physiological Zoology 52:113-121.

Greenfield, D. W. and G. D. Deckert. 1973. Introgressive hybridization between *Gila orcutti* and *Hesperoleucus symmetricus* (Pisces:Cyprinidae) in the Cuyama River basin, California. II. Ecological aspects. Copeia 1973:417-427.

Harrell, H. L. 1978. Response of the Devil's River (Texas) fish community to flooding. Copeia 1978: 60-68.

Hubbs, C. 1958. Geographic variations in egg complement of *Percina caprodes* and *Etheostoma spectabile*. Copeia 1958:102-105.

Hubbs, C. 1971. Competition and isolation mechanisms in the *Gambusia affinis* X *G. heterochir* hybrid swarm. Bulletin of the Texas Memorial Museum No. 19:1-46.

Hubbs, C. and W. F. Hettler. 1964. Observations on the toleration of high temperatures and low dissolved oxygen in natural waters by *Crenichthys baileyi*. Southwestern Naturalist 9:245-248.

Hubbs, C. and A. E. Peden. 1969. Gambusia georgei sp. nov. from San Marcos, Texas. Copeia 1969:357-364.

Hubbs, C., R. C. Baird and J. W. Gerald. 1967. Effects of dissolved oxygen concentration and light intensity on activity cycles of fishes inhabiting warm springs. American Midland Naturalist 77:104-115.

Hubbs, C., M. M. Stevenson and A. E. Peden. 1968. Fecundity and egg size in two central Texas darter populations. Southwestern Naturalist 13:301-324.

Hubbs, C. L., R. R. Miller and L. C. Hubbs. 1974. Hydrographic history and relict fishes of the north-central Great Basin. Memoirs of the California Academy of Sciences 7:1-259.

James, C. J. 1969. Aspects of the ecology of the Devil's Hole pupfish, Cyprinodon diabolis Wales. M. S. Thesis, University of Nevada, Las Vegas.

John, K. R. 1963. The effect of torrential rains on the reproductive cycle of Rhinichthys osculus in the Chiricahua Mountains, Arizona. Copeia 1963: 286-291.

John, K. R. 1964. Survival of fish in intermittent streams of the Chiricahua Mountains, Arizona. Ecology 45:112-119.

Keast, A. 1959. Australian birds: their zoogeography and adaptations to an arid continent. In: A. Keast, R. L. Crocker, C. S. Christian (Eds.), Biogeography and ecology in Australia. W. Junk, The Haag, pp. 89-114.

Kennedy, S. E. 1977. Life history of the Leon Springs pupfish, Cyprinodon bovinus. Copeia 1977:93-103.

Kinne, O. 1960. Growth, food intake, and food conversion in a euryplastic fish exposed to different temperatures and salinities. Physiological Zoology 33:288-317.

Kinne, O. and E. M. Kinne. 1962a. Rates of devlopment in embryos of a cyprinodont fish exposed to different temperature-salinity-oxygen combinations. Canadian Journal of Zoology 40:231-253.

Kinne, O. and E. M. Kinne. 1962b. Effects of salinity and oxygen on developmental rates in a cyprinodont fish. Nature 193:1097-1098.

Kodric-Brown, A. 1977. Reproductive success and the evolution of breeding territories in pupfish (Cyprinodon). Evolution 31:750-766.

Kopec, J. A. 1949. Ecology, breeding habits and young stages of Crenichthys baileyi, a cyprinodont fish of Nevada. Copeia 1949:56-61.

Koster, W. J. 1948. Notes on the spawning activities and the young stages of Plancterus kansae (Garman). Copeia 1948:25-33.

Koster, W. J. 1957. Guide to the fishes of New Mexico. University of New Mexico Press, Albuquerque.

Krebs, C. J., M. S. Gaines, B. L. Keller, J. H. Myers and R. H. Temarin. 1973. Population cycles in small rodents. Science 179:35-41.

Lagler, K. F., J. E. Bardach, and R. R. Miller. 1962. Ichthyology. Wiley, New York.

LaRivers, I. 1962. Fishes and fisheries of Nevada. Nevada State Fish and Game Commission, Carson City.

Lehtinen, S. and A. A. Echelle. 1979. Reproductive cycle of Notropis boops (Pisces:Cyprinidae) in Brier Creek, Marshall County, Oklahoma. American Midland Naturalist 102:237-243.

Lewis, M. A. 1978. Notes on the natural history of the longfin dace, Agosia chrysogaster, in a desert rheocrene. Copeia 1978:703-705.

Lowe, C. H., D. S. Hinds and E. A. Halpern. 1967. Experimental catastrophic selection and tolerances to low oxygen concentration in native Arizona freshwater fishes. Ecology 48:1013-1017.

MacArthur, R. H. and E. O. Wilson. 1967. The theory of island biogeography. Princeton University Press, Princeton, NJ.

McKay, F. E. 1971. Behavioral aspects of population dynamics in unisexual-bisexual Poeciliopsis (Pisces:Poeciliidae). Ecology 52:778-790.

Miller, R. R. 1948. The cyprinodont fishes of the Death Valley system of eastern California and southwestern Nevada. Miscellaneous Publications of the Museum of Zoology of the University of Michigan 68:1-155.

Miller, R. R. 1961. Man and the changing fish fauna of the American Southwest. Papers of the Michigan Academy of Science, Arts and Letters 46:365-404.

Miller, R. R. 1976. Four new pupfishes of the genus *Cyprinodon* from Mexico, with a key to *C. eximius* complex. Bulletin of the Southern California Academy of Sciences 75:68-75.

Minckley, C. O. and J. E. Deacon. 1973. Observations on the reproductive cycle of *Cyprinodon diabolis*. Copeia 1973:610-613.

Minckley, C. O. and J. E. Deacon. 1975. Foods of the Devil's Hole pupfish, *Cyprinodon diabolis* (Cyprinodontidae). Southwestern Naturalist 20:105-111.

Minckley, W. L. 1969a. Environments of the bolson of Cuatro Cienegas, Coahuila, Mexico. Texas Western Press, University of Texas at El Paso, Science Series No. 2:1-65.

Minckley, W. L. 1969b. Aquatic biota of the Sonoita Creek basin, Santa Cruz County, Arizona. Ecological Studies leaflet No. 15, Nature Conservancy.

Minckley, W. L. 1972. A survey of selected physico-chemical and biologic parameters of Aravaipa Creek, Arizona. Report to Defenders of Wildlife, and U.S. Bureau of Land Management, Safford, AZ.

Minckley, W. L. 1973. Fishes of Arizona. Arizona Game and Fish Department, Phoenix.

Minckley, W. L. and W. E. Barber. 1971. Some aspects of biology of the longfin dace, a cyprinid fish characteristic of streams in the Sonoran Desert. Southwestern Naturalist 15:459-464.

Minckley, W. L. and L. H. Carufel. 1967. The Little Colorado River spinedace, *Lepidomeda vittata*, in Arizona. Southwestern Naturalist 12:291-302.

Minckley, W. L. and J. E. Deacon. 1968. Southwestern fishes and the enigma of "endangered species." Science 159:1424-1432.

Minckley, W. L., J. N. Rinne and J. E. Johnson. 1977. Status of the Gila topminnow and its co-occurrence with mosquitofish. U. S. Forest Service Research paper RM-198.

Minshall, G. W. 1978. Autotrophy in stream ecosystems. BioScience 28:767-771.

Moore, W. S. and F. E. McKay. 1971. Coexistence in unisexual-bisexual species complexes of *Poeciliopsis* (Pisces:Poeciliidae). Ecology 52:791-799.

Moyle, P. B. 1976. Inland fishes of California. University of California Press, Berkeley.

Myers, G. S. 1952. Annual fishes. Aquarium Journal 23:125-141.

Naiman, R. J. 1974. Bioenergetics of a pupfish population and its algal food supply in a thermal stream. Ph.D. Dissertation, Arizona State University, Tempe.

Naiman, R. J. 1975. Food habits of the Amargosa pupfish in a thermal stream. Transactions of the American Fishes Society 104:536-538.

Naiman, R. J. 1976. Productivity of a herbivorous pupfish population (*Cyprinodon nevadensis*) in a warm desert stream. *Journal of Fish Biology* 9: 125-137.

Naiman, R. J. 1979. Preliminary food studies of *Cyprinodon macularius* and *Cyprinodon nevadensis* (Cyprinodontidae). Southwestern Naturalist 24: 538-541.

Noy-Meir, I. 1973. Desert ecosystems: environment and producers. Annual Review of Ecology and Systematics 4:25-51.

Ohmart, R. D. 1973. Observations on the breeding adaptations of the roadrunner. Condor 75:140-149.

Pianka, E. R. 1970. On *r*- and K-selection. American Naturalist 104:592-597.

Plantz, D. V., Jr. 1976. Size-selective predation by *Poeciliopsis occidentalis* (Baird and Girard). Ph.D. Dissertation, Arizona State University, Tempe.

Rinne, J. N. 1975. Changes in minnow populations in a small desert stream resulting from naturally and artificially induced factors. Southwestern Naturalist 20:185-195.

Sale, P. F. 1977. Maintenance of high diversity in coral reef fish communities. American Naturalist 111:337-359.

Schenck, J. R. and B. G. Whiteside. 1977. Reproduction, fecundity, sexual dimorphism and sex ratio of *Etheostoma fonticola* (Osteichthyes:Percidae). *American Midland Naturalist* 98:365-375.

Schoenherr, A. A. 1974. Life history of the topminnow *Poeciliopsis occidentalis* (Baird and Girard) in Arizona and an analysis of its interaction with the mosquitofish *Gambusia affinis* (Baird and Girard). Ph.D. Dissertation, Arizona State University, Tempe.

Schoenherr, A. A. 1977. Density dependent and density independent regulation of reproduction in the Gila topminnow, *Poeciliopsis occidentalis* (Baird and Girard). Ecology 58:438-444.

Schultz, R. J. 1969. Hybridization, unisexuality and polyploidy in the teleost *Poeciliopsis* (Poeciliidae) and other vertebrates. American Naturalist 103:605-619.

Scrimshaw, N. S. 1946. Egg size in poeciliid fishes. Copeia 1946:20-23.

Shrode, J. B. 1975. Developmental temperature tolerance of a Death Valley pupfish (*Cyprinodon nevadensis*). Physiological Zoology 48:378-389.

Shrode, J. B. and S. D. Gerking. 1977. Effects of constant and fluctuating temperatures on reproductive performance of a desert pupfish, *Cyprinodon n. nevadensis*. Physiological Zoology 50:1-10.

Sigler, W. F. and R. R. Miller. 1963. Fishes of Utah. Utah State Department of Fish and Game, Salt Lake City.

Smith, H. W. 1956. Kamango, the lungfish and the padre. Viking, London.

Smith, H. W. 1961. From fish to philosopher. Doubleday, London.

Soltz, D. L. 1974. Variation in life history and social organization of some populations of Nevada pupfish, *Cyprinodon nevadensis*. Ph.D. Dissertation, University of California, Los Angeles.

Soltz, D. L. and R. J. Naiman. 1978. The natural history of native fishes in the Death Valley system. Natural History Museum of Los Angeles County, Science Series 30:1-76.

Stearns, S. C. 1976. Life-history tactics: a review of the ideas. Quarterly Review of Biology 51:3-47.

Stevenson, M. M. and T. M. Buchanan. 1973. An analysis of hybridization between the cyprinodont fishes *Cyprinodon variegatus* and *C. elegans*. Copeia 1973:682-692.

Sumner, F. B. and U. N. Lanham. 1942. Studies on the respiratory metabolism of warm and cool spring fishes. Biological Bulletin 82:313-327.

Sumner, F. B. and M. C. Sargent. 1940. Some observations on the physiology of warm spring fishes. Ecology 21:45-54.

Sutton, M. and A. Sutton. 1966. The life of the desert. MacGraw-Hill, New York.

Thibault, R. E. 1974. Genetics of cannibalism in a viviparous fish and its relationship to population density. Nature 251:138-140.

Thibault, R. E. 1978. Ecological and evolutionary relationships among diploid and triploid unisexual fishes associated with the bisexual species, _Poeciliopsis lucida_ (Cyprinodontiformes:Poeciliidae). Evolution 32:613-623.

Thibault, R. E. and R. J. Schultz. 1978. Reproductive adaptations among viviparous fishes (Cyprinodontiformes:Poeciliidae). Evolution 32:320-333.

Tinkle, D. W. 1969. The concept of reproductive effort and its relation to the evolution of life histories of lizards. American Naturalist 103:501-516.

Turner, F. B., G. A. Hoddenback, P. A. Medica and J. R. Lannom. 1970. The demography of the lizard, _Uta stansburiana_ Baird and Girard, in southern Nevada. Journal of Animal Ecology 39:505-519.

Vanicek, C. D. and R. H. Kramer. 1969. Life history of the Colorado squawfish, _Ptychocheilus lucius_, and the Colorado chub, _Gila robusta_, in the Green River in Dinosaur National Monument, 1964-1966. Transactions of the American Fisheries Society 98: 193-208.

Vrijenhoek, R. C. 1978. Coexistence of clones in a heterogeneous environment. Science 199:549-552.

Welcomme, R. L. 1969. The biology and ecology of the fishes of a small tropical stream. Journal of Zoology 158:485-529.

Wilbur, H. M. and J. P. Collins. 1973. Ecological aspects of amphibian metamorphosis. Science 164: 1305-1314.

Williams, G. C. 1966a. Adaptation and natural selection. Princeton University Press, Princeton, NJ.

Williams, G. C. 1966b. Natural selection, the costs of reproduction, and refinement of Lack's principle. American Naturalist 100:687-692.

Wilson, E. O. 1975. Sociobiology: the new synthesis. Harvard University Press, Cambridge, MA.

Winkler, P. 1973. The ecology and thermal physiology of *Gambusia affinis* from a hot spring in southern Arizona. Ph.D. Dissertation, University of Arizona, Tucson.

Wourms, J. P. 1972. The developmental biology of annual fishes. III. Pre-embryonic and embryonic diapause of variable duration in the eggs of annual fishes. Journal of Experimental Zoology 182:389-414.

Zweifel, R. G. and C. H. Lowe. 1966. The ecology of a population of *Xantusia vigilis*, the desert night lizard. American Museum Novitates No. 2247:1-57.

TABLE 6. Data on the Natural History of Desert Fishes[a]

Species (1)	Geographical Range (2)	Habitat (3)	Morphology (4)	Food (5)	Breeding Season (6)	Body Size (7)	Age (8)
Stream Fishes							
Agosia chrysogaster	Throughout Sonoran Desert	Sandy-bottomed streams; ephemeral waters; not in spring heads	Fusiform; head rounded, mouth terminal	0	Jan-Nov, or Feb-Sept, depending on area	39 mm SL; max = 100 mm SL	1 yr; few live more than 2 yr
Catostomus clarki	Gila R. basin and Bill Williams tributaries, AZ	Streams: pools, riffles, rapids, rocky bottoms	Slightly depressed, cylindrical; mouth inferior	0	Late winter-early spring	$\leq$125 mm SL	
Catostomus insignis	Gila R. basin, NM and AZ; and Bill Williams basin, AZ	Creeks and small rivers; deep, quiet waters with gravelly bottoms	Fusiform, slightly compressed; mouth inferior, head large	0	Feb-July	Max = 0.8 m, 2.0 kg	
Gila robusta	Throughout Colorado R. basin	Creeks and rivers: pools, eddies behind boulders, mainstream; rarely in constant springs	Fusiform; body thick, caudal peduncle slender; mouth terminal	0	June-July	$\bar{x}$ = 300 mm TL	$\leq$5 yr

Meda fulgida	Formerly throughout Gila R. basin, AZ and NM	Streams: moving water, swift pools, eddies	Slightly compressed and attenuated; mouth terminal	C	May–June	39 mm SL	1 yr
Plagopterus argentissimus	Formerly throughout Gila and Virgin R. basins, AZ, NV, and UT	Streams: swift shallows, variable discharge, shifting sand bottom	Somewhat attenuated; caudal fin deeply forked; mouth inferior	O	Apr–Aug	Max = 71 mm SL	1 yr
Poeciliopsis lucida	Rio del Fuerte basin, NW Mexico	Streams: slow, fertile; not in headwaters	Somewhat elongate but rounded; mouth superior; caudal fin rounded			28 mm TL	
Poeciliopsis monacha	Rio del Fuerte basin, NW Mexico	Streams: small, headwater arroyos; dramatic fluctua tions in volume, food, and temperature	Somewhat elongate but but rounded; mouth superior; caudal fin rounded			28 mm TL	
Generalist Species							
Cyprinodon macularius	Formerly throughout deserts of AZ, CA, and NW Mexico	Diverse: constant-condition springs, stagnant pools, streams	Compressed; mouth superior; moderately truncated	O	Mar–Sept in variable environment	x = 45 mm SL, males mature at 15 mm SL	In warm water, mature at 2–3 mo; in variable environments mature at about 1 yr

Table 6. (Continued)

Species (1)	Geographical Range (2)	Habitat (3)	Morphology (4)	Food (5)	Breeding Season (6)	Body Size (7)	Age (8)
Cyprinodon nevadensis	Amargosa R. basin, CA and NV; widespread	Diverse: warm springs, temporary pools, streams	Compressed; mouth superior; moderately truncated; body deep	0	In warm springs; year-round; Mar–Oct in stream	19 mm SL in warm warm spring, 24 mm SL in stream	3–4 mo in warm spring; 1 yr in stream
Poeciliopsis occidentalis	Formerly throughout Gila R. system AZ, NM, and NW Mexico	Diverse: constant-condition springs, intermittent creeks, stream shallows	Somewhat elongate but rounded; mouth superior; caudal fin rounded	0	Year-round in warm springs; Apr–Aug in streams	20 mm SL	5 or 12 mo in warm springs, 1 yr in stream
Rhinichthys osculus	Widespread throughout W United States	Diverse: torrential creeks, warm springs, intermittent pools	Variable: fusiform; body slightly rounded; mouth subterminal	0	In N, late spring-early summer; in S, early spring and late summer	<70 mm SL	1 yr

Spring Fishes							
<u>Crenichthys baileyi</u>	Pluvial White R. basin, E and S NV; highly localized	Isolated, warm springs and their outflows; 30–38°C, relatively constant temperature	Body deep; head broad; posterior dorsal and anal fins	0	Year-round	24–28 mm SL	
<u>Crenichthys nevadae</u>	Railroad Valley drainage, NV	Isolated, warm springs; most warmer than 30°C	Head heavy; posterior dorsal and anal fins; pelvic fins absent				
<u>Cyprinodon diabolis</u>	Restricted to a single spring-cave called Devil's Hole, Nye Co., NV	Constant-condition (33–34°C) limestone spring	Compressed; mouth superior; head large; pelvic fins absent	0	Possibly year-round, distinct peak in spring	<26 mm SL	
<u>Relictus solitarius</u>	Pluvial lakes Franklin, Waring and Spring, E. NV	Isolated springs: relatively constant-condition; absent in canyon streams	Body chubby and deep, not streamlined; fins small and rounded; lateral line reduced; mouth terminal		June–Sept	66 mm SL	1 yr

Table 6. (Continued)

Fecundity (9)	Offspring Size (10)	Behavior (11)	Dispersal (12)	Population Characteristics (13)	References (14)
75–417 eggs	Egg diameter = 1.6 mm; fry length =4.0 mm SL	Aggregative spawning; no agonism among males; slight sexual dimorphism	Disperse into vast temporary reaches	Common; undergoes colonization and extinction regularly; high interannual variation in number of fry	Barber and Minckley 1966; Deacon and Minckley 1974; Koster 1957; Lewis 1978; Lowe et al. 1967; Miller 1961; Minckley 1969b, 1972, 1973; Minckley and Barber 1971; Rinne 1975; J. N. Rinne, personal communication
				Large interannual variation in reproductive success and local abundance	Koster 1957; Lowe et al. 1967; Minckley 1969b, 1972, 1973
		Two males spawn with each female		Frequent; high interannual variation in number of fry	Barber and Minckley 1966; Minckley 1969b, 1972, 1973
A female of 305 mm TL had 10,000 eggs	Egg diameter = 0.79 mm	Aggregative spawning; sexual dichromatism		Uncommon; no apparent relationship between annaul discharge and year-class strength	Barber and Minckley 1966; Cross 1978a; Deacon and Bradley 1972; Koster 1957; LaRivers 1962; Minckley 1969b, 1972, 1973; Sigler and Miller 1963; Vanicek and Kramer 1969; C. D. Vanicek, personal communication

121–219 eggs	Egg diameter = 1.65 mm	Aggregative spawning; sexual dichromatism; no territoriality		Locally common; large interannual variation in reproductive success and local abundance	Barber and Minckley 1966; Barber et al. 1970; Koster 1957; Minckley 1972, 1973
					Cross 1978b; LaRivers 1962 Minckley 1973; Sigler and Miller 1963; J. N. Cross, personal communication
11 embryos	Egg diameter = 1.4 mm; fry length = 6.8 mm TL	Territorial; not cannibalistic; sexually dimorphic and dichromatic			McKay 1971; Moore and McKay 1971; Schultz 1969; Thibault 1974, 1978; Thibault and Schultz 1978; R. E. Thibault, personal communication
12 embryos	Egg diameter = 2.1 mm; fry length = 8.5 mm TL	Territoral; cannibalistic; sexually dimorphic and dichromatic		Periodic local extinctions, large interannual variation in local abundance	McKay 1971; Schultz 1969; Thibault 1974; Thibault and Schultz 1978; Vrijenhoek 1978; R. E. Thibault, personal communication
50–800 eggs/ female/season	Egg diameter ≃ 2 mm; fry length = 3.7–5.3 mm TL, depending on temperature	Males intensely territorial; brightly colored; not cannibalistic; alternate male mating strategies	Occupies local flooded areas	Locally common; builds up large seasonal populations	Barber and Minckley 1966; Barlow 1958a, 1958b, 1961; Courtois and Hino 1979; Cowles 1934; Cox 1966; Crear and Haydock 1971; Deacon and Minckley 1974; Kinne 1960; Kinne and Kinne 1962a, 1962b; Lowe et al. 1967; Miller 1948; Minckley 1969b, 1973; Moyle 1976; Naiman 1979

Table 6. (Continued)

Fecundity (9)	Offspring Size (10)	Behavior (11)	Dispersal (12)	Population Characteristics (13)	References (14)
40 eggs in river, 10 in warm spring	Egg diameter = 1.39 mm in warm spring versus 1.34 mm in variable stream	Aggregative spawning in stream versus territoriality in spring; alternate male mating strategies	Moves into local inundated areas	Density and structure fluctuate more in stream than in spring; longevity shorter in warm springs	Brown 1971; Brown and Feldmeth 1971; Deacon 1968; Deacon and Deacon 1979; Deacon and Minckley 1974; Gerking et al. 1979; Kodric-Brown 1977; LaRivers 1962; Moyle 1976; Naiman 1974, 1975, 1976; Shrode 1975; Shrode and Gerking 1977; Soltz 1974; Soltz and Naiman 1978; M. F. Hirshfield, personal communication
5 embryos in warm spring, 14 in stream	Egg diameter = 1.9 mm; egg weight = 2.2 mg in warm spring, 1.6 mg in stream	Territorial; not cannibalistic; alternate male mating strategies	Invades local temporary waters	Locally dense; senescence may occur in springs; extirpated by violent flash floods	Barber and Minckley 1966; Constantz 1974, 1975, 1976, 1979, 1980, personal observation; Deacon and Minckley 1974; Minckley 1969b, 1972, 1973; Minckley and Deacon 1968; Minckley et al. 1977; Plantz 1976; Schoenherr 1974, 1977; Scrimshaw 1946

228–514 eggs	Egg diameter = 1.0 mm	Sexually dichromatic; several males fertilize eggs at each nest; minor aggression among males	Repopulates vast reaches of temporary water	Large interannual variation in local abundance; local extinction is rare	Barber and Minckley 1966; Deacon and Bradley 1972; Hubbs et al. 1974; John 1963, 1964; Koster 1957; LaRivers 1962; Lowe et al. 1967; Minckley 1969b, 1972, 1973; Moyle 1976; Rinne 1975; Sigler and Miller 1963; Soltz and Naiman 1978
5–19 eggs	Egg diameter = 1.9 mm; fry length = 4.3 - 5.3 mm TL	Males intensely colored; not cannibalistic; bimodal diel activity pattern	Precluded by surrounding desert	Locally common; lifespan shortened by hot water	Deacon and Bradley 1972; Deacon and Wilson 1967; Espinosa 1968; Hubbs and Hettler 1964; Hubbs et al. 1967; Kopec 1949; LaRivers 1962; Sumner and Lanham 1942; Sumner and Sargent 1940
		Emits sound during feeding and fighting	Precluded by surrounding desert		Cole 1968; Hubbs and Hettler 1964; Hubbs et al. 1967; Hubbs et al. 1974; LaRivers 1962
4–5 eggs	Fry length = 6.5 mm TL	Territoriality rare; sexually dichromatic; cannibalistic	Precluded by surrounding desert	Primary production, breeding, and population size cycle annually as function of solar input	Brown 1971; Deacon 1968; Deacon and Deacon 1979; Deacon and Minckley 1974; James 1969; LaRivers 1962; Minckley and Deacon 1973, 1975; Soltz and Naiman 1978

Table 6. (Continued)

Fecundity (9)	Offspring Size (10)	Behavior (11)	Dispersal (12)	Population Characteristics (13)	References (14)
		Sexually dimorphic fins, sluggish mid-water swimmer	Precluded by surrounding desert	Adult sex ratio favors females	Hubbs et al. 1974

[a]Geographical range = geographical distribution of the species; morphology = external morphology; food = food habits, where O = omnivory, C = carnivory; body size and age = for females at first breeding; fecundity = number of eggs carried by females.

9 Genetic Differentiation of Pupfishes (Genus *Cyprinodon*) in the American Southwest

DAVID L. SOLTZ
California State University, Los Angeles

MICHAEL F. HIRSHFIELD
Academy of Natural Sciences, Philadelphia

ABSTRACT

Pupfishes inhabit widely differing environments in the southwestern deserts. Some are extremely harsh and variable; others are constant thermal springs. The genus *Cyprinodon* has differentiated morphologically into 12 described species and subspecies in the Death Valley, Owens Valley, and Colorado River drainages. Despite this morphological differentiation after varying lengths of isolation, it has been generally thought that relatively little genetic differentiation has occurred in other portions of the pupfish genome. A major problem in the interpretation of pupfish evolution has been due to the strong direct effects of environmental factors on phenotypic characteristics. For this reason it is often difficult to determine whether observed differences between pupfish populations actually represent genetic differentiation.

We review here the extent of genetic differentiation of pupfish populations and the extent of direct environmental effects on the phenotypes of individuals. The most substantial differentiation is revealed in morphological characteristics; however, the magnitude of direct environmental effects is still unclear. In contrast, allozyme and karyotypic differences are minimal. Extremely plastic breeding behavior responds to the social and ecological environments. Physiological tolerances of populations in different environments appear to be similar, although slight genetic differences have been found between two populations. Life history characteristics of pupfish populations are strongly and directly affected by the environment. Genetic differences have been found between two populations; interspecific comparisons have not yet been made. Western pupfishes are highly interfertile, and quantitative hybridization studies must be done. To understand these patterns fully more geological information and expanded studies of critical features of pupfish biology are needed.

INTRODUCTION

The role of environmental factors in the differentiation of morphological, behavioral, physiological, and ecological characteristics of populations is a central focus of evolutionary biology. The basic questions have been asked for several decades, dating at least from the pioneering experimental studies on the nature of plant species by Clausen et al. (1940) and the coining of the term "ecophenotype" (Clausen et al. 1948; Clausen and Hiesey 1958).

Pupfishes, genus *Cyprinodon*, are particularly well suited to studies of environmental effects on adaptive traits. Pupfishes live in perhaps the widest range of aquatic environments of any vertebrate. Their habitats differ widely in temperature, salinity, dissolved oxygen, flow, and productivity. Habitat types include estuaries along the Atlantic and Gulf Coasts of North America, rivers, streams, and associated marshes in the southwestern United States, and isolated springs, often of constant temperature, concentrated in the Death Valley System of California and Nevada and in Cuatro Cienegas, Coahuila, Mexico (for a map see Turner and Liu 1977, Figure 1). In this review we have restricted our discussion to the six species of pupfishes that inhabit the Owens River, Death Valley, and Colorado River drainages of southwestern North America. Pupfishes in this region inhabit a wide range of environments. Their biology has been extensively studied, the hydrology of the area is well known, and it is the area with which we are most familiar. Between 25 and 30 isolated populations of pupfish live in this region.

The morphological divergence of these pupfishes (Miller 1948) in isolated, differing habitats has been used as an example of rapid evolution in harsh environments (Miller 1950). Studies of other aspects of pupfish biology (e.g., biochemistry, physiology, behavior, and the ability to hybridize), however, have shown little divergence or variation indicative of broad phenotypic plasticity (e.g., Turner 1974). Our recent studies of genetic differences in the life histories and physiological tolerances between two subspecies of *Cyprinodon nevadensis*, *C. n. amargosae* and *C. n. mionectes*, have shown that the two populations have diverged genetically in isolation in very different habitats (Hirshfield et al. 1980; unpublished data). The evidence of overall genetic differentiation between species and subspecies of pupfish is thus equivocal.

This chapter reviews the extent to which genetic differentiation has occurred among and within species of pupfish, the possibility of differential rates of evolution of different components of the pupfish genome, and the extent of direct environmental effects on the phenotypes of individuals within populations.

ENVIRONMENTAL SETTING

Present habitats, which consist of springs, streams (often intermittent), and associated marshes, have been described in detail by Soltz and Naiman (1978). The ancestral pupfish probably originated along the Atlantic coast of North America in estuaries and salt-water marshes, the habitat of C. variegatus today (Darling 1976). Pupfishes entered the Death Valley area during the early Pleistocene when it was part of a much larger interconnected system that included the Colorado River (Hubbs and Miller 1948; Miller, this volume). Since the end of the Pleistocene, as the climate of the southwestern United States became progressively more arid, pupfishes that had occupied rivers, large marshes, and shallow lakes (i.e., broadly variable environments) were isolated in the small streams, marshes, and constant temperature springs they now inhabit (Soltz and Naiman 1978). Pupfishes in this area never co-occur with congeners and occupy harsh and isolated environments that support only zero to four other species of native fishes (Echelle 1973; Soltz and Naiman 1978).

The Colorado River, source of the ancestral Cyprinodon, became separated from the Death Valley area at least 100,000 and perhaps as long as 1,000,000 years ago (Hubbs and Miller 1948; Miller, this volume). The Owens River was also isolated during that time. In the Death Valley area populations of C. nevadensis have been isolated for perhaps 400 to 5000 years, but C. salinus and C. diabolis have been isolated for at least 20,000 years and perhaps much longer (Miller 1948, 1950, this volume; Mehringer and Warren 1976). This system thus provides a series of fish populations that have been isolated in very different habitats for varying periods of time since the early Pleistocene.

The wide range of pupfish habitat variability in size, physiognomy, and physical and chemical characteristics is represented in Table 1. The range of environments for C. nevadensis approximtes that of all pupfishes in the Southwest. General descriptions of desert streams and springs are presented elsewhere in this volume (see Cole and Constantz). We stress the

TABLE 1. Characteristics of Habitats of Amargosa Pupfish, _Cyprinodon nevadensis_[a]

Locality	Subspecies	Habitat Type	Temperature	Total Desolved Solids	Volume or Flow
Amargosa River, near Tecopa	_C. n. amargosae_	Permanent stream; undergoes extensive desiccation in summer; periodically scoured by flash floods	Variable; 0 to 40°C	Variable; 5000–30,000 mg·liter^{-1}	3755 liter·s^{-1}, 17 yr av; 1.6 x 10^6 liter·s^{-1} maximum flood; no flow many days in summer
Big Spring	_C. n. mionectes_	Very large (20 m diameter 8 m deep) warm spring	Constant; 27.3°C (27.0–28.0)	Relatively constant; 1200 mg·liter^{-1}	10^5 liter·s^{-1}
Indian Springs	_C. n. pectoralis_	Two small, separate warm spring pools connected by a narrow channel 4 m long	Relatively constant; 30.4°C (29.7–31.5)	Relatively constant; <1000 mg·liter^{-1}	2750 liter
Marsh Spring	_C. n. pectoralis_	Shallow, warm spring with man-made marsh on outlet	Constant; 31.1°C (30.5–31.8)	Relatively constant; 1500 mg·liter^{-1}	1400 liter

[a]Modified from Soltz (1974).

wide differences between constant thermal springs and variable streams and marshes. Desert springs usually fluctuate little and in general exhibit warm temperatures (>20°C), low salinity, other chemical characteristics approximating those of potable freshwater (Table 2), and relatively high productivity (Brown 1971; Constantz 1976; Cox 1966; Deacon and Minckley 1974; Naiman 1976; Schoenherr 1974; Soltz 1974; Soltz and Naiman 1978). In contrast, temporal and spatial variation in temperature, salinity, ion concentrations, dissolved oxygen, and productivity in desert stream and marsh systems may have upper and lower extremes that far exceed those between desert spring ecosystems (Tables 1 and 2).

Desert stream and spring ecosystems provide a continuum in environmental variation, the extremes of which have been presented. It can be argued that such a spectrum of selective regimes should have led to relatively rapid differentiation of the often small and isolated pupfish populations, particularly in loss of variability and physiological tolerances of pupfishes that inhabit constant springs compared with those in variable environments (Brown and Feldmeth 1971; Hirshfield et al. 1980; Miller 1948, 1950; Soltz 1974; Turner 1974).

Some of the predicted differentiation and loss of variability have apparently occurred and are summarized below. Because pupfishes are extremely plastic phenotypically, the extent to which differences between populations are genetically based is unclear.

The central questions we address in this review are the following:

What characteristics of pupfish have been shaped by environmental factors since the isolation of the populations?

To what extent are the characteristics of individuals the result of direct effects of the environment on the phenotype?

To what extent have the species and populations of pupfish differentiated genetically?

TABLE 2. Comparative Water Analysis of Two Pupfish Habitats, Tapwater and Sea Water (in $mg \cdot liter^{-1}$).

	Big Spring	Tapwater, Los Angeles	Amargosa River	Sea Water[a] (35⁰/oo)
Total hardness ($CaCO_3$)	218	204	5050	
Ca	84	63	820	411
Mg	2	11	720	1293
Na	100	43	6700	10762
K	10	5	350	399
NH_4	0.2	0.0	0.9	
Total alkalinity ($CaCO_3$)	280	155	108	
HCO_3	342	159	131	142
SO_4	112	96	2550	2709
Cl	27	36	11183	19353
NO_3	1	10	50	
F	1.4	0.5	0.9	1.3
B	0.6	0.3	12.4	4.4
pH	7.8	8.2	7.1	
Conductivity[b]	840	610	31000	
Total dissolved solids		495	365	24660

[a]Values from Riley and Chester (1971).
[b]$\mu mho \cdot cm^{-3}$

To answer these questions we have reviewed studies of each set of adaptive characteristics critically. Particular attention has been focused on the role of environmental factors on both within and between population variation.

MORPHOLOGY

Five of the six species are clearly morphologically distinct (Table 3) and live specimens can be readily distinguished by an experienced investigator. The sixth species, _C. milleri_, a downstream isolate from Salt Creek in the Cottonball Marsh area (described recently by LaBounty and Deacon 1972), is very similar to _C. salinus_ (Miller, this volume). The morphological differences among the other five pupfishes involve characters presumably under polygenic control (e.g., dentition, meristics, and color pattern), which are considered indicative of significant evolutionary change (Kluge and Farris 1969).

The Death Valley and Owens Valley pupfishes are associated by a similar pattern of scale structure, which also distinguishes them from their presumptive ancestor in the Colorado River System, _C. macularius_ (Miller 1948, 1950). _C. radiosus_ from the Owens Valley is more like _C. macularius_ in other characters except in the presence of a truncate central cusp on the tricuspid teeth (which it shares with _C. nevadensis_). These patterns led Miller (1948) to the conclusion that _C. radiosus_ arose from a _C. macularius_-type ancestor at an early stage of the Death Valley System. The slender bodied _C. salinus_ has several unique morphological features; for example, many small, crowded scales on the body and a central ridge on the tricuspid teeth. Other characters, however, indicate a strong relationship with _C. nevadensis_, its presumed ancestor (Miller 1943, 1948). _C. nevadensis_, the characteristic morphological type of pupfish in the Death Valley area, is a widespread rather variable form with six named subspecies, all but one of which were isolated in one or more springs. _C. diabolis_, probably a morphologically unique isolate of _C. nevadensis_, is a small, slender, large-

TABLE 3. Some Major Morphological Differences Among Five Western Pupfish Species[a,b]

Character	*C. macularius*	*C. radiosus*	*C. nevadensis*	*C. salinus*	*C. diabolis*
Pelvic fins	+	+	+	+	–
Pelvic fin formula	7–7	7–7	6–6	6–6	NA
Number of scales around body	28	28	24	39–40	26
Projections on scale circuli	+	–	–	–	–
Vertical bars on female	+	+	+	+	–
Mean relative prehumeral length	1.5	1.5	1.8	1.8	1.5
Sexual dimorphism in median fin size	+	+	+	+	–
Mature male body shape	Deep	Deep	Deep	Slender	Slender
Median ridge on tricuspid teeth	–	–	–	+	–
Shape of central cusp on tricuspid teeth	Spatulate	Truncate	Truncate	Spatulate	Truncate
Relative size of central cusp	Broad	Broad	Narrow	Broad	Narrow

[a]Adapted from Miller (1948:Table 1).
[b](+) present or common; (–) absent or rare; (NA) comparison not applicable.

headed pupfish that has lost the pelvic fins and vertical barring of nuptial males, characteristic of the genus. It has been isolated for at least 20,000 years in the restricted high-temperature, low-productivity environment of Devils Hole, Ash Meadows, Nevada. It is morphologically most similar to *C. n. pectoralis*, which is geographically closest and occupies a series of the smallest and warmest springs in Ash Meadows (Miller 1948; Soltz and Naiman 1978).

To examine the relationship between morphological variability and environmental parameters coefficients of variation (CV) were calculated for 10 meristic characters for four of the six species; comparable data for *C. macularius* and *C. milleri* are not available (Table 4). The coefficients of variation were uniformly low for all 10 characters among the four species, with only one above 10 percent. The average CV for 10 characters was highest in *C. nevadensis* at 0.0637. *C. nevadensis* is the most ecologically generalized species; however, individual populations are isolated in a spectrum of habitats from the highly variable Amargosa River to small constant temperature springs in Ash Meadows. The CVs were generated from data gathered from more than 20 populations of *C. nevadensis*. *C. diabolis*, the species isolated in the most uniform and restricted habitat, had the lowest mean CV of 0.0500. The other two species, *C. salinus* and *C. radiosus*, had intermediate CVs of 0.0573 and 0.0537, respectively. These two species have been isolated in variable stream and associated marsh habitats for 20,000 to 100,000 years. Assuming that the phenotypic (meristic) variation used to calculate the CVs is genetically based, the species isolated in the most restricted regime, *C. diabolis*, is the least genetically variable. *C. diabolis* also has the smallest population size (200 to 500 individuals). It would be useful to gather data for an analysis of *C. macularius*, the presumptive ancestral pupfish in the Southwest.

An understanding of environmental effects can be further enhanced by examining a species with populations isolated in differing habitats, such as the Amargosa pupfish, *C. nevadensis*. Two of the subspecies, *C. n. amargosae* and *C. n. mionectes*, occupy different

TABLE 4. Coefficient of Variation of Morphological Characters in Four Species of *Cyprinodon* in Nature[a]

Character	*nevadensis*	*diabolis*	*salinus*	*radiosus*
Number of anal rays	0.043	0.040	0.037	0.051
Number of dorsal rays	0.044	0.050	0.064	0.060
Number of pectoral rays	0.032	0.027	0.033	0.047
Number of caudal rays	0.074	0.083	0.044	0.051
Lateral scale count	0.034	0.029	0.052	0.020
Predorsal scale count	0.071	0.073	0.050	0.064
Body scale count	0.106	0.039	0.061	0.046
Dorsal to pelvic scale count	0.078	–	0.079	0.069
Dorsal to anal scale count	0.087	0.056	0.091	0.063
Number of preopercular pores	0.068	0.053	0.062	0.066
Average of 10 characters	0.0637	0.050	0.0573	0.0537

[a]Calculated from data presented in Miller (1948).

habitats, the variable Amargosa River and the large, constant temperature springs in Ash Meadows, respectively. *C. n. pectoralis*, which inhabits a series of small warm springs in Ash Meadows, is unique in having a high number of pectoral rays. Otherwise it is similar to *C. n. mionectes*, except that the body is deeper and the head and mouth are wider. The fourth extant subspecies, *C. n. nevadensis*, occurs only in Saratoga Spring and marshy lakes along its outflow in the lower Amargosa River drainage.

CVs of 10 morphological characters were calculated for four populations of these subspecies (Table 5). The range of CVs within and between populations is extremely narrow. Of the 10 characters *C. n. amargosae* and *nevadensis* have the highest values for four (including one tie), *C. n. mionectes* is highest for two, and *C. n. pectoralis* is highest for only one. This parallels the trend in variability of the habitats; the two most variable forms occur in the Amargosa River and the fluctuating Saratoga Marsh and the other two, in constant temperature springs. Note that the average CVs for the latter two are as low as that of *C. diabolis*. The trend in CVs also parallels population size, the two most variable of which have by far the larger populations in nature.

The possibility thus exists that a genetic reduction in morphological variation has occurred in those small populations isolated in the most constant and restricted habitats. On the other hand, the data could reflect greater phenotypic, but not genotypic, variability in the more variable habitats that results from the direct environmental effects discussed below.

It is therefore critical to understand the extent on which morphological variation within and between species is actually genetically based. From a series of controlled rearing experiments of several species and subspecies of *Cyprinodon* Miller (1948) concluded that, for meristic characters at least, "a direct environmental influence is indicated, but it is thought that the observed differences are due in part to genetic adaptations that parallel those due to the direct effect of the surroundings."

TABLE 5. Coefficients of Variation of Morphological Characters in Four Natural Populations of *Cyprinodon nevadensis*[a]

Character	*amargosae*	*mionectes*	*pectoralis*	*nevadensis*
Number of anal rays	0.0483	0.0437	0.0399	0.0429
Number of dorsal rays	0.0581	0.0514	0.0526	0.0581
Number of pectoral rays	0.0434	0.0383	0.0331	0.0487
Number of caudal rays	0.0610	0.0567	0.0447	0.0589
Lateral scale count	0.0190	0.0235	0.0224	0.0148
Predorsal scale count	0.0690	0.0477	0.0635	0.0573
Body scale count	0.0570	0.0650	0.0724	0.0610
Dorsal to pelvic scale count	0.0586	0.0539	0.0572	0.0642
Dorsal to anal scale count	0.0536	0.0566	0.0395	0.0693
Number of Preopercular pores	0.0637	0.0742	0.0748	0.0595
Average of 10 characters	0.0532	0.0511	0.0500	0.0535

[a]Calculated from data presented in Miller (1948).

For example, the F_1 stocks of *C. n. amargosae* and *C. n. nevadensis* in Miller's (1948) experiments generally retained the meristic characters exhibited in nature. Pupfishes and fishes in general tend to exhibit negative correlations between meristic characters and temperature and positive correlations with salinity (Hubbs 1922; Lindsey and Harrington 1972; Miller 1948, 1950). The effects of these factors are not clear-cut; temperature and salinity often interact in affecting growth and development and different meristic characters are affected differently or not at all (Lindsey 1962). Many and perhaps all morphological characters used to distinguish the species and subspecies of *Cyprinodon* are subject to some direct environmental effects. Because the habitats of western pupfishes differ tremendously in temperature and salinity, the morphologies of the various populations may reflect these environmental differences directly; for example, in Miller's (1948) experiments scale number decreased significantly in offspring of *C. n. amargosae* when the F_1 stock was raised in freshwater rather than in their natural environment of higher salinity. The same result was observed in counts of predorsal scales and pelvic and caudal rays. Sweet and Kinne (1964) reared the eggs of *C. macularius* in different combinations of temperature and salinity and concluded that these two variables had significant effects on a number of morphometric variables, including the distance between the anal fin and caudal peduncle and body width (characteristics used to distinguish subspecies of *C. nevadensis*). Although we may presume that local genetic differentiation in morphology exists between the different populations of *C. nevadensis*, as Miller (1948) concluded, the extent of genetic divergence is equivocal.

BIOCHEMISTRY

Studies of allozyme variation of *Cyprinodon* species in the Death Valley and Colorado River area revealed relatively little variation within and between species (Turner 1973a,b, 1974). Among the 31 to 38 proteins surveyed only seven proteins were divergent in at least one of the five species sampled (Table 6). Protein polymorphisms were rare and ranged from zero to four polymorphic loci per population (Turner 1974). Average heterozygosity per locus per individual could not be calculated because of small sample sizes. In addition, karyotypes of the 16 *Cyprinodon* species studied to date are invariant, with a diploid number of 48 (Miller, this volume; Stevenson 1975).

C. macularius, presumably most similar to the ancestor of the other species, had the greatest number (five) of divergent proteins (Table 6). Only *C. nevadensis* showed no unique electrophoretic phenotypes. Proportional similarity indices (0.808 to 0.968), some of the highest reported for congeneric species (Dobzhansky 1972), indicate a high degree of genetic similarity among these five pupfish species. The divergent protein data do suggest the fixation of a few different alleles in the derivation of Death Valley pupfishes from a *C. macularius*-type ancestor. The Death Valley populations however, do not show the expected effects of population bottlenecking (periods of extremely small population size that certainly occur in *C. diabolis* and some *C. nevadensis* populations) on protein divergence and fixation of different alleles.

The proportion of polymorphic loci per population is a measure commonly used to compare genetic variation among taxa (Nevo 1978). The *C. n. amargosae* population from the highly variable Amargosa River had 14 percent polymorphic loci, whereas only one of the same 29 loci was polymorphic in stable spring populations of *C. n. nevadensis* and *C. n. mionectes* (Turner 1974). Although this suggests the expected positive correlation between genetic variability and ecological heterogeneity (Dobzhansky 1951; Levins 1968; Nevo 1978), the differences in proportion of polymorphic loci were not statistically significant. This niche-width/variation hypothesis is clearly not supported by the overall rarity of polymorphism in Death Valley pupfishes and the presence of only one polymorphic locus in *C. macularius* and *C. salinus*, or none in *C. radiosus*, the other habitat generalists from variable environments.

There is little agreement between the electrophoretic data, which suggest a high degree of genetic

TABLE 6. Divergent Proteins Between Species of Pupfish[a-d]

	G6PD-1	G6PD-2	AAT-1[b]	ES-1	ES-2	ES-5	ES-8
C. macularius	V_1	V_1	-	-	V_1	V_1	V_1[c]
C. radiosus	-	-	-	-	V_2	-	-
C. nevadensis	-	-	-	-	-	-	-
C. salinus	-	-	V_1	V_1	V_3	-	-
C. diabolis	-	-	-	-	-	-	V_1

[a]V indicates divergent form; subscripts assigned to different variants arbitrarily.

[b]Variation in liver tissue only.

[c]Only in Quitoboquito population; Salton Sea population had the same mobility as other species.

[d]Modified from Turner (1974).

similarity, and the morphological data, which suggest rather extensive divergence among the species of pupfish. Dendrograms based on 31 electrophoretic loci and 14 to 16 morphological and behavioral characters (Figure 1) produced two strikingly dissimilar arrays of

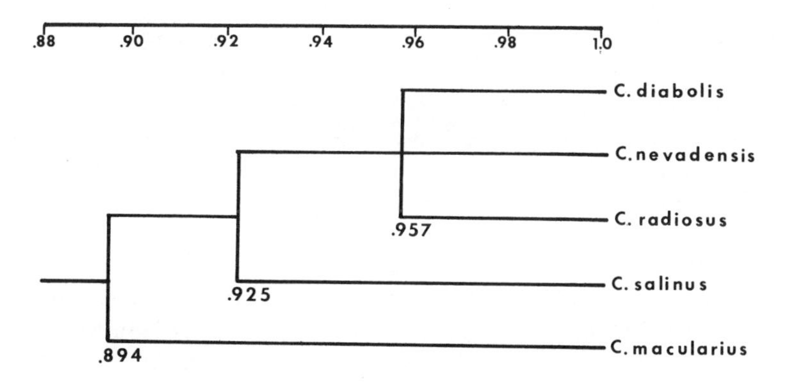

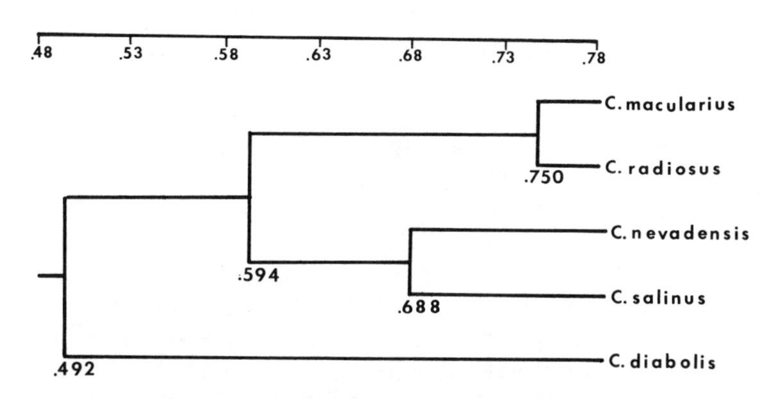

FIGURE 1. Dendrograms from unweighted pair group cluster analysis of electrophoretic (upper) and morphological-behavioral data (lower). Although the two arrays of species differ greatly, all five species cluster more closely based on electrophoretic characters, than do the two closest species based on morphological-behavioral characters. (Note the different scales.) From Turner (1974).

five species of Cyprinodon (Turner 1974). Although the lack of congruence between the morphological and allozyme data is striking, it is due in part to Turner's selection of "major morphological and behavioral differences among five western pupfish species," whereas the allozyme data is presumably a random sample of the genome. Recent comparative studies of vertebrate groups, in which larger numbers of morphological characters were used, have supported Turner's conclusion and established the general lack of concordance between morphological and allozyme data in phenetic analysis (e.g., Avise et al. 1975; Ferris et al. 1979; Mickevich and Johnson 1976; Schnell et al. 1978).

Turner (1974) proposed that

> the pupfish genome is essentially dichotomous with ... one group of genes (that) have been responsive to the selectional and stochastic factors, ... those genes which encode or mediate the expression of certain morphological and behavioral traits. The second group of genes have not been responsive to the ecological differences among the habitats. This group includes most of the loci surveyed by electrophoresis.

It it currently felt that speciation, as in the case of pupfishes in the Southwest, is often not accompanied by a "genetic revolution" (sensu Mayr 1963). Instead, speciation may involve evolutionary changes in morphology and behavior that may be attributed largely to changes in temporal and spatial expression of regulatory loci (Ferris et al. 1979). These changes may occur without concomitant changes in protein structure, which is more directly encoded by structural loci (Lewontin 1976). These mechanisms are consistent with Turner's (1974) hypothesis that in the ancestral estuarine pupfish there existed a "strongly coadapted, homeostatic, metabolic core of enzymes which evolved under strongly variable ecological conditions" and that "there was no selective pressure for substituting new alleles" in the stable environments that many populations of Cyprinodon subsequently invaded.

Presumably pupfish speciation has been associated with the differentiation of morphological characters rather than with electrophoretic adaptation (Miller 1948; Turner 1974). Because directly environmentally induced phenotypic variation in morphology is probably much greater than that in electrophoretic characters, the differences in morphology may represent a slight amount of genetic differentiation of essentially the same magnitude as the electrophoretic.

BEHAVIOR

Liu (1969) concluded that feeding, resting, agonistic, and spawning behavior, a total of 30 motor patterns, were uniform in all 13 species of *Cyprinodon* tested. Courtship behavior differed among the species complexes but was similar within a complex. In the western pupfishes Liu (1969) concluded that tactile stimulation was of primary importance in courtship. It was considered that the *C. nevadensis* complex (all Death Valley area forms) engage in minimal courtship with less visual stimulation than in the *C. macularius* complex. Recent studies suggest, however, that visual signals (perhaps not actual displays) are important in attracting females to territorial males in many western pupfishes (Echelle 1973; Kodric-Brown 1977, 1978; Soltz 1974). The basic breeding behavior of all western species of *Cyprinodon* is strikingly similar; the *macularius* and *nevadensis* complexes differ only in intensity of behavior.

Behavioral plasticity in the pupfish breeding system has already been reviewed (Kodric-Brown, this volume). *C. nevadensis* exhibits all four basic breeding systems: territoriality, dominance hierarchy, group spawning, and consort pair spawning (Soltz 1974). The environmental correlates of these breeding systems fit those generalized by Kodric-Brown with one exception. The associations between high primary productivity and territoriality and low productivity and consort pairing are not strong in *C. nevadensis* populations. Rather consort pairing appears to be part of the male mixed-breeding strategy within populations, for it occurs early in the breeding season when population densities are relatively low in habitats of large physical dimension; for example, Big Spring and Amargosa River (Soltz 1974). The various breeding systems appear to represent facultative responses of males to the social, ecological, and physical environments.

In a behavioral study of *C. n. amargosae* and *C. n. mionectes* in which density was varied in a series of outdoor pools, increasing population density revealed differences in the behavioral plasticity of the two subspecies (Banta and Soltz unpublished data). *C. n. amargosae* males established territories with or without the presence of females and maintained them at all densities (maximum density 80 males:80 females). In contrast, *C. n. mionectes* males did not begin territorial behavior until the sex ratios were between 5:1 and 5:3. The territorial system in the four *C. n. mionectes* pools began to be disrupted at densities of 25:25 or 20:30 and at 40:40 all territories were abandoned. At higher densities *C. n. mionectes* formed loose aggregations and only an occasional consort-pair spawning occurred. These observations suggest that, given a more stable and favorable environment than their natural habitat, *C. n. amargosae* adjusted rapidly and established a coherent territorial system rather than the group spawning system generally found in the field. The collapse of the *C. n. mionectes* breeding system at high densities in pools suggests that individuals derived from the low density Big Spring population may have lost some behavioral flexibility in coping with conditions outside their natural limits.

PHYSIOLOGICAL TOLERANCE

Pupfish habitats differ widely in temperature, salinity, and dissolved oxygen. The literature on temperature and salinity tolerance and effects in western species of *Cyprinodon* has been reviewed by C. R. Feldmeth, S. D. Gerking, and S. D. Hillyard (this volume). We therefore concentrate on the extent of differentiation in physiological tolerance.

Although it has been argued that fishes in constant springs should show reduced thermal tolerances compared with those experiencing variable temperatures, a survey of critical thermal maxima (CTM) at similar acclimation temperatures of various populations of Cyprinodon reveals little variation (Table 7). The uniformity of CTMs (40.7 to 42.7°C) is even more surprising considering the variation in prior field temperatures, acclimation temperatures, and rates of heating during the experiments. An attempt was made to select CTMs from similar acclimation temperatures within the range of pupfish reproductive activity.

Studies have revealed strong environmental effects on critical thermal maxima and minima. Lowe and Heath (1969) found a seasonal increase in CTM of C. macularius at all acclimation temperatures. The effect was strong enough that winter-acclimated fish could not tolerate high summer field temperatures. In a study of laboratory reared C. n. nevadensis, however, although the CTM of two-month-old larvae was directly correlated with temperature, differences in incubation and rearing temperature had no effect (Shrode 1975). Smaller younger fish had consistently higher CTMs than older larger fish. These results correlate with frequent field observations that juveniles of many pupfish species select warmer environments than the adults (Barlow 1958; Echelle 1973; Soltz 1974). The CTM of C. milleri, a species from relatively saline water, was higher at increased salinities than in freshwater (Otto and Gerking 1973). The effect was greatest at lower acclimation temperatures and maximal at 15 or 30 ppt, depending on acclimation temperature.

For two subspecies of C. nevadensis the CTM of fish acclimated to cycling temperatures (Table 7) was closer to that of fish acclimated to the higher temperature than to the mean temperature of the cycle (Feldmeth et al. 1974; Shrode and Gerking 1977). The critical thermal minimum of pupfishes apparently responds similarly, but more slowly, to acclimation (Feldmeth et al. 1974). Although caution should be used in drawing broad conclusions from the comparison of numerous populations of pupfishes using different experimental designs, the studies clearly suggest the importance of

TABLE 7. Thermal Tolerances of Popultions of *Cyprinodon* Acclimated to Similar Temperature Within Their Range of Reproductive Activity

	Acclimation Temperature (°C)	Critical Thermal Maximum (°C)	Critical Thermal Minimum (°C)	Habitat Temperature (°C)	Source of Fish Tested	References
C. nevadensis amargosae	27 ± 1.0	41.9	2.4	0 to > 40	Laboratory	Hirshfield et al. (1980)
C. n. amargosae	25 ± 1.0	40.8	NC	0 to > 40	Field	Brown and Feldmeth (1971)
C. n. amargosae	25 ± 1.0	41.8	2.3	0 to > 40	Field	Feldmeth et al. (1974)
C. n. amargosae	35 to 15[a]	42.6	2.4	0 to > 40	Field	Feldmeth et al. (1974)
C. n. mionectes	27 ± 1.0	41.6	2.7	27.3 ± 1.0	Laboratory	Hirshfield et al. (1980)
C. n. mionectes	> 20[b]	41.7	NC	27.3 ± 1.0	Field	Brown and Feldmeth (1971)
C. n. pectoralis	> 20[b]	42.1	6–7	32.5 ± 1.0	Field	Brown and Feldmeth (1971)
C. n. nevadensis	25 ± 1.0	41.7	NC	27.5 ± 1.0	Field	Brown and Feldmeth (1971)
C. n. nevadensis	28 ± 0.5	41.6	4.5	0 to > 40	Laboratory	Shrode and Gerking (1977)
C. n. nevadensis	32 to 28[a]	42.2		0 to > 40	Laboratory	Shrode and Gerking (1977)
C. diabolis	> 20[b]	42.7	6–7	33.8 ± 0.5	Field	Brown and Feldmeth (1971)
C. salinus	> 20[c]	41.6	NC	0 to > 40	Field	Brown and Feldmeth (1971)
C. milleri	< 20[d]	39.2		9 to 37	Field	Brown and Feldmeth (1971)
C. milleri	25 ± 0.5	40.7		9 to 37	Laboratory	Otto and Gerking (1973)

C. macularius	30 ± 0.5	42	0 to > 40	Field	Lowe and Heath (1969)

[a]Fish were acclimated to a 24-hr cycling temperature; see references for detailed methods.

[b]Fish from constant temperature springs were tested after temperature had converged to approximately 20°C and within 36 hrs of capture in the field.

[c]Fish were captured in April when habitat maximum temperature exceeds 30°C; sample was treated, as were spring fish, before testing.

[d]Fish were captured in January when maximum temperature is < 20°C; sample was treated, as were spring fish, before testing.

direct environmental effects in the phenotypic response of pupfish thermoregulatory physiology.

Brown and Feldmeth (1971) concluded that "all populations of *Cyprinodon* studied (11) are tolerant of a wide range (~40°C) of environmental temperatures and are able to shift within this range by means of thermal acclimation." Their results and those of most other thermal tolerance studies led them to conclude that "there is no evidence of genetic differences in short-term thermal tolerances between any of the populations tested." The fish they compared were collected from the field and tested after rapid or no laboratory acclimation to a fixed temperature. Their nonstatistical analysis of the data could not have detected slight, but real, differences. Studies of two populations of *C. nevadensis* reared under identical conditions in the laboratory for two generations produced different results (Hirshfield et al. 1980). *C. n. mionectes* from constant temperature Big Spring had a significantly narrower range of temperature tolerance than *C. n. amargosae* from a much more variable habitat. These results support the hypothesis that relatively rapid evolutionary loss of variability (in this case, physiological tolerances) should occur in organisms isolated in constant habitats. Environmental effects, however, can be as strong as and often stronger than genetic effects.

LIFE HISTORY CHARACTERISTICS

Egg Size

Despite the prediction that eggs produced in more productive habitats are expected to be smaller (Smith and Fretwell 1974; Bagenal 1971), analyses of eggs from pupfish inhabiting environments differing in productivity suggested that egg size was relatively constant; for example, Soltz (1974) found no differences in the diameter of preserved eggs from females collected in Big Spring and the Amargosa River. Turner and Liu (1977) stated that "so far as we are aware, *Cyprinodon* species are not significantly divergent in egg size or yolk content."

Shrode and Gerking (1977), however, found that yolk volume of *C. nevadensis* eggs produced at 18°C was 22.5 percent greater than that of eggs produced at 30°C. They argued that "changes in egg size may be a facultative response to the availability of food for fry, providing greater yolk reserve under less favorable conditions," because productivity should be positively correlated with temperature.

Because the productivity in Big Spring is apparently always lower than in the Amargosa River during the breeding season (Soltz 1974), we expected that the fish from Big Spring would be *genetically* programmed to produce larger eggs. Eggs from field-collected females from Big Spring were substantially larger than those from the Amargosa River (35 percent by volume).

This difference was not necessarily genetically based, however. Although the females from Big Spring were significantly smaller than those from the Amargosa River, they could have been growing more slowly, hence have been older. Repeated measurement of eggs laid by females in the F_1 generation showed that egg size increases with age in *C. nevadensis*, as in other species (e.g., *Oryzias latipes*, Hirshfield 1977). Therefore when the F_2 females became mature their eggs were measured at the first opportunity in crosses performed at the Claremont Colleges and California State University, Los Angeles (CSULA) in 1978. Comparison of newly laid eggs showed that *C. n. mionectes* produced significantly larger eggs than *C. n. amargosae*; the total egg volume differed by approximately 18 percent at Claremont.

The difference in volume was less at CSULA (approximately 7 percent). This resulted, in part, from a positive correlation between egg diameter and age (or size) at maturity in females at CSULA. This correlation, combined with the slightly greater age at maturity of *C. n. amargosae* females, reduced the difference between the two populations. Comparison of egg sizes between females maturing at the same age showed a difference of approximately 10 percent by volume.

Differences between eggs produced in the two laboratories were striking. Eggs produced at Claremont were substantially larger than those produced at CSULA;

the differences between the two labs were much greater than the differences between subspecies within labs.

Thus, although our experiments indicate significant differences between the two populations, differences consistent with predictions, direct environmental effects on egg size remain remarkably strong.

Developmental Rates, Hatching Success, Juvenile Viability

Gerking and coworkers (Gerking, this volume; Gerking et al. 1979; Shrode 1975; Shrode and Gerking 1977) and Kinne and Kinne (1962a,b) have reviewed the strong effects of temperature, salinity, and oxygen concentration on developmental rates and viability of *C. nevadensis* and *C. macularius* eggs. The hatching success of *C. n. nevadensis* studied by Gerking and coworkers declined during the course of their work (Table 8). In addition to differences in experimental technique, part of the decline may be attributed to the effects of

TABLE 8. Hatching Success (percent) of Eggs Laid and Reared at Various Temperatures During Three Studies by Gerking and Coworkers

Temperature (°C)				
Laid	Developed	1975[a]	1977[b]	1979[c]
28	24	>90		64
28	28	~95	~70	63
28	32	~85		32
34	24		~50	63
32	32		~15	51

[a]From Shrode (1975).
[b]From Shrode and Gerking (1977).
[c]From Gerking et al. (1979).

inbreeding during this period. Such variability between experiments indicates that comparisons must be made with caution if they are based on the work of a number of investigators. For this reason *C. nevadensis* cannot be compared directly with *C. macularius* because experimental techniques varied. Certainly no striking differences are apparent between the two species.

Some subjective evidence of differences in viability among pupfishes does exist; for example, Corzillius (1975) reported that *C. salinus* had a lower hatching rate than other species examined. *C. diabolis* is notoriously difficult to rear; eggs do not hatch or premature hatching with subsequent death of fry occur (Liu 1969). Successful reproduction has apparently taken place only in a refugium near Hoover Dame (Sharpe et al. 1978). We have observed lower hatching success and juvenile survival and greater frequency of abnormalities (albinism, jaw malformations, fin abnormalities) in *C. n. mionectes* than in *C. n. amargosae*.

These data suggest that the differences among pupfish in developmental tolerances may be substantial. Comparative studies of salinity and temperature tolerances of eggs and fry, in particular, of forms normally found in environments differing sharply in these parameters, should prove particularly rewarding.

Juvenile Growth Rate

Kinne (1960) showed that temperature and salinity have interactive effects on juvenile growth rates of *C. macularius*. In particular, juvenile growth was faster in freshwater than in sea water at low temperatures (15°, 20°C) but the reverse was true at high temperatures (25°, 30°, 35°C).

Kinne (1962) also concluded that the environment to which eggs were exposed during the first three to six hours after laying had permanent effects on growth and efficiency of food utilization in later life. Specifically, he claimed that individuals performed best in salinities that were the same as those in which they spent the first hours of life. Examination of the data does not support this claim, however; lower efficiencies can be attributed in all cases to greater size or to salinity in which the fish were reared; for example, at 25°C Kinne claimed that the performance of the fish spawned and reared at 15 ppt was superior, but in fact these fish were 2 mm smaller at the beginning of the experiments than fish transferred to 35 ppt and 2.2 mm smaller at the end. With growth rate as the criterion the fish at 35 ppt performed best, as expected by Kinne (1960).

Similarly, at 30°C fish transferred from freshwater to 35 ppt showed lower conversion efficiencies than those reared continuously in freshwater; but the fish in 35 ppt were 3.6 mm larger than those in freshwater at the start of the experiment and 5.6 mm larger by the end. Indeed, their growth rates were faster than those of fish spawned and reared at 35 ppt in a subsequent experiment. By focusing only on conversion efficiencies and ignoring growth invalid conclusions were drawn. All the data are consistent with direct effects of the environment in which the fish were reared. No evidence of "irreversible non-genetic adaptation" can be found.

Both genetic and environmental factors affect growth rates of juvenile Amargosa pupfish (*C. n. mionectes* and *C. n. amargosae*). Our experiments have shown that F_2 juvenile *C. n. amargosae* were approximately 5 percent larger and 10 percent heavier at seven weeks after laying than *C. n. mionectes* in both laboratories. Differences between laboratories were of approximately the same magnitude. No differences between the sexes were observed. Thus both genetic and environmental factors have strong effects on juvenile growth rates in this species.

Age and Size at Reproductive Maturity

Little information is available on age or size at maturity in pupfish. It is clear that *C. diabolis* matures at a smaller size than other species but whether this is a genetically fixed trait is unknown. *C. diabolis* in the Hoover Dam refugium are reported to grow substantially longer than those found in Devils Hole (Williams 1977), which indicates that the small sizes of individuals in natural conditions are due in part to food limitation. Whether individuals in the refugium mature at larger sizes is not known.

Soltz (1974) reported lengths at maturity for _C. nevadensis_ from the Amargosa River and Big Spring. Males from both populations matured at the same size (50 percent mature at 20 mm standard length). Females from the Amargosa River matured at a significantly larger size (50 percent mature at 24.5 mm SL) than females from Big Spring (50 percent mature at 20 mm SL).

Our laboratory experiments indicated that age and size at maturity are controlled by a number of factors. Males of both subspecies matured at earlier ages and smaller sizes than females in both laboratories. No significant differences between subspecies were found in age at maturity of males or females in either laboratory, and ages at maturity of both subspecies were significantly lower at CSULA than at Claremont. In both laboratories for males of both subspecies there was a significant negative correlation between size at seven weeks and age at maturity. The relationship for females was similar but weaker. This effect of growth rate presumably accounts for the difference between the two laboratories.

Size at maturity differed significantly between the sexes, subspecies, and laboratories. Males matured at smaller sizes than females, _C. n. mionectes_ of both sexes matured at smaller sizes than _C. n. amargosae_, and the fish at CSULA matured at larger sizes than those at Claremont, despite their younger age.

Size at maturity is also a function of juvenile growth rate and age at maturity. A positive relation exists between size at maturity and both seven-week size and age at maturity. The sizes at maturity of the two subspecies are very close after these two variables are accounted for. Similarly, the larger size of females at maturity simply reflects a greater age at maturity than males and not a greater juvenile growth rate.

Fecundity

Shrode and Gerking (1977) found strong effects of

temperature on the number of eggs per spawn and spawning frequency in *C. n. nevadensis*. Both parameters were highest at constant temperatures of 28° and 30°C, in relation to all other constant and fluctuating temperatures. Even taking into account the smaller egg size at these two temperatures, the total number of eggs produced was still substantially greater.

A similar set of experiments was run to test the effects of "irreversible non-genetic adaption" on reproduction (Gerking et al. 1979). Fish spawned and reared at 24°, 28°, and 32°C were transferred at maturity to the other two test temperatures or maintained at the temperature at which they were raised. No effect of rearing environment was observed, but temperature during reproduction again had significant effects; the greatest reproductive success occurred at 28°C. The differences between the two series of experiments were substantial; the performance of the fish in the (1977) series at 24° and 32°C was far superior to that of those in the (1979) series, whereas the reverse was true at 28°C (Table 9). In addition, the yolk diameters

TABLE 9. Comparison of Various Reproductive Parameters at Three Temperatures from Two Different Studies[a]

		24°	28°	32°
Eggs per spawn	1977	~5.5	~8.0	~6.0
	1979	2.1	11.5	2.7
Percent days when eggs laid	1977	~50	~75	~55
	1979	22.5	78	38
Eggs·d^{-1}	1977	~4.5	~6.0	~5.0
	1979	0.8	8.8	1.4
Mean yolk diameter (mm)	1977	1.06	1.01	1.02
	1979	1.05	1.05	1.02
GSI	1977	0.13	0.10	0.09
	1979	0.10	0.11	0.07

[a]Shrode and Gerking 1977; Gerking et al. 1979.

at 24° and 32°C were similar in (1977) and (1979) but much smaller in (1977) at 28°C (Table 9). The pattern for gonadosomatic index (GSI) was similar to that for the fecundity variables; 28°C fish were slightly higher in 1979 and 24° and 32°C fish, lower.

These differences may reflect genetic differences between the generations which result from progressive inbreeding or they may represent the inevitable differences between experimental series that must be considered before generalizations can be made.

Soltz (1974) presented the only fecundity data for field-collected pupfish. Comparisons of similar-sized females from Big Spring and the Amargosa River showed that female *C. n. amargosae* contained approximately 1.5 to 2 times as many mature ova. A female *C. n. mionectes* 32 mm SL would be expected to produce 25 eggs, whereas a female *C. n. amargosae* would produce nearly 40. These figures are in contrast to the maximum average of 12 eggs per spawn at 28°C, reported in the laboratory by Shrode and Gerking (1975) for *C. n. nevadensis*.

Our laboratory experiments demonstrated that the difference between *C. n. mionectes* and *C. n. amargosae* persisted even when the fish were reared under the same conditions. At CSULA the difference in eggs per day was highly significant; *C. n. amargosae* laid almost 40 percent more eggs. At Claremont the difference was not significant; the mean number of eggs laid per day by both subspecies was only half that of *C. n. mionectes* at CSULA.

Like Shrode and Gerking (1975), we found no correlation between fecundity and length and the pattern of reproduction was highly variable from day to day within and between individuals. The cause of such variability in pupfish remains an unsolved riddle, as indicated by Gerking et al. (1979). We found a negative correlation between age at maturity and average fecundity; females maturing earlier tended to lay more eggs per day during their adult lives.

Courtois and Hino (1979) reported significant effects of salinity and depth on spawning rates in desert pupfish (*C. macularius*). They found greater production at 5 than at 15 ppt and a preference for spawning at greater depths at 5 than at 15 ppt. Average rates were approximately 20 eggs per female per day during the one-month study conducted at temperatures fluctuating from 15° to 24°C at 5 ppt. Either *C. macularius* is innately much more fecund than *C. nevadensis* or housing 10 females with several males leads to greater fecundity than does isolation. Note also that 24°C is at the lower limit for reproduction reported by Shrode and Gerking (1977). Courtois and Hino (1979) also report that *C. macularius* customarily uses plants for egg deposition; *C. nevadensis* does not normally spawn on vegetation (Soltz 1974).

Adult Growth

Relatively little information is available on adult growth in pupfish. It is clearly slow in Devils Hole pupfish (*C. diabolis*), the maximum size of which is not much larger than 20 mm in nature. This apparently reflects low food availability in combination with high temperature because the fish grow more rapidly in the Hoover Dam refugium (Williams 1977). Soltz (1974) reported a maximum SL of 46 mm for male and 43 mm for female *C. n. amargosae*. No *C. n. mionectes* of more than 40 mm SL were ever collected. This result indicates faster adult growth rates or greater survival rates for the fish in the Amargosa River.

Our laboratory results indicated that fish at CSULA grew faster than fish at Claremont. Adult males grew faster than females in both laboratories. When growth rates of equivalent-sized fish were compared *C. n. mionectes* actually grew slightly faster than *C. n. amargosae* (although not significantly). The greater growth rates of *C. n. amargosae* as juveniles meant, however, that the *C. n. amargosae* were always larger than *C. n. mionectes* on any given date.

Summary

These studies indicate that life-history characters in pupfish are strongly affected by environmental parameters. Nevertheless, some apparently genetic differences exist, even between closely related subspecies, although the magnitude is not so great as might be expected. The remarkable variability seen by Gerking et al. (1979) may be part of the explanation: within-population variation, at least under laboratory conditions, is so extensive that few between-population differences are observed.

HYBRIDIZATION POTENTIAL

Species in the genus *Cyprinodon* are highly interfertile (Cokendolpher 1980; Corzillius 1975; Drewry 1967; Miller 1948; Turner and Liu 1977). Turner and Liu (1977), in the most extensive of these studies, produced 18 interspecific hybrids by using nine species from throughout the geographic range of the genus. The majority of the hybrids was fertile. Ten were carried to the F_2 or F_3 generation. One triple hybrid and two tetrahybrid crosses were produced; only one was known to be fertile.

All F_1 hybrids were qualitatively intermediate in shape and coloration. The F_2 that were produced closely resembled their parents, as expected. It is interesting that the courtship behavior of hybrids, scored for 10 motor patterns documented and described by Liu (1969), combined distinct elements of both parental species. Inheritance of behavior was not completely additive as it appeared to be for morphology.

All crosses (5 of 18) involving western pupfishes (*C. macularius*, *C. nevadensis*, *C. salinus*, and *C. radiosus* were used) produced fully fertile hybrids (Turner and Liu 1977). Their results agreed with those of Corzillius (1975) who produced fertile hybrids from three western species. In both studies aberrant sex ratios and unisexual (♂♂) sterility occurred in crosses between *C. variegatus*, a species widely distributed along the Atlantic and Gulf coasts of North America, and several western species. It is odd that Turner and Liu (1977) obtained sterile males in two crosses in which the maternal species was *C. variegatus* but not in the reciprocal crosses, whereas Corzillius (1975) obtained sterile males in two crosses in which the paternal species was *C. variegatus*. These data at best merely suggest a relationship between hybrid sterility and parental divergence in the genus as a whole.

The extreme interfertility of the *Cyprinodon* species in the Southwest casts doubt on the often presumed correlation between the ability of species to hybridize and their phylogenetic relationship (Hubbs 1963, 1970; Hubbs and Drewry 1959; Suzuki 1968). The results of hybridization experiments agree quite closely with the lack of allozyme divergence (Turner 1974) but are discordant with the morphological evidence (Miller 1948). The lack of success in hybridization of *C. variegatus* and western species may be due largely to the small egg and yolk diameters in *C. variegatus* (Corzillius 1975). In contrast to the presumption of Turner and Liu (1977), however, there may be considerable variation in yolk diameters of other species of *Cyprinodon* (see above). A more complete study of egg-size variation and its effect on the development of reciprocal crosses should therefore be conducted. More importantly, none of the hybridization studies has analyzed hybrid fitness quantitatively. Substantial reductions in hybrid viability that simply have not yet been seen may exist.

Although the highly interfertile *Cyprinodon* species are also largely invariant electrophoretically, sunfishes (genus *Lepomis*), which are perhaps even more interfertile, are strongly divergent electrophoretically (Avise and Smith 1974). These results suggest that morphological evolution, biochemical evolution, and hybridization potential are not strongly coupled. One explanation given for the apparent lack of genetically based differentiation of western pupfishes is that they are relatively new, recently isolated species (Corzillius 1975; Soltz 1974). Although the geological and paleoclimatological evidence is indirect, the recent postpluvial isolation of species of *Cyprinodon* in the Death Valley area has generally been assumed (Hubbs and Miller 1948; Mehringer and Warren 1976; Miller 1948; Soltz and Naiman 1978). Data presented in this volume, however, suggest that five of the six *Cyprinodon* species in the region have been isolated since mid-Pleistocene to late Pliocene times (Miller, this volume).

SUMMARY AND CONCLUSIONS

This review has pointed out the generally limited differentiation of the species of *Cyprinodon* in the American Southwest. Where significant variation or differentiation does occur there is often a striking lack of congruence between the data from different portions of the genome. The magnitude and nature of variation in those components are summarized below:

1. *Morphology*. Substantial differences in morphological characteristics are found between the species, greater than those found for other portions of the pupfish genome. Even within species the differences in body form between *C. n. mionectes* and *C. n. amargosae*, for example, are those expected between spring and river forms (see Constantz, this volume). Most differences between species, however, are in morphometric ratios and meristic characters. The magnitude of direct environmental effects on such characteristics, which may be substantial, is still unclear.

2. *Biochemical characters*. Allozymic variation among these species of *Cyprinodon* is of the order of that found for conspecific populations of other species, despite the substantial differences among habitats. This lack of differentiation is even more puzzling in the light of Darling's (1976) report of substantial geographic variation in allozymes *within C. variegatus*, variation as great as that seen *within* the Death Valley species examined by Turner (1974). The lack of differentiation of the pupfish of the Southwest thus remains an enigma.

3. *Behavior*. Breeding behavior of all western pupfish is strikingly similar. Behavioral plasticity in breeding systems is extensive and directly responsive to the social and ecological environments. Given this plasticity, and the absence of congeners or (frequently) other species of fishes, it is not surprising that little divergence has occurred in strictly programmed behavioral traits.

4. *Physiological tolerances*. Although small genetic differences in temperature and oxygen tolerances have been found between populations of *C.*

nevadensis, direct environmental effects tend to override these differences. In general, species of *Cyprinodon* are similarly able to tolerate extreme temperatures, salinities, and dissolved oxygen levels, regardless of the constancy or variability of their natural habitat. It should be noted that direct selection for reduced tolerances in constant environments should be weak. A most instructive comparison would be that between two populations from *constant* temperature springs that differed 5°C or more.

5. *Life histories*. Life-history characteristics of pupfishes are strongly and directly affected by environmental parameters. Nevertheless, some genetic differences were found between subspecies of *C. nevadensis*. Although genetic differences between species might be expected to be greater, the range of habitats encountered by these two subspecies is as great as that encountered by the six species in the area. Interspecific comparisons of life histories have still to be carried out.

6. *Hybridization*. Most species in the genus *Cyprinodon* and all western pupfishes examined to date are highly interfertile. Quantitative analysis of hybrid viabilities, however, as well as female choice experiments, have not been performed.

Although differentiation within this group of pupfishes is clearly slight, it is equally clear that many critical comparisons have not yet been made. Differentiation of these species may have been inhibited by the very attributes that have made the genus so successful in desert environments; that is, the extremely broad tolerance and flexibility derived from an estuarine ancestor similar to *C. variegatus* may slow the rate of local genetic adaptation. As argued by Levins (1969), "the greater the individual flexibility, the smaller the genetic differences required to meet a given degree of geographic and temporal long-term variation in the environment." Perhaps this argument is appropriate for these species of *Cyprinodon*.

One additional complication is the difficulty of determining the lengths of isolation of the populations and the long-term constancy of the habitats they currently occupy. Recent evidence (Smith, this volume) suggests that the environments currently occupied by pupfish may not be representative of those in the past. If selective pressures have reversed themselves frequently during periods of increasing and decreasing aridity, divergence of populations resulting from local adaptation would be further reduced.

Thus the view of pupfish differentiation as "less than expected" is colored by our notions of how different we perceive the environmental history of these populations to be. The evidence reviewed here and elsewhere in this volume suggests that more information is needed on the fishes and the environments that they have inhabited before we can fully understand the patterns that we perceive today.

ACKNOWLEDGMENTS

We thank J. A. Endler, M. A. Bell, and B. J. Turner for their helpful reviews and comments. Research was supported in part by Grant DEB 77-03897 from the National Science Foundation.

REFERENCES

Avise, J. C., J. J. Smith and F. J. Ayala. 1975. Adaptive differentiation with little genic change between two native California minnows. Evolution 29:411-426.

Avise, J. C. and M. H. Smith. 1974. Biochemical genetics of sunfish. II. Genic similarity between hybridizing species. American Naturalist 108:458-472.

Bagenal, T. 1971. The interrelationship of the size of fish eggs, the date of spawning, and the production cycle. Journal of Fish Biology 3:207-219.

Barlow, G. W. 1958. Daily movements of desert pupfish, *Cyprinodon macularius*, in shore pools of the Salton Sea, California. Ecology 39:580-587.

Brown, J. H. 1971. The desert pupfish. Scientific American 225:104-110.

Brown, J. H. and C. R. Feldmeth. 1971. Evolution in constant and fluctuating environments: thermal tolerances of desert pupfish (Cyprinodon). Evolution 25:390-398.

Clausen, J. and W. M. Hiesey. 1958. Experimental studies on the nature of species. IV. Genetic structure of local races. Carnegie Institute of Washington Publications 615.

Clausen, J., D. D. Keck and W. M. Hiesey. 1940. Experimental studies on the nature of species. I. Effect of varied environments on western North American plants. Carnegie Institute of Washington Publications 520.

Clausen, J., D. D. Keck and W. M. Hiesey. 1948. Experimental studies on the nature of species. III. Environmental responses of climatic races of Achillea. Carnegie Institute of Washington Publications 581.

Cokendolpher, J. C. 1980. Hybridization experiments with the genus Cyprinodon (Teleostei:Cyprinodontidae). Copeia 1980:173-176.

Constantz, G. D. 1976. Life history of the Gila topminnow, Poeciliopsis occidentalis: a field evaluation fo theory on the evolution of life histories. Ph.D. Dissertation, Arizona State University, Tempe.

Corzillius, B. 1975. Kreuzungsanalyse an vier arten der gattung Cyprinodon (Cyprinodontidae, Pisces) - ein beitrag zum speziationsproblem. Mitteilungen aus dem Hamburgischen Zoologischen Museum und Institut 72:229-240.

Courtois, L. A. and S. Hino. 1979. Egg deposition of the desert pupfish, Cyprinodon macularius, in relation to several physical parameters. California Fish and Game 65:100-105.

Cox, T. J. 1966. A behavioral and ecological study of the desert pupfish Cyprinodon macularius in Quitobaquito Springs, Organ Pipe Cactus National Monument, Arizona. Ph.D. Dissertation, University of Arizona, Tucson.

Darling, J. D. 1976. Electrophoretic variation in Cyprinodon variegatus and systematics of the subfamily Cyprinodontinae. Ph.D. Dissertation, Yale University, New Haven, Conn.

Deacon, J. E. and W. L. Minckley. 1974. Desert fishes. In: G. W. Brown, Jr. (Ed.), Desert Biology, Volume II, Academic, New York, pp. 385-487.

Dobzhansky, Th. 1951. Genetics and the Origin of Species. 3rd ed., rev., Columbia University Press, New York.

Dobzhansky, Th. 1972. Species of Drosphila. Science 177:664-669.

Drewry, G. E. 1967. Studies of the relationships within the family Cyprinodontidae. Ph.D. Dissertation, University of Texas, Austin.

Echelle, A. A. 1973. Behavior of the pupfish, Cyprinodon rubrofluviatilis. Copeia 1973:68-76.

Feldmeth, C. R., E. A. Stone and J. H. Brown. 1974. An increased scope for thermal tolerance upon acclimating pupfish (Cyprinodon) to cycling temperatures. Journal of Comparative Physiology 89: 39-44.

Ferris, S. O., S. L. Portnoy and G. S. Whitt. 1979. The roles of speciation and divergence time in the loss of duplicate gene expression. Theoretical Population Biology 15:114-139.

Gerking, S. D., R. Lee and J. B. Shrode. 1979. Effects of generation-long temperature acclimation on reproductive performance of the desert pupfish, Cyprinodon n. nevadensis. Physiological Zoology 52:113-121.

Hirshfield, M. F. 1977. The reproductive ecology and energetics of the Japanese medaka Oryzias latipes. Ph.D. Dissertation, University of Michigan, Ann Arbor.

Hirshfield, M. F., C. R. Feldmeth and D. L. Soltz. 1980. Genetic differences in physiological tolerances of Amargosa pupfish (Cyprinodon nevadensis) populations. Science 207:999-1001.

Hubbs, C. 1963. The use of hybridization in the determination of phylogenetic relationships. Proceedings of the 16th International Congress of Zoology 4:103-105.

Hubbs, C. 1970. Teleost hybridization studies. Proceedings of the California Academy of Science 38: 289-298.

Hubbs, C. and G. E. Drewry. 1959. Survival of F_1 hybrids between cyprinodont fishes, with a discussion of the correlation between hybridization and phylogenetic relationship. Publication of the Institute of Marine Science, University of Texas 6:81-91.

Hubbs, C. L. 1922. Variations in the number of vertebrae and other meristic characters of fishes correlated with the temperature of water during development. American Naturalist 56:360-372.

Hubbs, C. L. and R. R. Miller. 1948. The zoological evidence: correlation between fish distribution and hydrographic history in the desert basins of western United States. In: The Great Basin, with emphasis on glacial and postglacial times. Bulletin of the University of Utah 38(20), Biology Series 10:17-166.

Kinne, O. 1960. Growth, food intake, and food conversion in a euryplastic fish exposed to different temperatures and salinities. Physiological Zoology 33:288-317.

Kinne, O. 1962. Irreversible non-genetic adaptation. Comparative Biochemistry and Physiology 5:265-282.

Kinne, O. and E. M. Kinne. 1962a. Rates of development in embryos of a cyprinodont fish exposed to different temperature-salinity-oxygen combinations. Canadian Journal of Zoology 40:231-253.

Kinne, O. and E. M. Kinne. 1962b. Effects of salinity and oxygen on developmental rates in a cyprinodont fish. Nature 193:1097-1098.

Kluge, A. G. and J. S. Farris. 1969. Quantitative phyletics and the evolution of anurans. Systematic Zoology 18:1-32.

Kodric-Brown, A. 1977. Reproductive success and the evolution of breeding territoriality in pupfish (*Cyprinodon*). Evolution 31:750-766.

Kodric-Brown, A. 1978. Establishment and defense of breeding territories in a pupfish (Cyprinodontidae:*Cyprinodon*). Animal Behavior 26:818-834.

LaBounty, J. F. and J. E. Deacon. 1972. *Cyprinodon milleri*, a new species of pupfish (family Cyprinodontidae) from Death Valley, California. Copeia, 1972:769-780.

Levins, R. 1968. Evolution in Changing Environments. Princeton University Press, Princeton, NJ.

Levins, R. 1969. Thermal acclimation and heat resistance in Drosophila melanogaster. American Naturalist 103:483-499.

Lewontin, R. C. 1976. The Genetic Basis of Evolutionary Change. Columbia University Press, New York.

Lindsey, C. C. 1962. Experimental study of meristic variation in a population of threespine sticklebacks, Gasterosteus aculeatus. Canadian Journal of Zoology 40:271-321.

Lindsey, C. C. and R. W. Harrington, Jr. 1972. Extreme vertebral variation induced by temperature in a homozygous clone of the self-fertilizing cyprinodontid fish Rivulus marmoratus. Canadian Journal of Zoology 50:733-744.

Liu, R. K. 1969. The comparative behavior of allopatric species (Teleostei - Cyprinodontidae: Cyprinodon). Ph.D. Dissertation, University of California, Los Angeles.

Lowe, C. H. and W. G. Heath. 1969. Behavioral and physiological responses to temperature in the desert pupfish (Cyprinodon macularius). Physiological Zoology 42:53-59.

Mayr, E. 1963. Animal Species and Evolution. Harvard, Belknap Press.

Mehringer, P. J. and C. N. Warren. 1976. Marsh, dune and archaeological chronology, Ash Meadows, Amargosa Desert, Nevada. In: R. Elson (Ed.), Holocene Environmental Change in the Great Basin, Nevada Archaeological Survey, Research Paper 6, pp. 120-150.

Mickevich, M. F. and M. S. Johnson. 1976. Congruence between morphological and allozyme data in evolutionary inference and character evolution. Systematic Zoology 25:260-270.

Miller, R. R. 1943. Cyprinodon salinus, a new species of fish from Death Valley, California. Copeia 1943:69-78.

Miller, R. R. 1948. The cyprinodont fishes of the Death Valley System of eastern California and southwestern Nevada. Miscellaneous Publications of the Museum of Zoology, University of Michigan 68:1-155.

Miller, R. R. 1950. Speciation in fishes of the genera Cyprinodon and Empetrichthys inhabiting the Death Valley region. Evolution 4:155-163.

Naiman, R. J. 1976. Primary production, standing stock, and export of organic matter in a Mohave Desert thermal stream. Limnology and Oceanography 21:60-73.

Nevo, E. 1978. Genetic variation in natural populations: patterns and theory. Theoretical Population Biology 13:121-177.

Otto, G. and S. D. Gerking. 1973. Heat tolerance of a Death Valley pupfish (Genus Cyprinodon). Physiological Zoology 46:43-49.

Riley, J. P. and R. Chester. 1971. Introduction to Marine Chemistry. Academic, New York.

Schnell, G. D., T. L. Best and M. L. Kennedy. 1978. Interspecific morphologic variation in kangaroo rats (Dipodomys): degree of concordance with genic variation. Systematic Zoology 27:34-48.

Schoenherr, A. A. 1974. Life history of the topminnow, Poeciliopsis occidentalis (Baird and Girard) in Arizona and an analysis of its interaction with the mosquitofish Gambusia affinis (Baird and Girard). Ph.D. Dissertation, Arizona State University, Tempe.

Sharpe, F. P., H. R. Guenther and J. E. Deacon. 1978. Endangered desert pupfish at Hoover Dam. Reclamation Era 59:24-29.

Shrode, J. B. 1975. Developmental temperature tolerance of a Death Valley pupfish (Cyprinodon nevadensis). Physiological Zoology 48:378-389.

Shrode, J. B. and S. D. Gerking. 1977. Effects of constant and fluctuating temperatures on reproductive performance of a desert pupfish, Cyprinodon n. nevadensis. Physiological Zoology 59:1-10.

Smith, C. and S. Fretwell. 1974. The optimal balance between size and number of offspring. American Naturalist 108:499-506.

Sokal, R. R. and P. H. A. Sneath. 1963. Principles of Numerical Taxonomy. W. H. Freeman, San Francisco.

Soltz, D. L. 1974. Variation in life history and social organization of some populations of Nevada pupfish, *Cyprinodon nevadensis*. Ph.D. Dissertation, University of California, Los Angeles.

Soltz, D. L. and R. J. Naiman. 1978. The Natural History of Native Fishes in the Death Valley System. Natural History Museum of Los Angeles County, Science Series 30:1-76.

Stevenson, M. M. 1975. A comparative chromosome study of the pupfish genus *Cyprinodon* (Teleostei: Cyprinodontidae). Ph.D. Dissertation, University of Oklahoma, Norman.

Suzuki, R. 1968. Hybridization experiments in cyprinid fishes. II. Survival rates of F_1 hybrids with special reference to the closeness of taxonomic position of combined fishes. Bulletin of the Freshwater Fisheries Research Laboratory of Tokyo 18:113-156.

Sweet, J. G. and O. Kinne. 1964. The effects of various temperature-salinity combinations on the body form of newly hatched *Cyprinodon macularius* (Teleostei). Helgolander Wissenschaftliche Meeresuntersuchungen 11:49-69.

Turner, B. J. 1973a. Genetic variation of mitochondrial aspartate amino-transferase in the teleost *Cyprinodon nevadensis*. Comparative Biochemistry and Physiology 44B:89-92.

Turner, B. J. 1973b. Genetic divergence of Death Valley pupfish populations: species-specific esterases. Comparative Biochemistry and Physiology 46B:53-70.

Turner, B. J. 1974. Genetic divergence of Death Valley pupfish species: biochemical versus morphological evidence. Evolution 28:281-294.

Turner, B. J. and R. K. Liu. 1977. Extensive interspecific genetic compatibility in the New World killifish genus *Cyprinodon*. Copeia 1977:259-269.

Williams, J. E. 1977. Observations on the status of the Devils Hole pupfish in the Hoover Dam Refugium. REC-ERC-77-11.

10 The Contribution of the Desert Pupfish (*Cyprinodon nevadensis*) to Fish Reproduction in Stressful Environments

SHELBY D. GERKING
Arizona State University, Tempe

ABSTRACT

The reproductive performance of the desert pupfish *Cyprinodon n. nevadensis* was investigated in the laboratory in relation to temperature and pH. Reproductive temperature tolerance limits, based on egg production and egg viability, are narrow (25 to 30°C), compared with the critical thermal limits for survival (2 to 44°C). Reproduction is also more sensitive than survival to acidity. These studies and other supporting evidence suggest that reproduction is the most sensitive of the life functions to stress. Furthermore, oogenesis is apparently the most sensitive stage in the life history, manifested by incomplete oocyte maturation, noticeable egg abnormalities, and the failure of eggs to hatch in acid water. Reproductive performance shows no compensation to unaccustomed temperature or pH and may actually suffer lasting damage by acid exposure, even after the fish are transferred to their normal environment. Thus temperature and pH limits for reproduction are fixed and cannot be shifted by long-term exposure to marginal conditions. These findings with the pupfish contribute to our understanding of reproductive failure of fishes in waters affected by acidification and permanent man-made structures that alter natural thermal regimes.

INTRODUCTION

Our introduction to the pupfish (killifish family, Cyprinodontidae) grew out of thermal tolerance research. The work began with *Cyprinodon milleri* from Cottonball Marsh, Death Valley, California. The heat tolerance of this species, tested in the laboratory, was higher than any other teleost (Brett 1956), although it was not vastly superior in this respect (Otto and Gerking 1973). The critical thermal maximum (CTM), or the ability to survive high temperatures for short periods, was 43.0°C, after being acclimated to 35.0°C, and the upper incipient lethal tolerance limit, or long-term temperature tolerance, was 39.5°C at the same acclimation temperature. The CTM compares closely with other pupfish species (Brown and Feldmeth 1971; Lowe and Heath 1969; Strawn and Dunn 1967). Temperatures of Cottonball Marsh were recorded in August 1971 (Naiman et al. 1973) to compare them with the upper thermal limits. Water temperature varied from 30 to 37°C when the air temperature was 47°C, and it was concluded that Marsh conditions rarely if ever presented a serious challenge to the upper lethal temperature of this species.

The pupfish proved to be an excellent experimental animal during these initial studies. They can be handled without excessive mortality, they are not distracted by human presence, they are resistant to disease, and reproduction takes place almost year-round. Consequently the pupfish gave us an opportunity to broaden the scope of our work. We chose to study reproduction quantitatively by counting the eggs laid each day and following their viability under a variety of conditions. Studies of "reproductive tolerance," we reasoned, would yield information about the ability of a population in nature to sustain itself under conditions that permit survival but may impose limitations on reproduction. Fathead minnow, *Pimephales promelas* (Mount 1968), and two other killifishes, flagfish *Jordanella floridae* (Smith 1973), and sheepshead minnow *Cyprinodon variegatus* (Hansen et al. 1973; Schimmel et al. 1974), are in current use for this purpose.

The reproduction studies have utilized *Cyprinodon n. nevadensis* from the marsh at Saratoga Springs in the Death Valley region. The remainder of this report deals with the reproductive response of this species to temperature and pH stresses. Because reproduction is a fundamental life function and certain elements of the process are common to many species, the results can be used to evaluate the impact of permanent changes in environmental conditions brought about by man's activities. We have found that reproduction is the most sensitive life function to stress (sensu Brett 1958) and information on all aspects is urgently needed at this time when environmental alterations are rapid, massive, and widespread.

REPRODUCTIVE TEMPERATURE TOLERANCE

Egg Production

Our technique of measuring egg production consists of removing a divider used to separate the aggressive male from the female in a 20-liter tank and allowing the pair to spawn on a yarn mop for 30 min. The eggs are laid singly and can easily be seen against the black background of the yarn strands. They are removed from the mop with jewelers forceps and transferred to a petri dish for observation (for details see Shrode 1975; Shrode and Gerking 1977). The procedure is repeated for 21 days. The measurement of egg production is expressed as (1) percentage of days on which spawning occurred, (2) eggs laid per spawning, and (3) eggs laid per gram of body weight per day. All three criteria yield the same general result; therefore, only $eggs \cdot g^{-1} \cdot d^{-1}$ are presented here. Because the pairs are spawned separately, a measure of individual variation is available for statistical tests.

Experiments in the effects of temperature on reproduction exposed 7 to 10 pairs to the test temperature for three weeks before the 21-day test was performed. Constant temperatures were 18, 20, 22, 24, 26, 28, 30, 32, 34, and 36°C. Cycling (24-hr) temperatures were 24 to 16, 32 to 28, 36 to 28, 38 to 28, 39 to 30°C; the low temperature occurred at 0800 hr (spawning time) and the peak at 2000 hr. A photoperiod of 16L:8D was used in all experiments.

Egg production was statistically different at the constant temperatures, shown by an analysis of variance on log-transformed data. Less than 2 $eggs \cdot g^{-1} \cdot d^{-1}$ were laid at the lower temperatures of 18, 20, and 22°C and also at the higher 34 and 36°C (Figure 1). Peak egg production occurred at 30°C, 8 $eggs \cdot g^{-1} \cdot d^{-1}$, but this

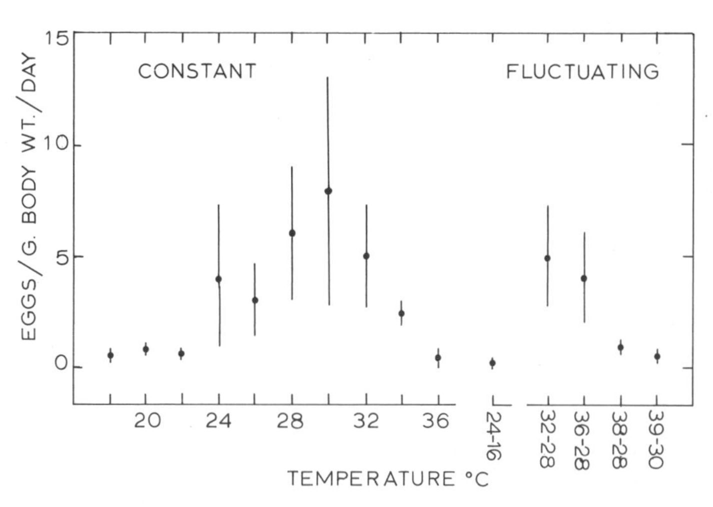

FIGURE 1. Reproductive performance of the desert pupfish, Cyprinodon n. nevadensis, under constant and fluctuating temperature conditions (from Shrode and Gerking, Physiological Zoology 50:6; copyright 1977 by the University of Chicago).

value could not be distinguished statistically from the results at temperatures ranging from 24 to 32°C. The response to cycling temperature conditions was near the mean of the cycle, as long as the extreme temperatures did not exceed the optimum range. When the extremes exceeded the upper or lower limits of the optimum range egg production was depressed.

Egg Viability

The temperatures at which at least 50 percent of the eggs hatched corresponded closely to those at which egg production was highest. The one exception occurred at 32°C, at which hatching was only 8 percent, although egg production was 5 eggs$\cdot g^{-1} \cdot d^{-1}$. These eggs possessed a soft chorion and some contained no yolk, a result observed in several other experiments performed since that time. Obviously the eggs are detrimentally affected at 32°C. We therefore adopted dual criteria for successful reproduction: (1) the temperature range over which 50 percent of the eggs hatch and (2) the temperature range over which at least 50 percent of peak egg production is experienced. These criteria are admittedly arbitrary, but smaller performance standards would extend the range very little because the curves that describe egg production and egg viability rise abruptly from low to high performance levels (Figure 2).

The egg viability tests exposed another feature of reproduction that should be taken into account. Shrode's (1975) original curve of developmental temperature tolerance was broadly domed and departed noticeably from a more sharply peaked response (see Figure 2) obtained in later studies; for example, an 80 percent hatch was obtained at 22°C, whereas only 2 percent hatched in the later series. The difference was attributed to experimental procedure. In the developmental temperature tolerance studies the eggs were laid at 28°C and transferred to other temperatures to test their ability to hatch. In the present set of experiments the eggs underwent maturation and incubation at the same temperature in a situation that would prevail in nature. Evidently the viability of the eggs is

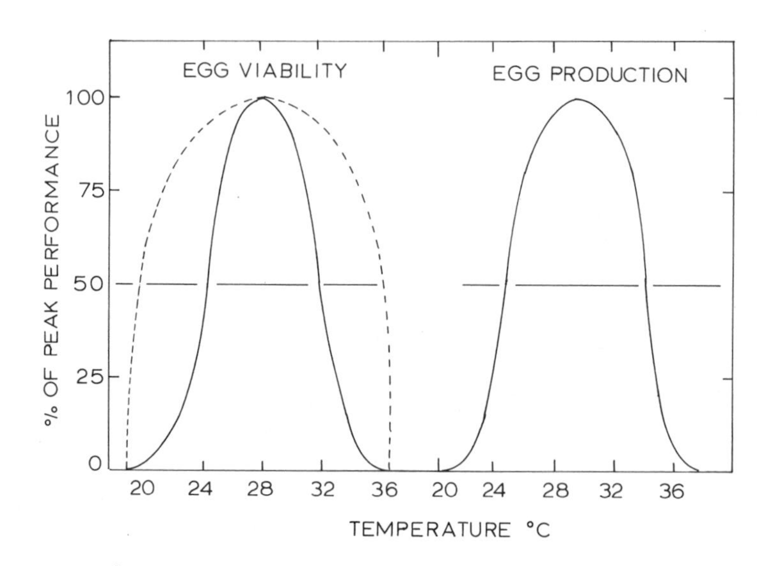

FIGURE 2. Egg viability and egg production as a percentage of peak performance at various temperatures. Peak performances were taken from raw data and percentages were calculated at each temperature. The smoothed curves shown here allow a direct comparison of the temperature range over which successful egg production and egg viability can be accomplished. The dashed line represents hatching success of eggs laid at 28°C and tested at various temperatures. The solid line is the hatching success to be expected when eggs undergo maturation and incubation at the same temperature. Modified from Gerking (1980).

affected by the exposure of the female to temperature extremes during the period in which the eggs are maturing inside the body or the eggs are detrimentally affected by temperature extremes immediately after being laid.

Reproductive Compensation to Temperature

The same stock of pupfish was subjected to generation-long (oogenesis to adult) exposure to two temperatures that produced marginal reproductive success, 24 and 32°C, and to 28°C in the optimum range. Each group was then divided into three subgroups, one to be tested at the rearing temperature and the other two at temperatures never before experienced (Gerking et al. 1979). The objective was to learn whether the limits of reproductive temperature tolerance could be broadened by long-term exposure to stressful conditions. If the limits could be broadened, fish might be able to reproduce under conditions that would otherwise prove to be detrimental if they were exposed long enough to the marginal conditions.

Briefly, no temperature compensation occurred (Figure 3). The response of egg production was typical of the test temperature, regardless of the temperature experience of the fish. Neither the optimum nor the reproductive tolerance limits were shifted upward or downward by generation-long exposure. Thus the tolerance limits are fixed and fish trapped in water with marginal temperatures cannot adapt their reproductive performance to a satisfactory level, even though they are exposed to these conditions for a whole generation.

Summary

The results of our thermal studies of *C. nevadensis* can be presented briefly in a diagram (Figure 4). The thermal limits determined by the CTM are 2 to 44°C, whereas the reproductive tolerance limits are 25 to 30°C, based on the dual criteria of 50 percent of peak performance in egg production and egg viability. Thus pupfish can reproduce successfully only over 10 percent of the temperature range that is favorable for survival for short periods of time. Recall that these reproductive tolerance limits are fixed and are not altered by exposure to temperatures beyond them. The narrowness of the reproductive tolerance limits in relation to the survival limits may indicate that a similar situation prevails in other species, although the absolute values of the temperature limits would differ.

Reproductive activity, such as male chasing the female, male vibration of the body against that of the female, and sexual coloration, cease below 12 and above 38°C. Between these temperatures, the fish may give

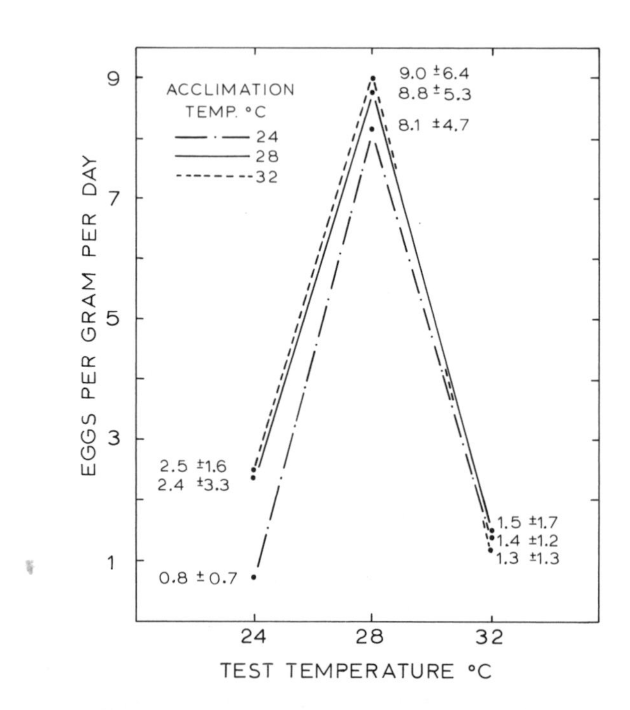

FIGURE 3. Reproductive performance of the desert pupfish, Cyprinodon n. nevadensis, under different temperature acclimation regimes. After generation-long exposure, each acclimation group was tested at its own and two other temperatures. Means and standard deviations are given for each point on the graph (from Gerking et al., Physiological Zoology 52:118; copyright 1979 by the University of Chicago).

the outward appearance of performing reproduction, but may lay no eggs in a significant portion of this temperature range. One can be easily misguided in nature by reproductive behavior that appears to be normal but never culminates in egg laying.

Eggs laid at an optimal temperature and tested for viability over a range of temperatures show developmental limits of 20 to 36°C. When rearing and incubation are performed at the same temperature, however, the limits are narrowed to 26 to 30°C.

The limits for growth, 18 to 30°C, are presented, although the experiments are not described in this

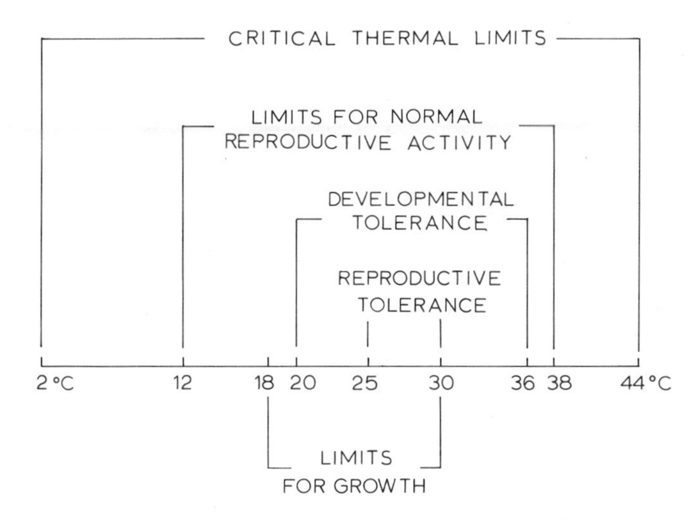

FIGURE 4. A comparison of critical thermal limits, various reproductive tolerance limits and thermal limits for growth of the desert pupfish, *Cyprinodon n. nevadensis* (from Gerking 1980, Fish Reproduction and Stress; In: M. Ali, Ed. *Environmental Physiology of Fishes*, Plenum Press).

report (Gerking and Lee unpublished data). They suggest that the pupfish can grow at temperatures several degrees below the lower limit for reproduction but the upper limit for both vital functions is the same.

REPRODUCTIVE TOLERANCE TO ACID WATERS

Egg Production

The techniques used in the thermal studies, described above, can also be applied to other stressful environmental conditions. For such an application we turned to a topic of current interest: the acidification of natural waters by the products of air pollution, SO_2 and NO_x. Acid rain has received considerable attention because, among other effects, low pH interferes with fish reproduction and often eliminates whole populations from lakes (Beamish and Harvey 1972,

Schofield 1976). Field investigations of this question conducted by examining ovaries of females collected from acid waters revealed that egg resorption is common. These field studies established pH values at which certain species disappeared, but field work is not capable of tracing the events that might have preceded their demise. We felt that the sensitivity of the pupfish model would provide additional details of reproductive failure.

The 21-day reproductive test was performed with 9 to 12 pairs of fish at pH 8.3 (control) and in each of five sulfuric acid solutions of pH 7.0, 6.5, 6.0, 5.5, and 5.0 (Lee and Gerking 1980a). The tapwater in which the tests were conducted had a conductivity of 842 $\mu mhos \cdot cm^{-1}$ (salinity = 0.5°/oo), a relatively high degree of hardness. The salinity analysis at Saratoga Springs is 3100 $\mu mhos \cdot cm^{-1}$ (salinity = 2.5°/oo). Because acidified natural waters typically have low conductivity, pupfish reproduction was tested in various salinities (Gerking and Lee 1980). Egg production was optimal over a broad range of 842 to 32,900 $\mu mhos \cdot cm^{-1}$ (salinities = 0.5 to 20.0°/oo), which is considerably beyond the survival tolerance limits of species in other teleost families. Therefore the results of the tests in tapwater should be representative of reproductive performance over a wide range of water hardness.

The effect of acid on reproduction was made readily apparent by examination of egg-laying frequency and egg production. The percentage of days on which eggs were laid remained relatively constant at pH 8.3 to 6.5 (60 to 67 percent) but was halved at pH 6.0 and 5.5 (36 to 38 percent) and virtually ceased at pH 5.0 (5 percent). Egg production paralleled this performance but revealed that any reduction in pH below that to which the fish were accustomed caused egg laying to fall off; for example, the reproductive output was 8.8 $eggs \cdot g^{-1} \cdot d^{-1}$ at the control pH and less than half this value at pH 7.0, 3.9 $eggs \cdot g^{-1} \cdot d^{-1}$. Below neutrality egg laying declined to 0.1 $egg \cdot g^{-1} \cdot d^{-1}$ at pH 5.0 (Table 1).

The fathead minnow provided the first quantitative measure of the effect of acidity on reproduction (Mount 1973). In replicate samples the mean number of eggs laid per female was 724 at pH 7.5 (control), 302 at pH

TABLE 1. Reproductive Output of the Desert Pupfish *Cyprinodon n. nevadensis* at Reduced pH Levels[a]

pH	Pairs	$Eggs \cdot g^{-1} \cdot d^{-1}$	Days Laid (percent)
8.3	10	8.8 ± 3.6	67 ± 15
7.0	8	3.9 ± 1.2	69 ± 12
6.5	12	3.4 ± 2.4	60 ± 23
6.0	9	1.8 ± 1.5	36 ± 22
5.5	9	2.0 ± 1.7	38 ± 20
5.0	9	0.1 ± 0.3	5 ± 11

[a]Means and standard deviations are given. Modified from Lee and Gerking (1980a).

6.6, and 84 at pH 5.9. Flagfish responded in a similar fashion; the mean number of eggs per female declined steadily from 16 at pH 6.8 to 0.2 at pH 4.5 (Craig and Baksi 1977). Brook trout (*Salvelinus fontinalis*), to the contrary, released about the same number of eggs over a wide (ph range of) 7.0 to 5.1 in laboratory experiments. Only very small numbers of females (1 to 4), however, actually spawned at each pH level (Menendez 1976). It remains to be seen how the reproduction of other species will react to lowering pH, but the results from the flagfish, fathead minnow, and pupfish imply that the reproductive potential of acid-stressed populations of some species is depressed in gradual fashion as acidity increases until they cannot reproduce at all.

Egg Viability

Considerably more research has been done on the effect of acid on egg viability than on egg production. Fewer fathead minnow eggs hatched at pH 5.2 (67 percent) than at the control of pH 7.5 (82 percent), but a pronounced drop to 1 percent was suffered at pH 4.5 (Mount 1973). Similar studies have been reported by Menendez (1976) on the eggs of brook trout, Johansson et al. (1973). On zebrafish (*Brachydanio rerio*), Johansson and Milbrinck (1976) on roach (*Rutilus rutilus*) and perch (*Perca fluviatilis*), and Johansson and Kihlstrom (1975) on pike (*Esox lucius*). These studies were undertaken with the purpose of establishing a minimum tolerance limit for egg survival. These values range from pH 4.8 to 6.5, depending on the species and the procedures used.

To illustrate how different procedures cause striking variances in egg viability our pupfish results (Lee and Gerking 1980b) can be contrasted with those of the flagfish (Craig and Baksi 1977). Flagfish eggs exhibited a surprisingly high hatching success after being transferred from pH 6.7 to 4.5 and 5.0, ca. 85 percent, whereas our experiments showed no hatching below pH 6.0. Flagfish eggs were transferred from the control pH of 6.7 to the test solutions within 24 hr after being laid, whereas the pupfish eggs were incubated at the same pH at which they were fertilized (continuous exposure). Results of continuous exposure by Craig and Baksi exhibited considerable variability between tests and the difference among the pH levels were not clear-cut.

Because the egg becomes resistant to osmotic changes soon after it is laid, we decided to test the hypothesis that the delayed transfer of the eggs to acid solutions was responsible for the high flagfish hatching success at low pH. In a series of experiments with the pupfish continuous exposure of eggs produced statistically lower hatching percentages (Table 2) at pH 7.0, 6.5, 6.0, and 5.5 than the 1- or 12-hr transfer. The delayed transfer of eggs to test solutions obviously has a profound effect on egg viability. A standard procedure must be adopted if egg viability is to be used to establish minimum tolerance limits. This conclusion is voiced by Kellog et al. (1978), who noticed an increase in the resistance of alewife (*Alosa pseudoharengus*) eggs that corresponded to the time when they were transferred from control temperatures to thermal shocks, and also by Daye and Garside (1979), who tested the survival of Atlantic salmon (*Salmo salar*) eggs in waters of various levels of acidity. A continuous exposure procedure mimics more closely the conditions that the fish experience in nature.

TABLE 2. Egg Viability of Pupfish, *Cyprinodon n. nevadensis*. Eggs Transferred to Acid Conditions at Various Times After Being Laid

pH	Percentage hatch		
	Continuous Exposure	Transfer (1-hr)	Transfer (12-hr)
8.3 (control)	51	57	51[a]
7.0	32	44	--
6.5	9	21	56
6.0	3	15	32
5.5	0	0	30

[a]The continuous exposure and 12-hr transfer experiment was conducted by dividing a batch of eggs into two groups. The control values therefore are the same. Taken from Lee and Gerking (1980b).

Oogenesis Effect

Not only is the quantity of eggs affected by acid exposure but the quality as well. An "oogenesis effect" is described in considerable detail by Ruby et al. (1977), who examined ovaries of flagfish exposed for 20 days to pH 6.7 (control), 5.5, 5.0, and 4.5. They classified the oogonia and oocytes into six stages. The virtual absence of the final stage of oocyte development was interpreted as a disturbance of yolk deposition and protein synthesis.

During the time when Ruby and his associates were finishing their work we were making independent observations of acid-stressed pupfish ovaries. The results of a less detailed study of immature eggs led us to the same conclusion. By pooling the data from several females a total of 343 oocytes was examined in the control group (pH 8.3), of which 18 were mature. The groups at pH 5.5 and 5.0 had 269 and 265 oocytes, of which 12 and 8, respectively, were mature. This observation was coupled with an increased frequency of abnormal eggs (opaqueness and unusual yolk diameter) immediately after deposition, which was associated with increased acidity. Abnormalities amounted to 12 percent at control pH 8.3, whereas 24 and 60 percent were abnormal at pH 5.5 and 5.0. The evidence is strongly in favor of the detrimental effect of acid on oogenesis.

Ruby et al. (1978) carried their studies further by examining spermatogenesis under the same pH conditions mentioned above. Gradually declining volumes of spermatogonia, spermatocytes, and spermatids were observed. They calculated an index of reproductive impairment and compared the male and female (Table 3). Both sexes manifested the impairment clearly, although the female was by far the more sensitive.

TABLE 3. Reproductive Impairment Based on Histological Examination of Oogenesis and Spermatogenesis in the Flagfish, *Jordanella floridae*[a]

pH	Percent Reproductive Impairment	
	Male	Female
6.7 (control)	0	0
6.0	24	79
5.5	27	84
5.0	37	98
4.5	35	92

[a]Modified from Ruby et al. (1978).

Comparison of Survival and Reproductive Tolerances

Survival tolerance of pupfish to sulfuric acid solutions was tested by a standard 96-hr bioassay technique that uses resistance time to 50 percent mortality. In keeping with other studies (Lloyd and Jordan 1964; Robinson et al. 1976), young fish had a lower tolerance to acid conditions (pH 4.72) than adults (pH 4.56). These limits do not depart greatly from a lower limit of pH 5.0, at which mortality may be expected and productivity reduced in European (EIFAC 1969) as well as American fish populations (Cooper and Wagner 1973).

Pupfish survival, however, is much less sensitive than reproduction to acid conditions. The 50 percent hatching rate and 50 percent of peak egg production were reached at the surprisingly high pH of 7.0 (Lee and Gerking 1980a). The same criteria applied to the fathead minnow reproductive performance places the limit for that species at about pH 6.6 (Mount 1973) and at pH 6.0 for the flagfish (Craig and Baksi 1977), although the data are not so clear for the latter species. Based on 50 percent egg viability alone, the reproductive tolerance limit for brook trout is between pH 5.0 and 6.0 (Menendez 1976) and for zebrafish, about pH 5.5 (Johansson et al. 1973).

The sensitivity of the pupfish to an increase in acidity is probably due to physiological adaptation on an evolutionary time scale to the alkaline environment in Death Valley. If our results can be extrapolated to nature, _C. nevadensis_ populations would be expected to survive in waters below pH 7.0 but their reproductive potential would be impaired. Generally speaking, other species have the same relationship between the limits for reproduction and survival, although both are shifted to more acid conditions, compared with the pupfish.

Reproductive Compensation to Acidity

The question of acclimation to acid exposure has not been settled, despite its obvious application to the acidification of lakes and streams. EIFAC (1969) admits that evidence of physiological compensation to acid conditions is conflicting. At the time their review was written they adopted the view that fish were not able to compensate to low pH but also offered a qualifying statement, "...that some species of fish can become acclimated to pH values lower than those found to be lethal in the laboratory." Some evidence of compensation by cardinal (_Cheirodon axelrodi_), neon (_Hyphessobrycon innesi_), and emperor (_Nematobrycon palmeri_) tetras is claimed (Dunson et al. 1977). These fish normally live in tropical acid "blackwaters."

When they were transferred every two to five weeks to successively lower pH levels one specimen lived for more than five weeks at pH 3.1 and the lower incipient lethal pH was tentatively established at 3.35 to 3.40. The number of specimens in the test was small, however, and no comparison of survival was made with fish that had not been exposed to the stepwise transfer.

The evidence of failure to compensate to acid environments is stronger. Fathead minnows kept for 12 months at pH 4.5 occasionally experienced pH 4.2 as the acidity in the tank fluctuated. They showed marked distress each time the fluctuation occurred, as if they had made no long-term physiological adjustment (Mount 1973). Brook trout appeared to be harmed by 168 hr of exposure to pH 3.75. On transfer to pH 2.5 and 3.0, these fish suffered a 20 to 25 percent decrease in survival time, compared with controls held at pH 7.5 before transfer to the low pH solutions (Robinson et al. 1976).

Pupfish held at pH 6.0 for six weeks and transferred to pH 6.5 had only half the egg production (1.7 eggs$\cdot$g$^{-1}\cdot$d^{-1}) of those held at pH 8.3 and transferred to the pH 6.5 (3.9 eggs$\cdot$g$^{-1}\cdot$d^{-1}). In addition, the acid-exposed pupfish spawned less frequently (49 percent of the days of the 21-day test) than those originally maintained at tapwater pH 8.3 (70 percent) (Lee and Gerking 1980a). Some lasting damage of acid exposure on egg production is suggested by these results. The damage found in these tests corresponds to the results of survival tests conducted by Robinson et al. (1976). A more elaborate experiment with rigorous controls recently completed in our laboratory confirms this conclusion. The results also demonstrate that reproduction in the pupfish does not exhibit compensation to acid exposure.

Summary

Egg production, egg-laying frequency, and egg viability are more sensitive to acidic conditions than survival. The decrease in these parameters with increasing hydrogen-ion concentration can be traced to the effect of acid on the eggs while they are in the body of the female, the "oogenesis effect." A similar effect on spermatogenesis has also been observed. No compensation of survival or the reproductive function to acid waters has been convincingly demonstrated. Bioassay tests of egg survival should be examined critically because the egg becomes more resistant to exposure to stress conditions during the first 12 hr after being laid.

GENERAL CONCLUSIONS

The following conclusions, some of which are modified from Gerking (1980), are supported by research on temperature and pH effects with the pupfish and a few other species. Like any group of conclusions, it is expected that they will be modified or reinterpreted by further work.

1. Reproduction is the most sensitive to stress among all the life functions. As a result, successful reproductive performance should be assigned as a priority criterion in describing conditions that are suitable to maintaining populations in nature.
2. Oogenesis is the most sensitive stage in the reproductive process; it thereby affects the quantity and quality of eggs. A consequence of this conclusion is that success or failure of hatching may be the result of stress that the female experiences while the eggs are maturing inside her body. Additional work is required to establish this event in seasonal spawners. Also, less than complete information is available on the timing of meiosis in relation to spawning to appreciate the possible impact of prespawning stress on reproductive performance.
3. Reproductive performance shows no compensation to temperature or pH stress. This means that the tolerance limits for successful reproduction cannot be broadened by exposure to conditions outside the optimal range.
4. The egg becomes resistant to environmental stresses within hours after being laid. Egg-survival tests should be designed with this factor in mind.

Most procedures have delayed for various periods the transfer of eggs from "normal" or optimal environments to test solutions, whereas a test that continuously exposes the egg to stress from the time of fertilization more closely approaches conditions that prevail in nature.

DIRECTIONS FOR FUTURE RESEARCH

The environmental effects on oogenesis suggested here have far-reaching implications. The fundamental effect may be on the mechanism of meiotic division, or chromosomal distribution, and to the formative stages of the ovum, such as yolk deposition and protein synthesis mentioned earlier. Environmental effects on spermatogenesis have only been scratched and need much more research. So far attention has been directed at the female because eggs are larger and more easily handled, whereas spermatogenesis is more difficult to quantify and must necessarily be studied at the microscopic level.

At present, fishery biology lacks a comprehensive view of the relation of food intake to fecundity, egg size, sexual maturity, and ultimate reproductive success. Because egg production can be quantified, and we have successfully conducted growth experiments with the pupfish in the laboratory, these problems are subject to direct experimentation by bioenergetic analysis.

Finally, the pupfish reproductive test can be used as a powerful bioassay procedure. The effects of toxic substances on reproduction can be assayed more sensitively with the pupfish than with any other fish species now being used for this purpose.

ACKNOWLEDGMENTS

This research was performed under contract EY-76-5-02-2498 sponsored by the U.S. Department of Energy. I wish to thank a number of former graduate students and a parade of hourly workers for their interest and dedication to the objectives of the research. Dr. J.

B. Shrode and Mr. R. M. Lee deserve special mention because they established and have kept the pupfish colony thriving since its inception.

REFERENCES

Beamish, R. J. and H. H. Harvey. 1972. Acidification of the La Cloche mountain lakes, Ontario, and the resulting fish mortalities. Journal of the Fisheries Research Board of Canada 29:1131-1143.

Brett, J. R. 1956. Some principles in the thermal requirements of fishes. Quarterly Review of Biology 31:75-87.

Brett, J. R. 1958. Implications and assessments of environmental stress. _In_: P. A. Larkin (Ed.), The Investigation of Fish-power Problems. University of British Columbia, Vancouver, pp. 69-83.

Brown, J. H. and C. R. Feldmeth. 1971. Evolution in constant and fluctuating environments: thermal tolerances of desert pupfish (_Cyprinodon_). Evolution 25:390-398.

Cooper, E. L. and C. C. Wagner. 1973. The effects of acid mine drainage on fish populations. Ecological Research Series, Environmental Protection Agency R3-73-032:73-124.

Craig, G. R. and W. F. Baksi. 1977. The effects of depressed pH on flagfish reproduction, growth and survival. Water Research 11:621-626.

Daye, P. G. and E. T. Garside. 1979. Development and survival of embryos and alevins of the Atlantic salmon, _Salmo salar_ L., continuously exposed to acidic levels of pH from fertilization. Canadian Journal of Zoology 57:1713-1718.

Dunson, W. A., F. Swarts and M. Silvestri. 1977. Exceptional tolerance to low pH of some tropical blackwater fish. Journal of Experimental Zoology 201: 157-162.

European Inland Fisheries Advisory Commission (EIFAC). 1969. Water quality criteria for European freshwater fish:extreme pH values and inland fisheries. Water Research 3:593-611.

Gerking, S. D. 1980. Fish reproduction and stress. In: M. Ali (Ed.), Environmental Physiology of Fishes. Plenum, New York, pp. 569-587.

Gerking, S. D. and R. M. Lee. 1980. Reproductive performance of the desert pupfish (Cyprinodon n. nevadensis) in relation to salinity. Environmental Biology of Fishes 5:375-378.

Gerking, S. D., R. M. Lee and J. B. Shrode. 1979. Effects of generation-long temperature acclimation on reproductive performance of the desert pupfish, Cyprinodon n. nevadensis. Physiological Zoology 52:113-120.

Hansen, D. J., S. C. Schimmel and J. Forester. 1973. Arochlor 1254 in eggs of sheepshead minnows: effect on fertilization success and survival of embryos and fry. In: 27th Annual Conference Southeastern Association of Game and Fish Commissioners, pp. 420-426.

Johansson, N. and J. E. Kihlstrom. 1975. Pikes (Esox lucius L.) shown to be affected by low pH values during first weeks after hatching. Environmental Research 9:12-17.

Johansson, N., J. E. Kihlstrom and A. Wahlberg. 1973. Low pH values shown to affect developing fish eggs (Brachydanio rerio Ham.-Buch.). Ambio 2:42-43.

Johansson, N. and G. Milbrinck. 1976. Some effects of acidified water on the early development of the roach (Rutilus rutilus L.) and perch (Perca fluviatilis L.). Water Research Bulletin 12:39-48.

Kellog, R. L., J. J. Salerno and D. L. Latimer. 1978. Effects of acute and chronic thermal exposures on the eggs of three Hudson River anadromous fishes. In: J. H. Thorp and J. W. Gibbons (Ed.), Energy and Environmental Stress in Aquatic Systems. United States Department of Energy. pp. 714-725.

Lee, R. M. and S. D. Gerking. 1980a. Survival and reproductive performance of the desert pupfish, Cyprinodon n. nevadensis, in acid waters. Journal of Fish Biology 17:507-515.

Lee, R. M. and S. D. Gerking. 1980b. Sensitivity of fish eggs to acid stress. Water Research 14:1679-1681.

Lloyd, R. and D. H. M. Jordan. 1964. Some factors affecting the resistance of rainbow trout, *Salmo gairdneri*, in acid waters. International Journal of Air and Water Pollution 8:393-403.

Lowe, C. H. and W. G. Heath. 1969. Behavioral and physiological responses to temperature in the desert pupfish, *Cyprinodon macularius*. Physiological Zoology 42:53-59.

Menendez, R. 1976. Chronic effects of reduced pH on brook trout (*Salvelinus fontinalis*). Journal of the Fisheries Research Board of Canada 33:118-123.

Mount, D. I. 1968. Chronic toxicity of copper to fathead minnows (*Pimephales promelas* Rafinesque). Water Research 2:215-223.

Mount, D. I. 1973. Chronic effect of low pH on fathead minnow survival, growth and reproduction. Water Research 7:987-993.

Naiman, R. J., S. D. Gerking and T. D. Ratcliff. 1973. Thermal environment of a Death Valley pupfish. Copeia 1973:366-369.

Otto, R. G. and S. D. Gerking. 1973. Heat tolerance of a Death Valley pupfish (Genus *Cyprinodon*). Physiological Zoology 46:43-49.

Robinson, G. D., W. A. Dunson, J. E. Wright and G. E. Mamolito. 1976. Differences in low pH tolerance among strains of brook trout (*Salvelinus fontinalis*). Journal of Fish Biology 8:5-17.

Ruby, S. M., J. Aczel and G. R. Craig. 1977. The effects of depressed pH on oogenesis in flagfish, *Jordanella floridae*. Water Research 11:757-762.

Ruby, S. M., J. Aczel and G. R. Craig. 1978. The effects of depressed pH on spermatogenesis in flagfish, *Jordanella floridae*. Water Research 12:621-626.

Schimmel, S. C., D. J. Hansen and J. Forester. 1974. Effects of Arochlor 1254 on laboratory reared embryos and fry of sheepshead minnows (*Cyprinodon variegatus*). Transactions of the American Fisheries Society 103:582-586.

Schofield, C. L. 1976. Acid precipitation: effects on fish. Ambio 5:228-230.

Shrode, J. B. 1975. Developmental temperature tolerance of a Death Valley pupfish (*Cyprinodon nevadensis*). Physiological Zoology 48:378-389.

Shrode, J. B. and S. D. Gerking. 1977. Effects of constant and fluctuating temperatures on reproductive performance of a desert pupfish, *Cyprinodon n. nevadensis*. Physiological Zoology 50:1-10.

Smith, W. E. 1973. A cyprinodontid fish, *Jordanella floridae*, as a laboratory animal for rapid chronic bioassays. Journal of the Fisheries Research Board of Canada 30:329-330.

Strawn, K. and J. E. Dunn. 1967. Resistance of Texas salt and freshwater marsh fishes to heat death at various salinities. Texas Journal of Science 19:57-76.

11 The Evolution of Thermal Tolerance in Desert Pupfish (Genus *Cyprinodon*)

C. ROBERT FELDMETH
The Claremont Colleges
Claremont, California

ABSTRACT

Thermal tolerance in desert pupfish includes four main evolutionary strategies: behavioral thermoregulation, developmental plasticity, temperature acclimation, and evolutionary adaptation. Behavioral thermoregulation involves precise selection of preferred water temperatures in thermal gradients common to many desert aquatic habitats. Young pupfish are more tolerant of high temperature when compared with adults and eggs of the same species. This ontogenetic plasticity appears to allow very young fish to survive in warmer waters of pools and stream margins. When thermal gradients are not present, or when seasonal temperature changes occur, physiological compensation to warmer conditions can take place. Pupfish from thermally fluctuating environments adjust their critical thermal maximum and minimum as much as 4°C from summer to winter. Amargosa pupfish have also been known to increase their thermal scope (measured by the difference between critical thermal maximum and minimum) when acclimated to a diel fluctuating thermal cycle. Finally, a long-term evolutionary adjustment of thermal tolerance for pupfish from a constant temperature thermal spring was found when compared with a closely related population from a nearby desert stream with a seasonally fluctuating thermal regime. Thus the extremes of the thermal niche, bounded by upper and lower lethal limits, have been reduced for the pupfish population living in the constant temperature habitat.

INTRODUCTION

Among North American fishes there is wide variation in the ability to survive high temperatures, as indicated by laboratory determinations of thermal tolerance. Pupfish (*Cyprinodon*) found in desert habitats of North America tolerate natural environmental temperatures as high as 42°C (Brown 1971) and a critical thermal maximum[1] tolerance as high as 44.6°C has been recorded in the laboratory (Lowe and Heath 1969). In contrast, salmon (*Oncorhynchus*) of the cool streams and rivers of the Pacific Northwest have upper lethal temperatures[2] as low as 21.3°C (Brett 1956). All other fishes of temperate North American freshwater habitats probably have tolerances between these extremes, with correlation between the thermal regime of the environment and thermal tolerance (Brett 1956).

---

[1]Critical thermal maximum is an index of thermal tolerance determined by heating the water at a specific rate and measuring the temperature at which fish lose equilibrium or die.

[2]Upper lethal temperature is a mean thermal tolerance determined by transfer of fish from an acclimation temperature to a warmer test temperature and measuring time until loss of equilibrium or death.

The ecological significance of thermal tolerance in fishes has been questioned because relatively few reports of heat death under natural conditions have been made for most North American fishes, and the temperature tolerances of fishes are usually well above their habitat temperatures (Brett 1959). Species of the genus Cyprinodon, however, occur in numerous habitats in the deserts of the American southwest and northern Mexico, where they not only are known to experience frequent and drastic natural heat kill (Lowe and Heath 1969) but can also be found living at temperatures just below their critical thermal maximum (Brown and Feldmeth 1971).

Why a salmon dies when water temperature reaches 21.3°C and a pupfish tolerates temperatures as high as 44.6°C is apparently determined by the effect of thermal energy on the integrity of chemical bonds, termed weak bonds (Hochachka and Somero 1971). Biochemically functional structures formed by van der Waals forces, hydrogen bonds, and ionic bonds have bond energies at least one order of magnitude below covalent chemical bonds. Because proteins, membranes, and nucleic acids have tertiary and quartenary structures that are dependent on these weak bonds, environmental temperature can increase to a point of which these complex structures become deranged or perturbed in a manner that is lethal to an organism (Hochachka and Somero 1973; Somero 1978).

Hence a probable physiological sequence as temperature increases to a lethal level is deactivation of enzymes by structural rearrangement, failure to form metabolic products, and decline of cell metabolism. As a result, organ systems fail and loss of equilibrium and respiratory failure ensues. Thermal death thus occurs because life depends on biochemical structure and integrity; environmental temperature provides enough energy to break bonds and perturb essential structures, hence disrupting metabolism.

Because relative lethal limits of various fishes are correlated with environmental temperatures, differences in thermal tolerance must have developed over a long period of evolutionary history. Much comparative research, however, has focused on fish from widely different taxa; comparisons among closely related fishes should provide more information on the evolution of thermal tolerances.

Desert pupfish evolved from ancestors that occupied thermally fluctuating coastal bays and salt marshes of the Gulf of Mexico (Miller 1948). Thus their ability to tolerate extreme temperatures is not surprising. Because _Cyprinodon_ has occupied the southwestern deserts of the United States for more than a million years and has become isolated into a number of species (Hubbs and Miller 1948; Miller 1950, this volume) that occur in a variety of thermal regimes, perhaps some indication of the evolution of different thermal tolerances can be obtained by a careful examination of the lethal temperatures of desert pupfish species.

Some species of _Cyprinodon_ live in habitats that fluctuate thermally on a seasonal and a daily basis, perhaps with even greater extremes than those experienced by ancestral pupfish. Others occur in constant temperature pools, where tolerance to temperature extremes is no longer required (Brown and Feldmeth 1971). Because nonfunctional characteristics tend to be eliminated by natural selection (Regal 1977), one would expect that pupfish which presumably inhabited fluctuating environments would lose thermal tolerance if they subsequently evolved in less variable environments (Brattstrom 1968; Snyder and Weathers 1975).

Careful examination of the thermal tolerances of pupfish from habitats of various constant temperatures as well as those from daily and seasonally fluctuating environments sheds light on the question of the evolution of thermal tolerance of North American fishes. The purpose of this chapter is to review the studies of thermal tolerance of _Cyprinodon_ species living in desert habitats to determine to what extent their environments have influenced the evolution of behavioral and physiological characteristics related to temperature.

THERMAL TOLERANCE

Four main strategies for dealing with temperature are available to ectotherms: behavioral thermoregulation, developmental plasticity, temperature acclimation, and evolutionary adaptation. In aquatic environments behavioral thermoregulation involves seeking out thermally appropriate habitats by preference and avoidance responses along thermal gradients. In desert aquatic environments in particular these behavioral responses can be critical to survival. Very young fish may have a higher thermal tolerance than adults, hence can tolerate the warmer temperatures at stream or pond margins. In a thermally homogeneous environment temporary phenotypic adjustments such as temperature acclimation allow fish to cope with changing temperatures. A final strategy uses evolutionary adaptation of physiological tolerance to temperature, which requires adjustment of the thermal niche.

Behavioral Thermoregulation

Thermoregulation can be viewed physiologically and ecologically. From a physiological perspective thermoregulatory behavior is a homeostatic mechanism that tends to maintain internal temperatures favorable to metabolism. From an ecological viewpoint behavioral thermoregulation is an aspect of habitat selection, hence constitutes a dimension of the species niche (Magnuson et al. 1979; Reynolds 1979). Precise control of body temperature by behavioral means is well known in reptiles (Bogert 1949; Cowles and Bogert 1944; Huey 1974; Huey and Slatkin 1976) and also occurs to some extent in amphibians (Brattstrom 1979; Brattstrom and Warren 1955; Lillywhite 1970; Lillywhite et al. 1973). Fishes aggregate in a preferred temperature (Norris 1963) and definitely show some behavioral thermoregulatory capability (Beitinger and Fitzpatrick 1979; Crawshaw 1977; Crawshaw and Hammel 1974; Neill 1979; Reynolds 1977).

In the Salton Sea of southeastern California pupfish (_C. macularius_) show what appears to be true behavioral thermoregulation as they move into warm shallow pools (up to 37°C) in the morning and remain there feeding throughout the day. They return to cooler areas in deeper water in the evening (Barlow 1958). This apparent diel thermoregulatory cycle indicates that pupfish have developed the ability to sense and respond rapidly to natural thermal gradients. In Quitobaquito Spring, southern Arizona, this same species swims into very warm shallow pools (40 to 41°C) when much cooler water (30°C) is nearby. The lethal temperature of these fish is somewhere between 43 and 44°C; hence *Cyprinodon* regulates its body temperature behaviorally within 2 to 3°C of thermal death (Lowe and Heath 1969).

In a small artesian spring in the Mojave Desert (Tecopa Bore) near Death Valley Amargosa pupfish (*C. nevadensis amargosae*) live in temperatures as high as 42°C (Brown 1971). Water issues from the ground at 47.5°C and cools 8 to 12°C over a distance of about 300 m. A dense mat of bluegreen algae forms along the stream margins and pupfish are the dominant herbivores in the ecosystem (Naiman 1976). Pupfish occur throughout the stream below the 42°C isotherm. On cold, windy winter days cooling of the hot stream water occurs more rapidly and the 42°C isotherm moves much closer to the stream source (Figure 1). This allows pupfish to exploit an area of the stream that in summer is warmer than their critical thermal maximum. As air temperatures warm, the length of the stream, which is 42°C or greater, increases and the pupfish are forced back downstream (Brown 1971). At times fish become isolated in small, shallow side pools which are slightly cooler than the main stream. Precise temperature measurements in such a portion of the stream indicate that the pupfish are selecting temperatures just below their thermal death point. If startled, fish will dart out of these side pools onto the warmer main stream, where they instantly lose equilibrium and die.

Precise behavioral temperature selection is not restricted to desert pupfish. Laboratory studies of several freshwater fish species indicate that temperature differentials as small as 0.05°C can be detected (Bardach and Bjorklund 1957). However, the pupfish are living by choice at the extreme upper limit of their zone of thermal tolerance. Thus they are different

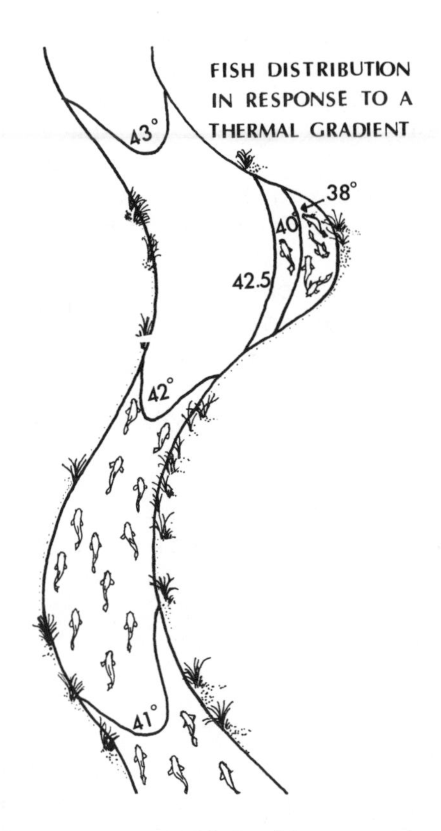

FIGURE 1. Amargosa pupfish (C. nevadensis amargosae) are able to respond precisely to temperature differentials in an artesian thermal spring. Fish swim up to the 42°C isotherm, which is their upper thermal tolerance limit (from Soltz and Naiman 1978).

from many other North American fishes that normally never encounter environments near their lethal temperature. Precise temperature selection is an important ecological factor in a thermal artesian spring habitat; however, the physiological and behavioral adaptations demonstrated by the Amargosa River pupfish are also a definite advantage in coping with the numerous other types of desert habitat in which these remarkable fish survive.

Another behavioral aspect of the thermal ecology of desert pupfish concerns segregation of Comanche

Springs pupfish (*C. elegans*) as a result of apparent competition (Gehlbach et al. 1978). In the laboratory *C. elegans* shifts to cooler water in a horizontal temperature gradient when in the presence of the Pecos gambusia (*Gambusia nobilis*). When alone, each species prefers a temperature range of 26 to 30°C; when placed together in a thermal gradient the pupfish chooses a temperature area of 21 to 25°C, whereas the gambusia selects the 26 to 30°C portion of the gradient. In a sense this thermal segregation is equivalent to a niche shift (Magnuson et al. 1979).

The Effect of Temperature on Developmental Stages

The ability of eggs and juvenile animals to survive environmental conditions has often been neglected but is of considerable ecological significance. Survival of pupfish populations in isolated desert habitats can occur only if all life stages can tolerate existing environmental factors; hence the effect of temperature on the embryological development of desert pupfish has been rather thoroughly examined. Kinne and Kinne (1962) observed that eggs of *C. macularius* from the Salton Sea, California, develop normally and produced greater than 50 percent hatch between 17 and 34°C at 35 ppt salinity. Shrode (1975) found that the eggs of *C. nevadensis nevadensis* (Saratoga Spring, Death Valley) developed normally with greater than 50 percent hatch from 20 to 36°C; when exposed to daily temperature fluctuations this level of hatch success occurred up to 38°C. Although the developmental tolerance range of pupfish is broad, it is narrower than the range for normal adult activity. From their experiments with thermal tolerance limits for oogenesis and egg development in this pupfish Shrode and Gerking (1977) estimate that thermal reproductive limits are approximately one-fifth the adult activity range and one-seventh the critical thermal range.

Shrode (1975) also determined critical thermal maxima for young pupfish and found that they were consistently higher than those of older fish acclimated under the same conditions. Shrode concluded that tolerance to high temperatures is lowest during the egg stage, highest in young fish, and intermediate in larger adults. Lowe and Heath (1969) observed that young C. macularius were regularly observed in temperatures significantly higher than adults in the spring-fed pond at Quitobaquito in southern Arizona. Similar observations have been documented in laboratory selection studies on juvenile yellow perch (McCauley and Read 1973) and juvenile rainbow trout (Kwain and McCauley 1978). Feldmeth and Eriksen (1978) also observed marked differences in temperature tolerance and behavioral temperature selection for grayling fry, compared with adults, and suggested that the higher thermal tolerance of young fish is indicative of physiological and ecological adaptations that allow them to remain in warmer shallow water to avoid predation.

Acclimation and Thermal Tolerance

The hypothesis first presented by Brown and Feldmeth (1971), that there should be an evolutionary loss of the ability to withstand temperature extremes among fishes isolated in constant temperature environments is critical to a determination of the mechanisms and rates of evolution of thermal tolerance in fishes.

In a survey of 11 populations of Cyprinodon from the Death Valley area Brown and Feldmeth found that upper and lower lethal temperatures were closely related to environmental water temperatures. Fish from constant temperature springs had upper lethal temperatures ranging from 41.2°C for Forest Spring (water temperature, 26.5°C) to 42.7°C for Devils Hole (water temperature, 33.9°C; see Figure 2). Pupfish from the Amargosa River, which fluctuates in temperature seasonally, had an upper lethal temperature of 38.8°C, when collected on January 19, when the environmental temperature was about 10°C (Figure 3), whereas the same subspecies (C. nevadensis amargosae) had an upper lethal temperature of 43.5°C when collected from the 42°C region of the Tecopa Bore (Figure 4). Field acclimatized fish from Salt Creek and the Amargosa River had a lower lethal temperature of less than 1.0°C when collected during January (about 10°C), whereas pupfish from

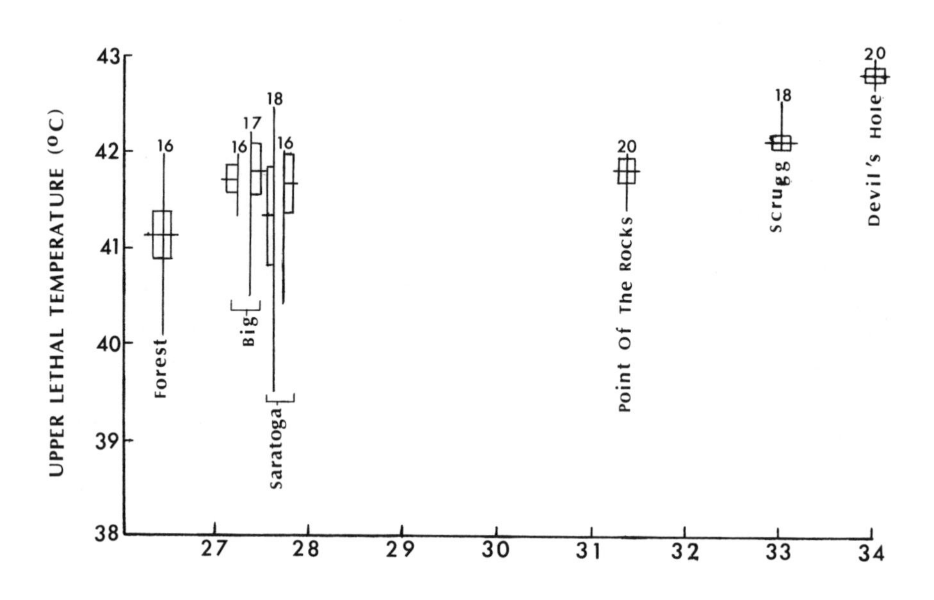

FIGURE 2. Relationship between upper lethal temperature and temperature of spring for populations of Cyprinodon which are inhabiting thermally constant springs. Vertical lines represent ranges; horizontal lines, means; rectangles, 95 percent confidence intervals; numbers above each diagram, sample size. Replicate samples for Big and Saratoga Springs were tested in January and April (from Brown and Feldmeth 1971).

constant temperature warm springs showed higher and definitely habitat-correlated lower lethal temperatures which ranged from 2 to 7°C (Table 1).

The fact that fish can adjust their thermal tolerance levels to seasonal temperature changes is well known (Brett 1944; Brett 1956; Fry 1947). From the data for Amargosa pupfish Brown and Feldmeth (1971) suspected that the differences in lethal temperatures they had measured for various warm spring pupfishes might be due entirely to thermal acclimation. To test this hypothesis fish from the Amargosa River, a seasonally fluctuating habitat, were compared with a closely

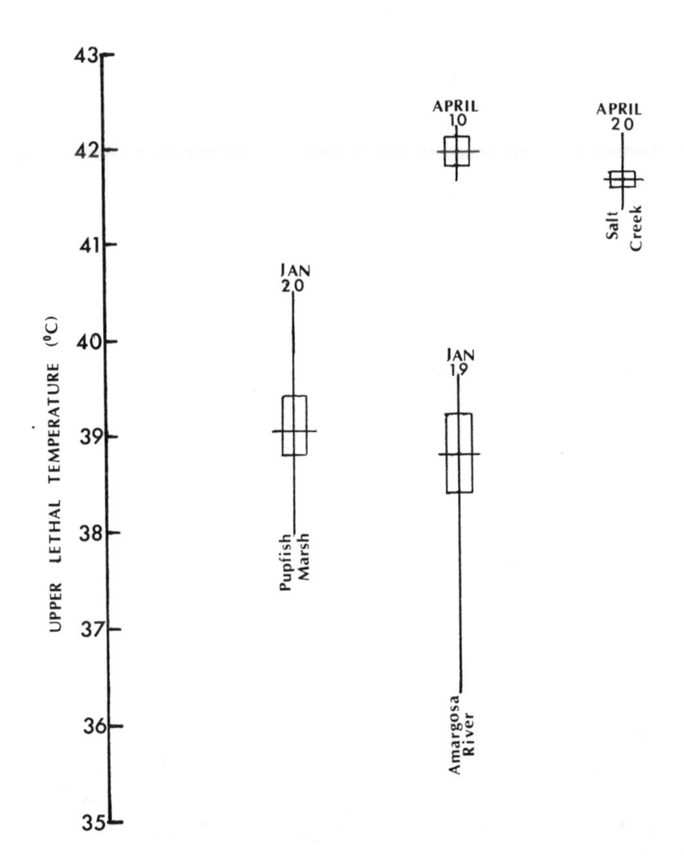

FIGURE 3. Seasonal variation in the upper lethal temperatures of populations of Cyprinodon that inhabit thermally fluctuating waters. Symbols are those in Figure 2. All three localities have similar elevations on the floor of Death Valley and their inhabitants are exposed to similar temperature regimes (from Brown and Feldmeth 1971).

related subspecies from a warm, constant temperature spring (C. nevadensis nevadensis from Saratoga Spring; 27.5°C). Fish from both habitats were held in the laboratory at 15 and 25°C for a period of three weeks. The two populations acclimated at 15°C had virtually identical upper lethal temperatures of 39°C, whereas the two groups acclimated to 25°C had upper lethal temperatures of about 41°C, although the fish from

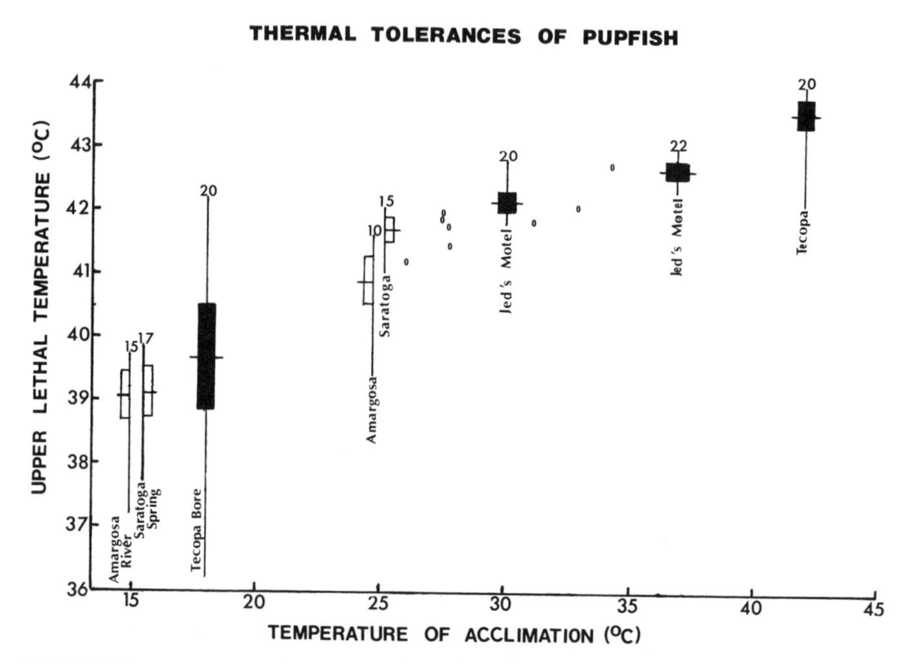

FIGURE 4. Effects of thermal acclimation on the upper lethal temperatures of several populations of _Cyprinodon_. Circles indicate populations naturally acclimated in thermally constant springs. Shaded diagrams illustrate the results of "natural" experiment; unshaded diagrams, laboratory experiments. Symbols are those in Figure 2 (from Brown and Feldmeth 1971).

Saratoga Spring tolerated slightly higher temperatures (Figure 4).

Brown and Feldmeth then performed a series of additional temperature tolerance experiments on Amargosa pupfish from two thermal artesian streams, the Tecopa Bore and a nearby stream issuing from a well at Jed's Motel. Fish were collected from Tecopa Bore at 18 and 42°C and from the stream at Jed's Motel at temperatures of 30 and 37°C.

The upper lethal temperatures for these field-acclimatized groups of pupfish (Figure 4) correlate well with the acclimation data and seem to indicate that differences in thermal tolerances for the pupfish

TABLE 1. Tolerance of Death Valley Area *Cyprinodon* to Low Ambient Temperatures

Species	Habitat	Thermal History	Loss of Equilibrium
C. diabolis	Devils Hole	Field acclimatized; 39.9°C	6–7°C
C. nevadensis pectoralis	Scrugg's Spring	Field acclimatized; 32.5°C	6–7°C
C. n. mionectes	Point-of-Rocks-Spring	Field acclimatized; 31.4°C	4–5°C
C. n. nevadensis	Saratoga Spring	Field acclimatized; 27.5°C	2–3°C
C. n. nevadensis	Saratoga Spring	Laboratory acclimatized; 15°C	1°C
C. n. amargosae	Amargosa River	Laboratory acclimatized; 15°C	1°C
C. n. amargosae	Amargosa River	Field acclimatized in winter ca. 10°C	1°C
C. salinus	Salt Creek	Field acclimatized in winter ca. 10°C	1°C

of the Death Valley region may be explained as a phenotypic physiological adjustment. Slight differences in thermal tolerances were observed, however, and more precise temperature tolerance tests on these populations may show distinct differences. Amargosa pupfish collected in winter (January 19) have upper lethal temperatures below 39°C, whereas the same population had an upper lethal temperature of about 42°C when collected in the spring (April 10) (Figure 3). Otto and Gerking (1973) found similar results when Cottonball Marsh pupfish (C. milleri) were acclimated to a range of temperatures. Clearly thermal acclimation is another important adaptation pupfish used to tolerate desert environments.

In carrying out the numerous experiments described above we observed that pupfish from habitats with a daily thermal cycle such as the Amargosa River seemed to have more resistance to cold than fish from a constant temperature habitat. To test this observation a thermally cycling experimental aquarium was designed to produce a diel temperature regime of 15°C at night and 35°C during the day (Feldmeth et al. 1974). Warming and cooling rates were equal, as were the constant high (day) and low (night) temperature period. Amargosa pupfish were placed in the thermally cycling aquarium and in constant temperature aquaria of 15, 25, and 35°C. All fish were exposed to a 12 to 12, dark-light photoperiod.

Amargosa pupfish had critical thermal maxima and minima about 39°C apart when acclimated to the three constant temperature regimes (Figure 5). This thermal scope for tolerance shifted upward as the acclimation temperature increased and a loss of cold tolerance to gain heat tolerance has been observed in other North America fishes (Brett 1944; Fry 1967; Heath 1963).

When pupfish were acclimated to a fluctuating temperature cycle of 15 to 35°C the critical thermal maximum was equal to that of fish acclimated to a constant temperature of 35°C; however, the critical thermal minimum was equal to that of fish held in 25°C. Thus the thermal tolerance scope for this pupfish living in the fluctuating temperature extremes of the Amargosa River expands when acclimated to a cycling thermal regime

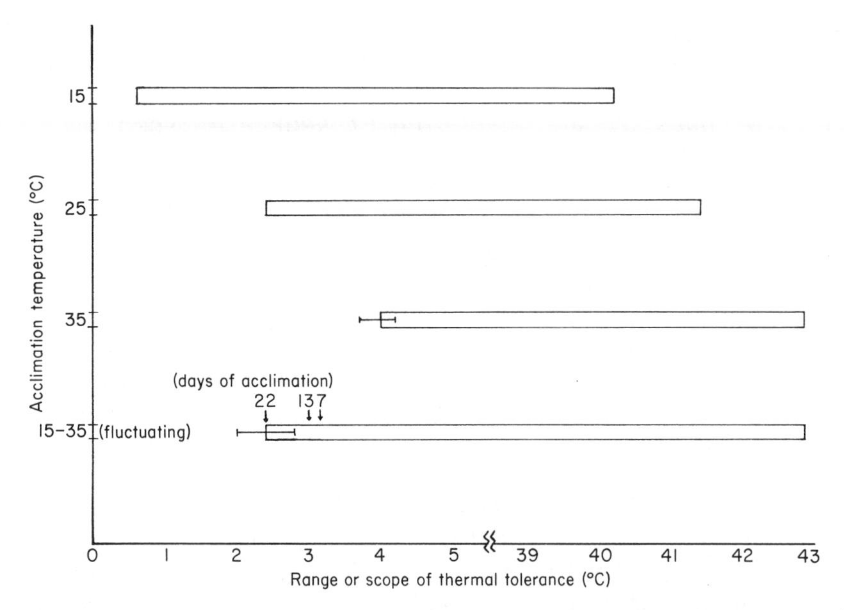

FIGURE 5. Range of thermal tolerance for pupfish acclimated to various constant temperatures and a 15 to 35°C daily temperature cycle. Bars indicate 95 percent confidence intervals. Critical thermal maxima and minima experiments were performed in duplicate (approximately 18 fish per experiment). For fish exposed to cycling temperatures acclimation to low temperature occurs more slowly than to high temperature, as indicated by critical thermal minima after 7, 13, and 22 days acclimation (from Feldmeth et al. 1974).

(Figure 5); note also that it requires 21 days of acclimation for the full extension of scope.

Thermal acclimation is a physiological compensation in ectotherms to allow some homeostatic control of metabolism to take place. The colder water temperatures of winter would greatly limit the behavior of fish were it not for the fact that these metabolic adjustments have evolved. The mechanism involved in the relatively slow process described here (21 days) is probably by control of the level of enzyme variants (isoenzymes) uniquely suited to function at particular environmental temperatures (Hochachka and Somero 1973).

It is remarkable that our experiments on acclimation to fluctuating temperatures indicate that it is possible for fish to acclimate to daily changes in temperature, perhaps by the induction of separate isoenzymes, which allows compensation for warm and cold temperatures to occur simultaneously.

Genetic Aspects of Thermal Tolerance

The results of Brown and Feldmeth (1971) indicate that the evolutionary rate of thermal tolerance alterations in desert pupfish may be slow. Even _C. diabolis_, which has been isolated in a thermally constant spring for at least 20,000 years, appears to be capable of tolerating a range of temperatures as wide as _C. salinas_ (Salt Creek, Death Valley) and populations of _C. nevadensis_ which encounter temperature fluctuations of 0 to 40°C in their present habitats. There seems to be a strong tendency to retain homeostatic capabilities with respect to temperature. Turner's (1974) work with metabolic enzymes, hemoglobin, and muscle proteins for five morphologically distinct pupfish from Death Valley, Owens Valley, and the Colorado River drainages indicates that minimal genetic divergence of these biochemical characteristics has occurred in desert pupfishes. Turner presented a hypothesis that states that a portion of the evolving pupfish genome must have encoded a strongly coadapted, homeostatic, metabolic core of enzymes appeared initially in widely fluctuating thermal habitats. Although the evolution of morphological characteristics occurred rapidly for the various developing species, rates of evolution of metabolic enzymes and other proteins were apparently much slower (Turner 1974).

Perhaps, however, small differences in enzymes or physiological or ecological adaptation to temperature exist in populations of pupfish in habitats with different thermal characteristics that were not detected in the experiments designed by Brown and Feldmeth (1971) or the biochemical genetic studies of Turner (1974). To approach this question Hirshfield et al. (1980) examined two closely related subspecies from different thermal environments. The populations chosen were *C. nevadensis mionectes* Miller from Big Spring, a constant temperature habitat (27.3°C) and *C. n. amargosae* from the Amargosa River, which varies considerably in temperature (0 to 40°C). Brown and Feldmeth (1971) concluded that no genetic differences occurred in short-term thermal tolerances; however, fish were collected directly from the field and subjected to a tolerance test that required a relatively slow change in temperature ($2°C \cdot h^{-1}$). This method may not have been precise enough to detect slight differences in thermal tolerance that might nevertheless exist. Also, because fish were tested immediately after collection, other environmental variables may have affected the results.

We (Hirshfield et al. 1980) therefore collected eggs from 50 females (fertilized by 10 males) for each of the two populations. Offspring of both species were isolated before hatching and raised under identical conditions (temperature, water, and feeding regimes). Hybrids of the two subspecies were also reared.

Two thermal tolerance tests were used: thermal shock and critical thermal maxima and minima experiments (CTM). In the first, resistance time to mortality (cessation of opercular movement) was measured when the fish were transferred from 25 ± 1°C (temperature of acclimation since hatching) to test aquaria at 8°C to determine cold tolerance and by transfer to 38 or 39°C to measure heat tolerance. Critical thermal maxima and minima were determined with fish maintained at 27 ± 1°C since the egg stage. Individual fish were heated or cooled in their containers. Water temperatures were lowered approximately 10°C the first hour and 5°C an hour thereafter. The temperature at loss of equilibrium was recorded for each fish with a recording thermograph to 0.1°C. At the conclusion of the experiment the temperature was returned slowly to 27°C; two days later the temperature was raised approximately 8°C the first hour and 5°C an hour thereafter. Temperature at cessation of respiratory ventilation was recorded.

The CTM experiments indicated small but significant differences in heat and cold tolerance for the two pupfish subspecies ($P < .05$; two-tailed W test). The thermal scope for *C. n. mionectes* from constant temperature Big Spring was 38.9, whereas C. n. amargosae had a thermal scope of 39.5°C (Figure 6). These slight differences in thermal tolerance between the two subspecies are even more pronounced in the time to mortality or thermal shock experiments and indicate the delicately balanced nature of temperature tolerances in pupfish. At the 38°C test temperature 8 to 10 C. n. mionectes died within 6 min, whereas 2 lived longer than 15 min, at which point the experiments were terminated. None of 10 C. n. amargosae died within 15 min. These proportions are different at $P < .01$, based on binomial confidence limits for proportions. Three of 10 C. mionectes amargosae hybrids died within 15 min.

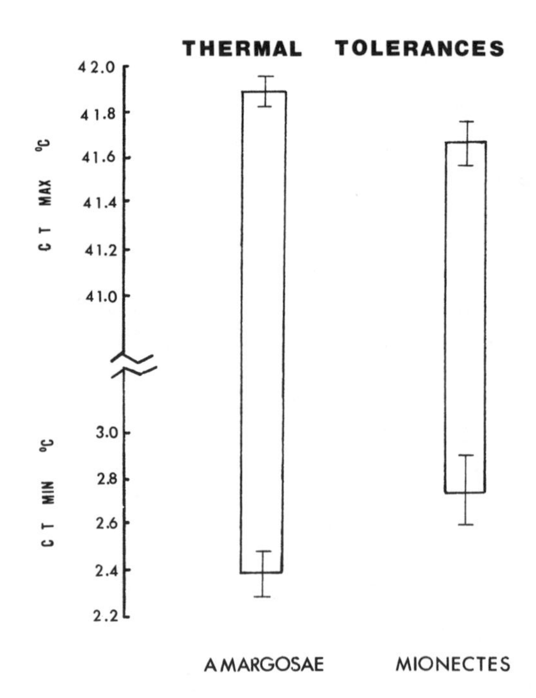

FIGURE 6. Critical thermal maxima and minima (means ± SE) for pupfish reared at 27°C. For C. n. amargosae n = 23; for C. n. mionectes, n = 12. Values adjacent to vertical lines are average thermal ranges of individual pupfish (from Hirshfield et al. 1980). Copyright 1980 by the American Association for the Advancement of Science.

At 39°C 8 of 10 _C. n. amargosae_ died within 1 min of entering the water, whereas the _C. n. mionectes_ appeared to die instantly.

The cold shock tests showed that _C. n. amargosae_ survived significantly longer than _C. n. mionectes_ (_n_ = 20, _P_ < .002; two-tailed W test of the median time of survival). Hybrids again showed intermediate tolerances; the mean time to loss of equilibrium was 1.26 min for _C. n. mionectes_, 1.88 min for hybrids, and more than 3.68 min for _C. n. amargosae_ (Figure 7).

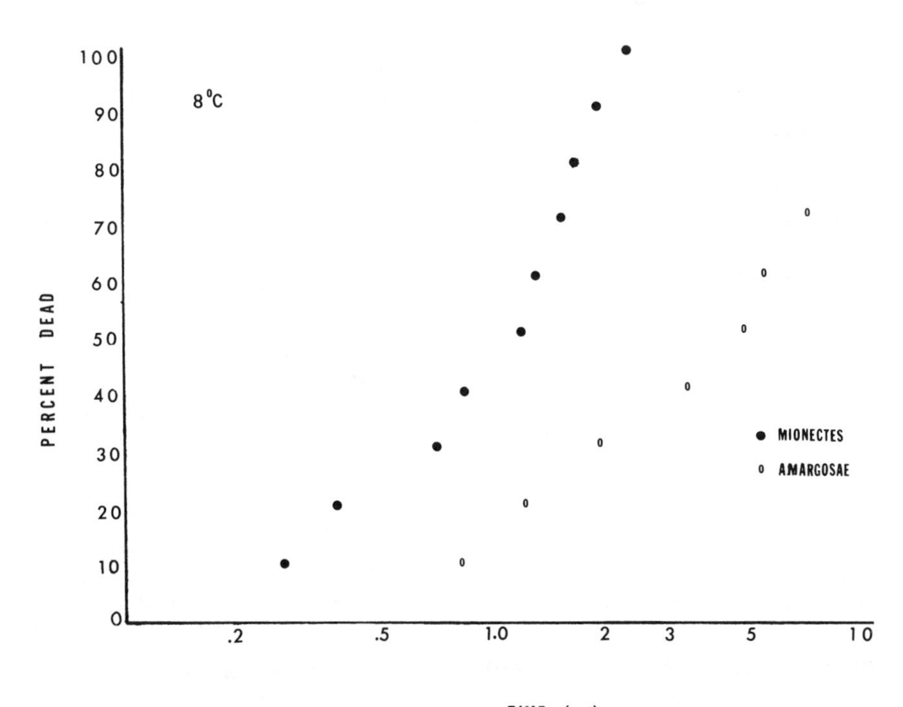

FIGURE 7. Cold tolerance for pupfish reared at 25°C determined by resistance time method. Both subspecies (_C. n. mionectes_, solid circles and _C. n. amargosae_, open circles) were transferred to an 8°C aquarium and time until loss of equilibrium was determined (from Hirshfield et al. 1980). Copyright 1980 by the American Association for the Advancement of Science.

These results demonstrate that the two populations have indeed diverged genetically since their isolation. The length of time that this population pair has been isolated cannot, however, be exactly established. Big Spring is located in Ash Meadows, Nevada, 10 km and 66 m higher in elevation than the dried bed of the Amargosa River at Death Valley Junction, California (Figure 8). During periods of heavy rainfall (as in 1936-1937 and 1969) this portion of the river carries water to a section of permanent flow (inhabited by *C. n. amargosae*), approximately 65 km to the south. Somewhat greater flooding would be necessary to carry pupfish from Big Spring to the river. Mehringer and Warren (1976) document a period of peat formation in Ash Meadows during the last 5000 years and infer that rainfall and spring discharge have been greater or less than present for several lengthy periods. *C. n. amargosae* almost certainly could not have reached Big Spring during this period. The last more or less continuous connection between the two subspecies was at least 10,000 to 12,000 years ago, at the end of the last pluvial period. Downstream gene flow for *C. n. mionectes* to the *C. n. amargosae* habitat is more probable.

Although the amount of genetic differentiation in the thermal tolerances of these subspecies is not great, these data indicate that it is possible to measure the evolutionary rate of a physiological adaptation. Perhaps evolutionary rates of other ecologically important characteristics can now be measured for these two subspecies. It is certainly true that thermal acclimation plays a more important role in determining the tolerances of these fish than genetic differences. Nevertheless, the relatively slight differences should be important to fish under natural conditions. Because pupfish have frequently been observed swimming within 1 or 2°C of their lethal temperature, the reduced thermal tolerance of *C. n. mionectes* could therefore lead to their elimination if they were moved by flood waters into the Amargosa River and downstream into the permanent portion of the river. Thermal tolerance is only one of many important ecological characteristics involved in the evolutionary adaptation

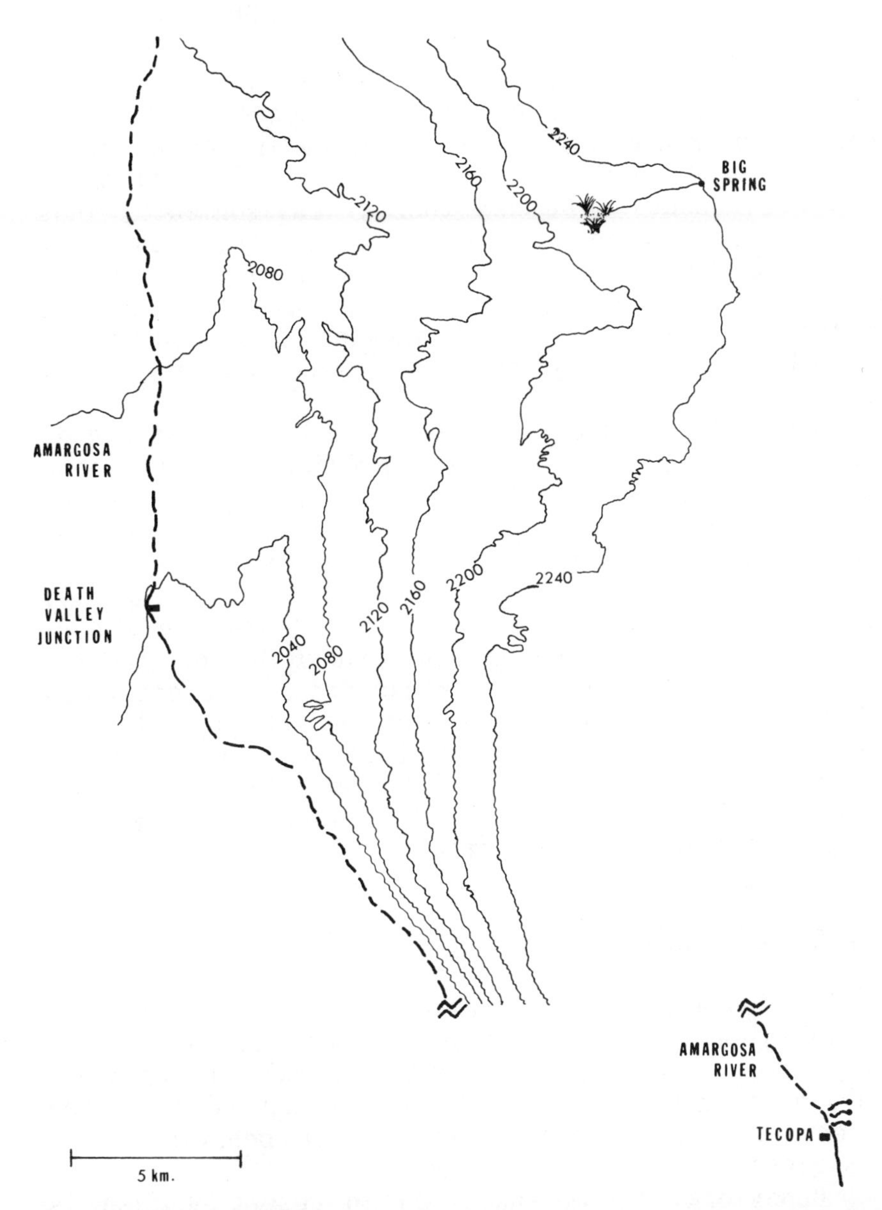

FIGURE 8. Topographic and geographic illustration of Big Spring (habitat for C. n. mionectes) and the Amargosa River. The thick broken line indicates normally dry river bed; the solid thick line south of Tecopa denotes permanent flowing portion of the river (habitat for C. n. amargosae). Thin lines are topographic contours; numbers indicate elevation in feet. Big Spring and Tecopa are approximately 65 km apart.

of an ectotherm to its environment. Genetic changes in thermal tolerance are probably indicative of an array of other ecologically important alterations that occur as a fish adapts to a new thermal environment. Thus the thermal tolerance range for a pupfish, bounded by upper and lower lethal limits, is a measure of the ultimate dimension of the thermal niche. Other aspects of this thermal niche might include physiological or metabolic optima, behavioral performance optima, or behavioral preferences (Magnuson et al. 1979).

Not only are the results obtained by Hirshfield et al. (1980) consistent with the prediction that organisms from constant environments should show reduced ability to deal with environmental variability or have a narrower thermal niche they also indicate that such evolutionary changes can take place fairly rapidly. They also present an important first step in understanding the mechanism of the evolution of thermal tolerance in other North American fishes. Closely related species and subspecies of desert pupfishes in isolated waters of different thermal characteristics may yield considerably more information on the nature of the evolution of ecologically important behavioral and physiological characteristics.

SUMMARY AND CONCLUSIONS

It appears that an examination of behavioral and physiological characteristics of desert pupfish populations from markedly different thermal environments could be a valuable key to understanding the evolution of ecologically important aspects of behavioral and physiological responses to temperature.

Behavioral thermoregulation in desert pupfish is an important response, physiologically and ecologically, because it allows the regulation of internal temperature and makes up a significant dimension of the species niche. Pupfish show the ability to select temperature differentials precisely and to live at the very limit of their thermal tolerance.

Egg production and embryonic development, although effected by environmental temperature, occur over a wide temperature range that reflects the eurythermal nature of pupfish.

Pupfish, along with most other fishes from temperate environments, have developed the ability to alter thermal tolerance by temperature acclimation which is significantly important in habitats with fluctuating temperatures. The Amargosa River pupfish, and probably others from thermally unstable environments, has also developed the ability to expand its thermal niche by acclimating to a diel fluctuating temperature regime.

Perhaps the most significant result of thermal tolerance studies has been experimental support of the hypothesis that animals from constant thermal environments should show reduced ability to cope with environmental variability. A clear divergence in thermal tolerance range has been demonstrated among two populations of *C. nevadensis* that probably have been separated for only a few thousand years.

FUTURE RESEARCH ON DESERT PUPFISH

Uncovering the mechanisms involved in the evolution of ecological and physiological characteristics related to temperature in fishes will require numerous investigations, and desert pupfish and their habitats will continue to be an ideal subject of these evolutionary studies. If small differences in thermal tolerance and thermal niche width exist in closely related forms, perhaps considerably greater differences exist for more distantly related pupfish. Genetic differences may exist and be detected by precise physiological tests for allopatric populations of the same subspecies; for example, a population of *C. nevadensis amargosae* has lived in the warm thermal gradient of the Tecopa Bore (an artesian well drilled near Tecopa, California) for approximately 15 years and another population of the same subspecies is present in the lower Amargosa River in Death Valley National Monument, some distance to the west.

Thermal tolerance is one means of measuring an evolutionary adaptation such as thermal niche width and indicates the extreme thermal limits for life. Perhaps more sensitive ecological tests could also be devised to elucidate genetically based differences related to the different thermal histories of pupfish populations. Preferred temperature studies for pupfishes from constant and fluctuating environments may yield even more marked genetic differences. The effect of temperature on cruising speed, maximum swimming velocity, and metabolic rate of Cyprinodon may give an even clearer view of the adaptative evolution of behavior and physiology to the remarkable environments that these small fishes inhabit.

REFERENCES

Bardach, J. E. and R. G. Bjorklund. 1957. The temperature sensitivity of some American freshwater fishes. American Naturalist 91:233-251.

Barlow, G. W. 1958. Daily movements of desert pupfish, Cyprinodon macularius, in shore pools of the Salton Sea, California. Ecology 39:580-587.

Beitinger, T. L. and L. C. Fitzpatrick. 1979. Physiological and ecological correlates of preferred temperature in fish. American Zoologist 19:319-329.

Bogert, C. M. 1949. How reptiles regulate their body temperature. Scientific American 200:105-120.

Brattstrom, B. H. 1968. Thermal acclimation in anuran amphibians as a function of latitude and altitude. Comparative Biochemistry and Physiology 24:93-111.

Brattstrom, B. H. 1979. Amphibian temperature regulation studies in the field and laboratory. American Zoologist 19:345-356.

Brattstrom, B. H. and J. W. Warren. 1955. Observations on the ecology and behavior of the Pacific tree frog, Hyla regilla. Copeia 1955:181-191.

Brett, J. R. 1944. Some lethal temperature relations of Algonquin Park fishes. University of Toronto Studies in Biology Series 52: Publication Ontario Fisheries Research Laboratory 63:1-49.

Brett, J. R. 1956. Some principles in the thermal requirements of fishes. Quarterly Review of Biology 31:75-87.

Brett, J. R. 1959. Thermal requirements of fish: three decades of study, 1940-1970. U. S. Public Health Second Seminar in Biological Problems in Water Pollution, Transactions, Cincinnati, Ohio, April 20-24.

Brown, J. H. 1971. The desert pupfish. Scientific American 225:104-110.

Brown, J. H. and C. R. Feldmeth. 1971. Evolution in constant and fluctuating environments: thermal tolerances of desert pupfish (*Cyprinodon*). Evolution 25:390-398.

Cowles, R. B. and C. M. Bogert. 1944. A preliminary study of the thermal requirements of desert reptiles. American Museum of Natural History, Bulletin 83, Article 5:261-296.

Crawshaw, L. I. and H. T. Hammel. 1977. Physiological and behavioral reactions of fishes to temperature change. Journal Fisheries Research Board of Canada 34:730-734.

Crawshaw, L. I. 1974. Behavioral regulation of internal temperature in the brown bullhead *Ictolurus nebulosus*. Comparative Biochemistry and Physiology 47A:51-60.

Echelle, A. A., C. Hubbs and A. F. Echelle. 1972. Developmental rates and tolerances of the Red River pupfish, *C. rubrofluviatilis*. Southwestern Naturalist 17:55-60.

Feldmeth, C. R. and C. H. Eriksen. 1978. A hypothesis to explain the distribution of native trout in a drainage of Montana's Big Hole River. Verhandlungen Internationale Vereinigung Limnologie 20: 2040-2044.

Fry, F. E. J. 1947. Effects of the environment on animal activity. University of Toronto Studies in Biology Series, No. 55; Publication, Ontario Fisheries Research Laboratory, No. 68:1-62.

Fry, F. E. J. 1967. Responses of vertebrate poikilotherms to temperature. *In*: A. H. Rose (Ed.), Thermobiology. Academic, New York, pp. 352-409.

Gehlbach, F. R., C. L. Bryan and H. A. Reno. 1978. Thermal ecological features of *Cyprinodon elegans* and *Gambusia nobilis*, endangered Texas fishes. The Texas Journal of Science 30:99-101.

Heath, W. G. 1963. Thermoperiodism in sea-run cutthroat trout (Salmo clarki clarki). Science 142: 486-488.

Hirshfield, M. F., C. R. Feldmeth and D. L. Soltz. 1980. Genetic differences in physiological tolerances of Amargosa pupfish (Cyprinodon nevadensis) populations. Science 207:999-1001.

Hochachka, P. W. and G. N. Someo. 1973. Strategies of Biochemical Adaptation. Saunders, Philadelphia.

Hubbs, C. L. and R. R. Miller. 1948. The zoological evidence: correlation between fish distribution and hydrographic history in the desert basins of western United States. In: The Great Basin, with emphasis on glacial and postglacial times. Bulletin University of Utah, 38(20), Biological Series 10(7):17-166.

Huey, R. B. 1974. Behavioral thermoregulation in lizards. Importance of associated costs. Science 184:1001-1003.

Huey, R. B. and M. Slatkin. 1976. Costs and benefits of lizard thermoregulation. Quarterly Review of Biology 51:363-384.

Kinne, O. and E. M. Kinne. 1962. Rates of development in embryos of a cyprinodont fish exposed to different temperature-salinity-oxygen combinations. Canadian Journal of Zoology 40:231-253.

Kwain, W. and R. W. McCauley. 1978. Effects of age and overhead illumination on temperatures preferred by underyearling rainbow trout, Salmo gairdneri, in a vertical temperature gradient. Journal Fisheries Research Board of Canada 35: 1430-1433.

Lillywhite, H. B. 1970. Behavioral thermoregulation in the bullfrog, Rana catesbeiana. Copeia 1970: 158-168.

Lillywhite, H. B., P. Licht and P. Chelgren. 1973. The role of behavioral thermoregulation in the growth energetics of the toad, Bufo boreas. Ecology 54:375-383.

Lowe, C. H. and W. G. Heath. 1969. Behavioral and physiological responses to temperature in the desert pupfish Cyprinodon macularius. Physiological Zoology 42:53-59.

McCauley, R. W. and L. A. A. Read. 1973. Temperature selection by juvenile and adult yellow perch (*Perca flavescens*) acclimated to 24 C. Journal of Fisheries Research Board of Canada 30:1253-1255.

Magnuson, J. J., L. B. Crowder and P. A. Medvick. 1979. Temperature as an ecological resource. American Zoologist 19:331-343.

Mehringer, P. J. and C. N. Warren. 1976. Marsh, dune and archeological chronology, Ash Meadows, Amargosa Desert, Nevada. *In*: R. Elson (Ed.), Holocene environmental change *in* the Great Basin, Nevada Archaeological Survey, Research Paper 6, pp. 120-150.

Miller, R. R. 1948. The cyprinodont fishes of the Death Valley System of eastern California and southwestern Nevada. Miscellaneous Publications Museum of Zoology, University of Michigan 68:1-155.

Miller, R. R. 1950. Speciation in fishes of the genera *Cyprinodon* and *Empetrichthys* inhabiting the Death Valley region. Evolution 4:155-163.

Naiman, R. J. 1976. Productivity of a herbivorous pupfish population (*Cyprinodon nevadensis*) in a warm desert stream. Journal of Fish Biology 9: 125-137.

Neill, W. H. 1979. Mechanisms of fish distribution in heterothermal environments. American Zoologist 19:305-317.

Norris, K. S. 1963. The functions of temperature in the ecology of the percoid fish Girella nigricans (Ayres). Ecological Monographs 33:23-62.

Otto, R. G. and S. D. Gerking. 1973. Heat tolerance of a Death Valley pupfish (Genus *Cyprinodon*). Physiological Zoology 46:43-49.

Regal, P. J. 1977. Evolutionary loss of useless features: is it molecular noise suppression? American Naturalist 111:123-133.

Reynolds, W. W. 1979. Perspective and introduction to the symposium: thermoregulation in ectotherms. American Zoologist 19:193-194.

Shrode, J. B. 1975. Developmental temperature tolerance of a Death Valley pupfish (*Cyprinodon nevadensis*). Physiological Zoology 48:378-389.

Shrode, J. B. and S. D. Gerking. 1977. Effects of constant and fluctuating temperatures on reproductive performance of a desert pupfish, _Cyprinodon n. nevadensis_. Physiological Zoology 50:1-10.

Soltz, D. L. and R. J. Naiman. 1978. The Natural History of Native Fishes in The Death Valley System. Natural History Museum of Los Angeles, Science Series 30:1-76.

Snyder, G. K. and W. W. Weathers. 1975. Temperature adaptations in amphibians. American Naturalist 109:93-101.

Turner, B. J. 1974. Genetic divergence of Death Valley pupfish species: biochemical versus morphological evidence. Evolution 28:281-294.

12 Energy Metabolism and Osmoregulation in Desert Fishes

STANLEY D. HILLYARD
University of Nevada, Las Vegas

ABSTRACT

Desert fishes are often found in habitats in which extremes in temperature, dissolved oxygen, and salinity occur. In some cases, such as the pupfish *Cyprinodon salinus*, a single species may be confined to a highly restricted habitat with fluctuating environmental parameters, whereas other species, such as the springfish *Crenichthys baileyi*, may be found in several isolated springs, each having widely different but relatively stable environmental conditions. A basic physiological measurement used to assess environmental demands on an animal is energy metabolism which is usually determined as oxygen consumption (VO_2) either as routine metabolism (RMR) or as standard metabolism (SMR). Standard metabolism values (ml $O_2 \cdot g^{-1} \cdot hr^{-1}$) for

Cyprinodon salinus and those for the roundtailed chub *Gila robusta* were considerably lower than RMR values from other desert species, such as the Utah chub *Gila atraria* and *Crenichthys baileyi*, primarily because care was taken to minimize activity in the respiratory chambers. Acclimation or acclimatization of desert fishes to progressively lower temperatures produces levels of RMR and SMR that are lower than those observed at higher acclimation temperatures, thus indicating an incomplete compensation of energy metabolism with changing temperatures. An exception to this pattern is seen in *Crenichthys baileyi*, among which populations acclimatized to 37°C springs have the same RMR at 21°C as populations acclimatized to 21°C springs. Changes in salinity have marked effects on SMR in the euryhaline species *C. salinus*, in which SMR increased as the fish were acclimated to increasing salinity. It would appear that a greater level of tissue metabolism is required for the fish to extrude salt at higher salinities than to take in salt at lower salinities. This is supported by the observation that plasma osmolality at higher salinities climbs rapidly with decreasing acclimation temperature, whereas in freshwater the osmolality declines little with reduced acclimation temperature. At the tissue level the specific activity of the enzyme Na^+-K^+ ATPase in the gills is lowest in fish acclimated to 1/2 SW and increases as salinity decreases to FW or rises to that of SW or 1 1/2 SW. Thus the level of this transport enzyme correlates with the osmotic gradient between the fish and its surroundings.

INTRODUCTION

Desert fishes occur in habitats characterized by extremes in temperature, dissolved oxygen, and salinity, all of which place physiological demands on aquatic organisms (Prosser 1973). Death Valley pupfish (*Cyprinodon* spp.), for example, may encounter ambient temperatures (T_a) as low as 0°C or as high as 40°C and salinity ranging from freshwater to that of greater than two times sea water (Brown 1971). The killifish *Crenichthys baileyi*, on the other hand, inhabits springs in which temperature and salinity are constant but dissolved oxygen (DO) often falls below one part per million (ppm) (Hubbs et al. 1967). Thus a wide range of habitats exist where physiological adaptation to T_a, DO, and salinity can be studied singly or in combination. The purpose of this review is to examine the effects of these parameters on the metabolic rates of selected desert fishes and to assess the metabolic cost of the physiological mechanisms that allow particular species to survive. When appropriate, respiratory adjustments associated with changes in metabolic rate are discussed. Finally, changes in the osmotic concentration of the body fluids produced by extremes in temperature and salinity are evaluated in terms of possible mechanisms involved in osmoregulation.

ENERGY METABOLISM

Definition of Terms

Energy metabolism in fishes is measured in a variety of ways and any discussion of metabolic adjustments to changing physical factors requires a definition of experimental methods. The most common method of measuring metabolic rate in fishes is to place an animal in a closed chamber and allow time for it to consume measurable amounts of oxygen; DO is determined initially and at the end of the experimental period and metabolic rate is calculated as ml O_2 consumed per gram body weight per hour (ml $O_2 \cdot g^{-1} \cdot hr^{-1}$; VO_2). This measure of VO_2 is termed routine metabolism (RMR) and the caloric equivalent of oxygen (4.8 kcal·liter) can be used to assess the energetic demands of the fish. Measurements of RMR are eaily accomplished in the field with a minimum of equipment; this technique, however, often gives variable results because fish may not become fully accustomed to the chamber and will increase their activity in an effort to escape. To overcome these problems a "flow-through" system (Fromm 1957) may be used with aerated water continuously perfusing the chamber. VO_2 is then calculated from the difference in DO between the incurrent and excurrent flow and the flow rate. The lowest VO_2 measured over a period of several hours is termed the standard metabolic rate (SMR) and is presumed to approximate the metabolic cost of homeostasis under minimally active conditions. The caloric demands of increasing activity are measured by creating a current against which the fish must work. VO_2 under these conditions is termed activity metabolism (AMR) and reaches a maximum when the respiratory and circulatory systems can no longer provide oxygen to the tissues. The difference between SMR and maximal VO_2, termed the metabolic scope (Bartholomew 1977), provides an index of the aerobic capacity of the animal. More useful is the critical swimming speed [see Beamish (1978) for a review of swimming performance in fishes] which is the maximum speed that a fish can maintain for a prescribed time in a water tunnel. Ideally, the critical swimming speed is measured under conditions where rectilinear flow can be maintained in the experimental chamber so that the work load on the fish is known (Beamish 1978). To date no detailed studies have been conducted on swimming performance or maximal VO_2 in desert fishes. As with other poikilotherms, the level of SMR, RMR, and AMR is temperature dependent and measurements of VO_2 in fishes are specified according to the method used and the temperature of determination.

Figure 1 summarizes VO_2 measurements on four species of desert fishes for which the temperature of determination was the same as the acclimation temperature. The term acclimation is used in the context of this review as the compensation in physiological function in response to changes in a single environmental variable (Bartholomew 1977). VO_2 measurements in Figure 1 were made on groups of fish of each species which had been maintained continuously at the test temperature for two to four weeks. VO_2 in the Utah chub (*Gila atraria*) increases progressively under conditions of SMR, RMR, and AMR (Rajagopal and Kramer 1974), although swimming speed was not controlled during the AMR measurements. Note that the effect of temperature acclimation on VO_2 varies with the method of determination. In comparison the SMR of the roundtail chub (*Gila robusta*) is considerably lower than that of *G.*

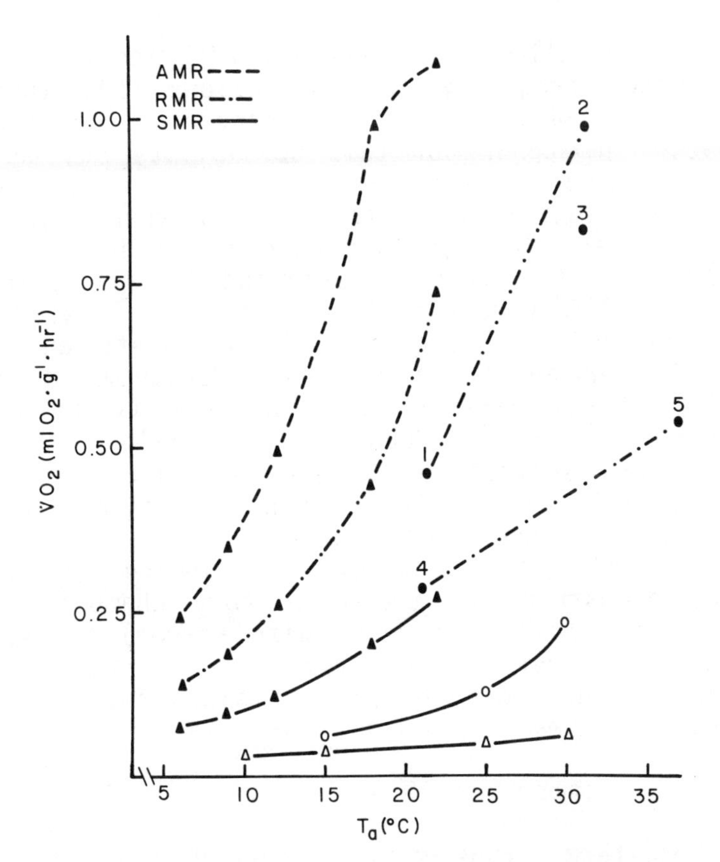

FIGURE 1. A comparison of standard metabolic rate (SMR), routine metabolic rate (RMR), and activity metabolism (AMR) in *Gila atraria* (solid triangles, Rajagopal and R. H. Kramer 1974), *Gila robusta* (open triangles, Schumann 1978), *Crenichthys baileyi* (closed circles 1, 2 and 3 from Hubbs et al. 1967; 4 and 5 from Sumner and Lanham 1942) and *Cyprinodon salinus* (open circles, Stuenkel and Hillyard 1981). In all cases VO_2 was measured at an ambient temperature (T_a) equal to the acclimation or acclimatization temperature.

atraria, even at higher T_a (Schumann 1978). This may be due in part to the larger size of *G. robusta* (average weight 53 g versus 4 to 7 g for *G. atraria*); however, Schumann (1978) used an electrode system (Davey

1966) which allowed continuous monitoring of ventilation rate (frequency of respiration, FR); thus VO_2 was measured when the fish was observed to be minimally active. In support of this possibility Stuenkel and Hillyard (1981) found the SMR of *Cyprinodon salinus* (average weight 0.66 g) to be lower than that of *G. atraria* when VO_2 was measured in conjunction with minimal FR. In both *G. robusta* and *C. salinus* the SMR values at 25°C were near those predicted in studies of other teleosts of comparable size (Brett and Grove 1979; Kayser and Heusner 1964). Measurements of VO_2 in *Crenichthys baileyi* show that RMR is lower when measured nocturnally (Hubbs et al. (1967). Correlated with these observations, Hubbs et al. (1967) found that *C. baileyi* is less active at night in its native habitat. Sumner and Lanham (1942) obtained even lower levels of RMR for *C. baileyi* by allowing the fish to become accustomed to the respirtory chambers (glass tubes that could be covered with screen) for several hours or more before the chambers were sealed for VO_2 determination. Note that these values for *C. baileyi* are close to SMR values for *G. atraria*.

Effects of Temperature

Long-term exposure to varying T_a may result from seasonal variation or, alternatively, different populations of a particular species may inhabit a series of isolated springs with quite different temperatures, although the temperature of a given spring is constant. Pupfish are a good example of the former and have been studied in the laboratory and field. Stuenkel and Hillyard (in press a) acclimated *C. salinus* to 15 and 30°C for two to four weeks in the laboratory and then measured SMR at 15, 20, 25, and 30°C (Figure 2). It was found that the SMR of the 15°C group was greater than that of the 30°C group at all T_a values. The SMR of the 15°C group at 15° was considerably lower than that of the 30°C group at 30°C, which indicates an incomplete compensation in SMR. Sumner and Sargent (1940), on the other hand, measured metabolic rate indirectly by observing how long pupfish (now *C. nevadensis*), removed immediately from their habitat,

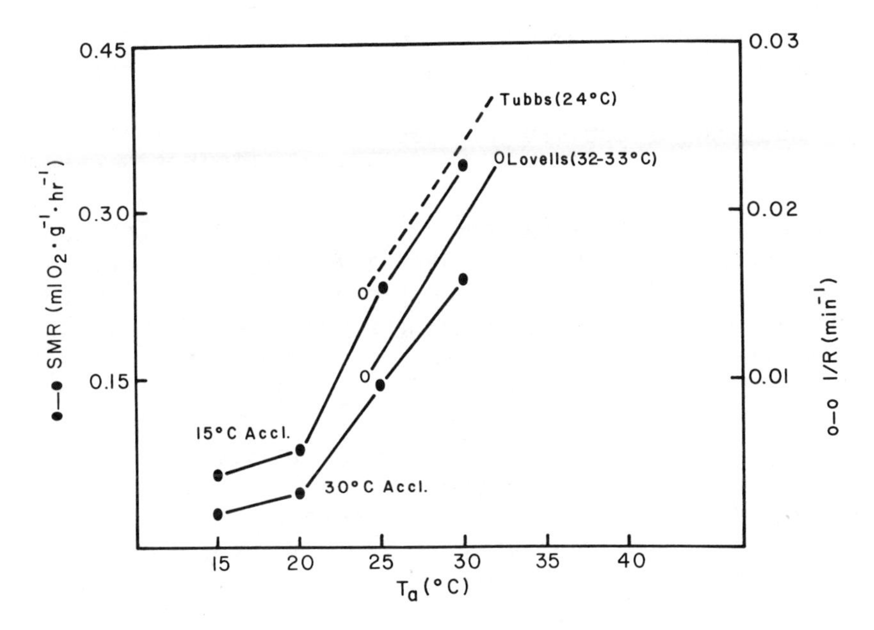

FIGURE 2. The effect of temperature acclimation on standard metabolism in Cyprinodon salinus (solid circles, data from Hillyard and Stuenkel 1981), compared with the inverse of the survival time in 1 mM KCN which was observed for C. nevadensis resident in warm and cool springs (open circles; data from Sumner and Sargent 1940).

could survive in a 1 mM KCN solution. The inverse of the survival time in the KCN was taken to be proportional to the oxygen demand by the fish. Comparison of a population from a warm habitat (Lovell's spring, 32 to 33°C) with one from a cooler habitat (Tubb's spring, 24°C) showed that the metabolic rate of the warmer acclimatized[2] fish was lower at 24°C (Lovell's in Tubb's) than that of the fish native to the cooler

---

[2]Acclimatization refers to physiological compensation that occurs in response to all the environmental variables that may fluctuate in the fish's habitat (Bartholomew 1977). In hot and cool spring populations it is assumed that temperature is the predominant variable.

spring (Tubb's in Tubb's) (Figure 2). Unfortunately Tubb's spring population was not tested in Lovell's spring.

Routine metabolism indicated by KCN survival time was also tested by Sumner and Sargent (1940) for populations of *Crenichthys bai1eyi*, which inhabit warm (Mormon Springs, 37°C) and cooler (Preston Springs, 21°C) habitats. It was found that the KCN survival times at 37°C indicated a lower metabolic rate for the fish resident in the warmer spring; however, the apparent metabolic rate for the two populations at 21°C was not significantly different (Figure 3). Hubbs et al. (1967) confirmed this observation by comparing the RMR of the 21 and 37°C populations at 21 and 32°C (the 21°C fish do not survive long enough at 37°C for RMR determinations). The pattern of temperature compensation in metabolic rate processes is thus quite different in pupfish and *C. baileyi*.

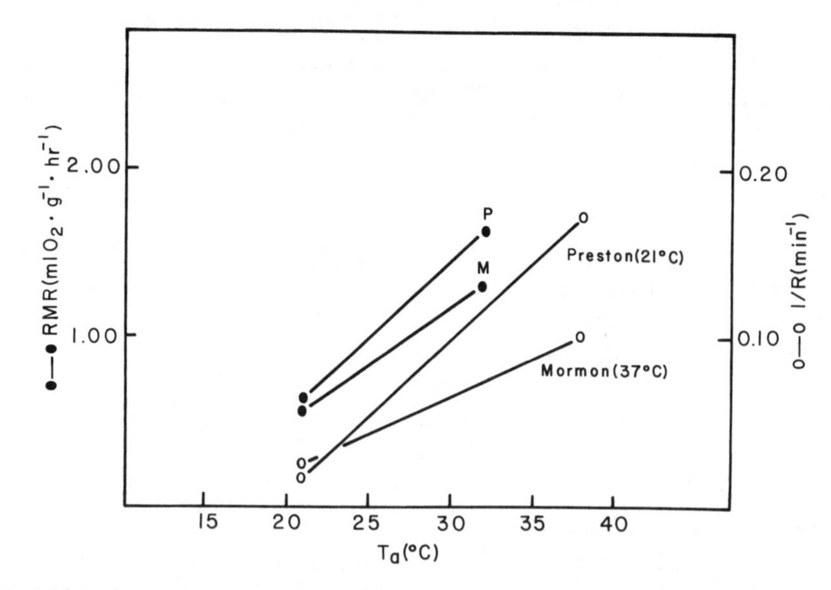

FIGURE 3. Routine metabolism of *Crenichthys baileyi* from warm and cool springs (solid circles, data from Hubbs et al. 1967), compared with the inverse of the survival time in 1 mM KCN for the same two populations (open circles data from Sumner and Sargent 1940).

Effects of Dissolved Oxygen

Dissolved oxygen levels in aquatic habitats are usually measured in (ppm) or ml O_2 ·liter H_2O. A more useful measure in physiological studies, however, is the partial pressure (PO_2) which represents the driving force for oxygen loading by the blood. A systematic study on the effects of PO_2 on VO_2 in desert fishes remains to be done, although a recent study by Lomholt and Johansen (1979) on hypoxia acclimation in carp (*Cyprinus carpio*) provides insight on a species that is tolerant of low PO_2. Using closed system respirometry on 80 to 200 g *C. carpio*, the authors observed that the hypoxia-acclimated (PO_2 = 30 Torr) fish had a lower VO_2 than normoxia-acclimated (PO_2 = 120 Torr) fish when the ambient PO_2 was above 80 Torr. When PO_2 was lowered to 10 Torr, however, VO_2 was greater in the hypoxia-acclimated group. They concluded that the predominant compensatory mechanism for sustaining oxygen transport at low external PO_2 must involve a high O_2-hemoglobin affinity that can be adjusted for loading at the gills and unloading at the tissues over a wide range of environmental PO_2. In this regard Greaney and Powers (1977) have found that ATP is the major organic phosphate required for the reduction in O_2 affinity (Bohr effect) of *Fundulus heteroclitus* blood when the pH is lowered. Furthermore, specific lactate dehydrogenase (LDH) isozymes have been associated with different ATP/Hb ratios (Powers et al. 1979), which suggests that the capacity to modulate O_2-hemoglobin affinity may be genetically determined by the properties of metabolic enzymes in the red blood cells. It is also important to realize that fish blood contains multiple hemoglobins (Powers 1980) and that changes in O_2-Hb affinity may develop from changes in the relative amounts of high and low affinity hemoglobin varieties present at a given time.

An interesting example of modification in oxygen binding by different hemoglobins in the blood of desert fishes is seen in the suckers *Catostomus insignis* and *Pantosteus clarki* (Powers 1972, 1980). The blood of both species contains hemoglobin varieties that migrate to the positive pole (cathodal Hb) and negative pole (anodal Hb) when subjected to starch gell electrophoresis. In *C. insignis* the anodal Hb and cathodal Hb varieties demonstrated a Bohr effect, whereas the cathodal Hb from *P. clarki* did not. It was suggested that the lack of a Bohr effect in the cathodal Hb of *P. clarki* is related to its preference for swift water habitats in which bursts of activity would generate lactate accumulation and consequent acidification of the blood. If the Bohr effect were present in all the Hb of these fishes, this acidification could reduce the affinity of the blood to the extent that adequate O_2 loading could not occur at the gills and the fish could not sustain activity. *C. insignis*, on the other hand, prefers areas of lower flow and can reduce its activity after bursts of rapid swimming. The greater amount of Hb with a Bohr effect in *C. insignis*, however, might be expected to allow for a more effective delivery of O_2 to the tissues under normoxic or hypoxic conditions.

These observations may relate to results in a recent study by Schumann et al. (submitted) that *P. clarki* acclimated to progressively lower temperatures select lower temperatures when placed in a continuous thermal gradient (Figure 4). *Catostomus latipinnis*, on the other hand, selects a relatively constant temperature, regardless of the acclimation temperature. If we assume that anodal and cathodal hemoglobins from *C. latipinnis* demonstrate a Bohr effect like those of *C. insignis* and that *C. latipinnis* and *P. clarki* have higher metabolic rates when acclimated to lower temperatures, the following could occur: cold and warm acclimated *C. latipinnis*, in selecting similar temperatures, will have elevated and reduced metabolic rates, respectively. The ability to modulate oxygen-hemoglobin affinity would then allow for oxygen delivery under conditions in which oxygen demand by the tissues is altered and thus allow for a narrow range of behavioral thermoregulation. In *P. clarki* acclimated to progressively lower temperatures the selection of a lower temperature will reduce oxygen demand by the tissues. Apparently thermally acclimated *P. clarki* regulate the oxygen demand of their tissues by selecting higher or lower temperatures. It would be interesting to measure VO_2 of thermally acclimated *C. latipinnis* and *P.*

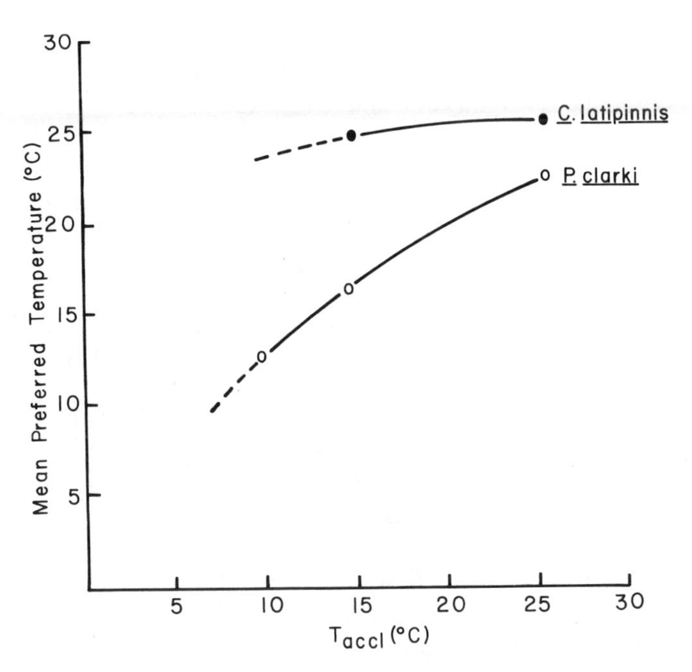

FIGURE 4. The effect of acclimation temperature (T_{accl}) on the mean preferred temperature selected by Catostomus latipinnis and Pantosteus clarki placed in a thermal gradient (data from Schumann et al., personal communication).

clarki at the temperature they selected in a thermal gradient. Also important would be the determination of the temperature effect on oxygen-hemoglobin affinity. Increasing temperature is known to reduce oxygen affinity in most species (Stryer 1974; Wood 1980) and the absence of a Bohr effect in P. clarki does not rule out the capacity to modulate oxygen-hemoglobin affinity with changes in VO_2 that arise from temperature changes. Dill et al. (1932) found that oxygen affinity was reduced considerably when skate (Raja oscillata) blood was warmed from 0.2 to 37.5°C. Note that skate hemoglobin shows a reduced Bohr effect to demonstrate that the temperature-induced changes in oxygen affinity can be independent of the Bohr effect.

Effects of Salinity

Increasing salinity is known to cause an increase in active salt extrusion and in passive salt and water fluxes across the gills of euryhaline teleosts (Motais and Isaia 1972). In addition, increasing salinity reduces the DO in the water and alters the composition of the body fluids (Houston 1973), both of which may alter the energy demands for homeostasis. Stuenkel and Hillyard (1981) have measured SMR and FR in C. salinus acclimated to salinities that approximate freshwater (FW, 80 $mosm \cdot kg^{-1}$), one half sea water (1/2 SW, 541 $mosm \cdot kg^{-1}$) and sea water (SW, 1168 $mosm \cdot kg^{-1}$). SMR increased significantly with increasing salinity at each temperature examined, except at 30°C, where SMR was the same for both 1/2 SW and SW acclimated fish (Figure 5a). Living in elevated salinities would therefore appear to increase the energetic demand of the tissues in these fish. Kinne (1960) has, in fact, shown increases in assimilation with increasing temperature in C. macularius, although assimilative efficiency is lower. The highest salinities encountered by desert species usually occur during the summer when T_a is also highest and primary production is maximal; therefore food supply may not be a limiting factor in the adjustment to higher salt concentration.

The reason for the leveling off in SMR with increasing salinity in the 30°C acclimated C. salinus is not clear. It could be due to metabolic compensation for a reduced oxygen demand by the tissues or to a reduced ability to deliver oxygen to the tissues. The FR was the same in the SW-acclimated fish at 25 and 30°C (Figure 5b), which indicates comparable levels of ventilation. The 30°C fish apparently were not encountering respiratory difficulties. It was noted that FR is lower or unchanged in 1/2 SW when compared with FW, even though SMR is higher in the 1/2 SW fish. This may be related to changes in gill permeability or branchial blood flow which are known to occur with increasing salinity and temperature (Maetz and Evans 1972; Motais and Isaia 1972). It is evident that T_a, DO, and salinity are interacting factors that affect oxygen demand by and delivery to the tissues; thus they

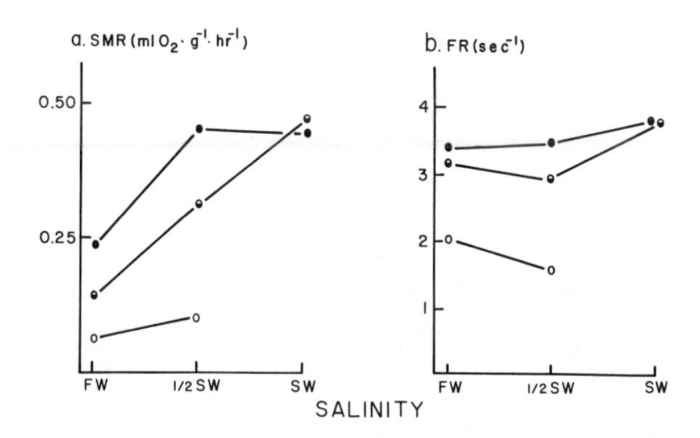

FIGURE 5. The effect of salinity acclimation on (a) standard metabolism and (b) ventilation rate (FR) in Cyprinodon salinus acclimated to 15°C (open circles), 25°C (half-open circles), and 30°C (solid circles). Data from Stuenkel and Hillyard (in press, a).

should be studied together in assessing energy metabolism in desert fishes.

OSMOREGULATION

Definition of Terms

Studies of osmoregulation are described in terms that relate to the amount of salt in solution and its effect on the osmotic pressure of body fluids and external medium. A common term is salinity, the weight of salt in a liter of solution ($g \cdot liter^{-1}$ or ppt) with sea water having a salinity of 35 ppt. As salinity increases, however, the behavior of salts in solution becomes less ideal and the effective osmotic concentration becomes progressively lower than the actual molar concentration. For this reason osmotic concentration is measured in terms of the concentration of an ideal solute which corresponds to the osmotic pressure of a given sample. Osmotic concentration is given as milliosmoles per liter of solution (osmolarity, $mOsm \cdot liter^{-1}$), or as milliosmoles per kg of water (osmolarity, $mosm \cdot kg^{-1}$). Osmolality is a more accurate term from a thermodynamic standpoint because the number of water molecules added to the solution remains constant at all solute concentrations (Marshall 1978).

Effects of Temperature and Salinity on Body Fluid Composition

Desert fishes encounter varying extremes in salinity. The woundfin (Plagopterus argentissimus, Cyprinidae) for example, inhabits the Virgin River, where the osmotic concentration is about 57 $mosm \cdot kg^{-1}$ (Williams 1977). The upper distribution of P. argentissimus in the Virgin River is at LaVerkin Hot Springs in southern Utah, where warm water with an osmotic concentration of 261 $mosm \cdot kg^{-1}$ enters the river (Cross 1975). Williams (1977) found that P. argentissimus could tolerate 100 prcent LaVerkin Springs water at 25°C in the laboratory, although plasma osmalility increased from 303 to 327 $mosm \cdot kg^{-1}$ after 42 days' exposure. It was concluded that salinity alone would not limit P. argentissimus above LaVerkin Springs, although it might contribute to other factors such as increased T_a and decreased DO which are also produced by inflow from the springs.

In contrast to the woundfin, pupfish inhabit areas in which salinity may vary from that of FW to that in excess of four times sea water (Gunter 1956). This characteristic of pupfish has promoted several studies of the effect of increased environmental salinity on plasma osmolality. Renfro (1970) studied osmotic and ionic regulation in the Red River pupfish (Cyprinodon rubrofluviatilis), which may be found in waters with a salinity of 50 ppt or greater. In this study fish were acclimated to FW (65 to 70 $mosm \cdot kg^{-1}$) or to salt water (996 to 1,130 $mosm \cdot kg^{-1}$) for two weeks, and blood samples were analyzed for osmolality, Na^+, and K^+ concentrations. It was found that on transfer of FW-acclimated fish to salt water the plasma osmolality increased from 321 to 599 $mosm \cdot kg^{-1}$ after 4 hr but dropped to 367 $mosm \cdot kg^{-1}$ after 8 hr and remained fairly constant thereafter. This increase in osmolality was accompanied by an increase in plasma Na^+ from 180 to 204 milliequivalents per liter ($meq \cdot liter^{-1}$), whereas K^+ increased from 4.3 to 6.5 $meq \cdot liter^{-1}$). In addition, the rate of Na^+ exchange between the fish and their surroundings, measured isotopically, increased from 9.6 $meq \cdot kg^{-1} \cdot h^{-1}$ in FW to 44.9 $meq \cdot kg^{-1} \cdot h^{-1}$ in salt water. In conjunction with the increased Na^+ turnover the drinking rate in salt water acclimated fish amounted to 1.4 percent of the body weight per hour. In FW, on the other hand, drinking amounted to only 0.2 percent of the body weight per hour.

Naiman et al. (1976) found that the Death Valley species, *Cyprinodon milleri*, could withstand salinities up to 87.5 pt (about 2500 $mOsm \cdot liter^{-1}$) with little mortality by gradually increasing the acclimation salinity; a few individuals were able to tolerate a salinity of 105 ppt (about 3000 $mOsm \cdot liter^{-1}$). It was found that plasma osmolarity increased linearly from 293 $mOsm \cdot liter^{-1}$ in FW to 503 $mOsm \cdot liter^{-1}$ at 87.5 ppt but was not further elevated at 105 ppt.

Stuenkel and Hillyard (1981) examined the effect of temperature on osmoregulation in the Salt Creek pupfish *Cyprinodon salinus*, which was initially acclimated at 25°C in Salt Creek water which had an osmotic concentration of 560 $mosm \cdot kg^{-1}$. After this initial acclimation fish were transferred to salinities of 80, 641, 1168, and 1772 $mosm \cdot kg^{-1}$ with acclimation temperatures of 15, 25, and 30°C maintained for each salinity. After two weeks' acclimation to these temperature-salinity combinations plasma osmolaity was determined. In Figure 6 it can be seen that *C. salinus* was unable to survive the full acclimation period after transfer to a 1168 $mosm \cdot kg^{-1}$ medium at 15°C. At 25°C the fish could tolerate acclimation to 1168 $mosm \cdot kg^{-1}$, although plasma osmolality was signficantly greater than that of 30°C acclimated fish at this salinity. Only the 30°C acclimated fish were able to survive the full acclimation period at all osmotic concentrations tested. Thus it would appear that pupfish can osmoregulate more effectively in hypersaline media when T_a is elevated; this coincides with the field observations that maximum salinity occurs during the summer when high air temperature causes increased evaporation (Soltz and Naiman 1978). Survival in FW,

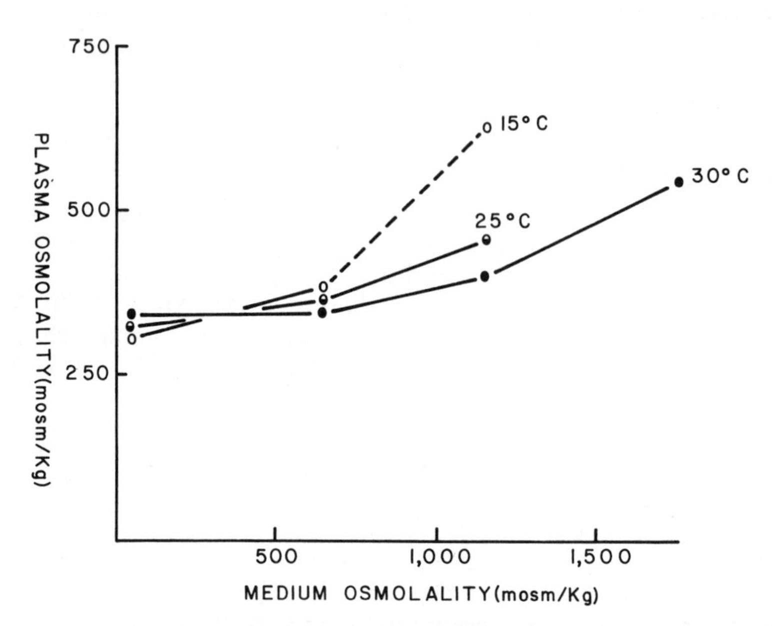

FIGURE 6. The osmolality of plasma obtained from Cyprinodon salinus acclimated to increasing salinity at 15°C (open circles), 25°C (half-open circles), and 30°C (closed circles). Data from Stuenkel and Hillyard (1981).

on the other hand, is not dependent on elevated T_a which may relate to the observation by Deacon and Minckley (1974) that flooding occurs in Salt Creek during winter rains and that abrupt transfer to FW in nature would occur when T_a is low.

The Role of the Gills in Osmoregulation

The gills are believed to be the primary site of Na^+ and Cl^- uptake from FW and their active extrusion when fish are exposed to hypersaline media [see Prosser (1973) for a general discussion and list of references on gill function in teleosts]. The increase in Na^+ exchange which Renfro (1970) observed in salt water-acclimated C. rubrofluviatilis was then produced by an increase in the active extrusion of this ion in the face of an increase in Na^+ uptake caused by drinking and an increase in the passive permeability of the gills to Na^+. Similar findings have been documented for a variety of euryhaline teleosts, including the eel *Anguilla anguilla* (Motais and Isaia 1972) and the flounder *Platichthys flesus* (Maetz and Evans 1972). These studies also showed that water turnover rates increased with increasing salinity and that increasing temperature produced an elevated turnover of Na^+ and H_2O in these species. To date a detailed study of the combined effects of temperature and salinity on salt and water turnover rates in desert species has not been conducted.

Salt transport across the gill epithelium involves the chloride cell, a specialized cell whose function has been reviewed (see Schmidt-Nielsen 1980). In the present discussion I describe some of the cellular mechanisms that have been suggested for branchial Na^+ and Cl^- transport and review data that do exist for desert species.

In FW (Figure 7a) Na^+ entry from the external medium is thought to be coupled to H^+ extrusion across the apical plasma membrane. Cellular Na^+ is then actively transported across the basal-lateral membrane into the extracellular fluid by the Na^+ pump (Na^+-K^+ ATPase). H^+ arises from carbonic acid which is rapidly formed from CO_2 and water by the action of carbonic anhydrase. There is also thought to be an HCO_3^--Cl^- exchange across the apical plasma membrane which is important for regulating pH; ammonium ion exchange for Na^+ has also been suggested (Maetz 1973).

When fish are transferred to hypersaline media Na^+ and Cl^- are actively transported out of the fish. Silva et al. (1977) have proposed a mechanism to account for this reversal of direction (Figure 7b). The Na^+ pump continues to function as before to create a gradient for passive Na^+ reentry into the chloride cell. This passive reentry is thought to be coupled to Cl^- entry into the cell, across the basal-lateral membrane, and the negative cellular potential promotes Cl^- extrusion across the apical membrane. An important investigation that lead to these models, especially for Cl^- extrusion into hypersaline media,

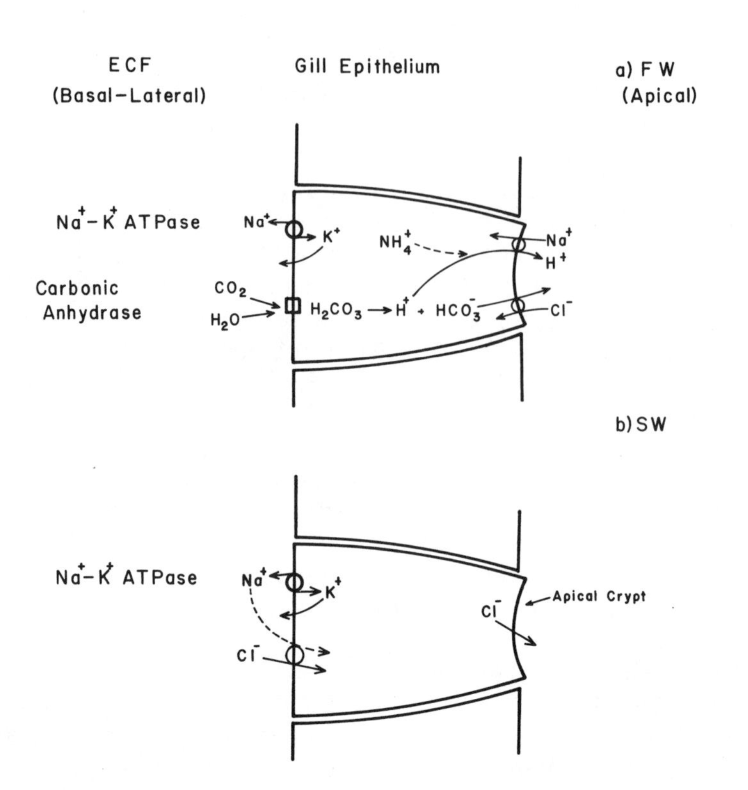

FIGURE 7. Schematic drawings of mechanisms proposed to account for Na^+ and Cl^- uptake from freshwater (a) and Cl^- extrusion when fish are exposed to elevated salinities (b).

was conducted by Karnaky et al. (1976a, b) on _Cyprinodon variegatus_ and _Fundulus heteroclitus_, which localized Na^+-K^+ ATPase on the basal-lateral membranes of chloride cells and correlated the increase in activity of this enzyme in saline media with the increase in surface area of the membranes. In addition to increased development of basal infoldings, the chloride cells of saline-acclimated fish demonstrated apparent increases in secretory activity at the apical membrane (e.g., the apical crypt region) which suggests that exocytosis may be involved in active salt extrusion.

The Na^+ pump is a component of cellular models for both Na^+ uptake (Figure 7a) and Cl^- extrusion (Figure 7b). The specific activity of Na^+-K^+ ATPase has been measured in gill tissue from a variety of euryhaline teleosts and the general observation is an increase in activity with increasing acclimation salinity, which is consistent with a need to extrude salt actively. In FW, on the other hand, the role of Na^+-K^+ ATPase is not so clearly defined, although several studies have found that the specific activity of this enzyme is lowest in salinities near that of the extracellular fluid and increases when fish are transferred to FW or a more saline environment. Stuenkel and Hillyard (1980) confirmed this finding in the gills of *C. salinus* transferred from a 560 $mosm \cdot kg^{-1}$ medium to FW or media with concentrations of 1168 and 1772 $mosm \cdot kg^{-1}$ (Figure 8). Similar results have been obtained in *Anguilla* (Butler and Carmichael 1972) and in *Fundulus heteroclitus* (Towle et al. 1976; Figure 8). The study by Towle et al. (1976) is of particular interest because it was observed that Na^+-K^+ ATPase activity was fully adjusted to changes in salinity within 1 hr after transfer. Thus, euryhaline fish seem to be able to adapt rapidly to changes in the salinity of their environment. It should be noted, however, that other investigators who studied *Fundulus* and *Anguilla* (Epstein et al. 1967; Karnaky et al. 1976b; Utida et al. 1971) have observed that Na^+-K^+ ATPase activity increases progressively with increasing salinity. The role of the Na^+ pump in salt transport therefore needs further investigation and euryhaline desert fishes may be useful animals for study.

CONCLUSIONS

Relatively few studies have been made on the mechanisms by which desert fishes regulate their energy metabolism and internal environment under varying conditions of T_a, DO and salinity. The levels of SMR and RMR in the desert species that have been studied are similar to those of other teleosts when VO_2 is measured under comparable conditions. Activity metabolism and swimming capacity in desert species have not been investigated to date. Studies of this kind would be particularly useful for comparing adaptations of

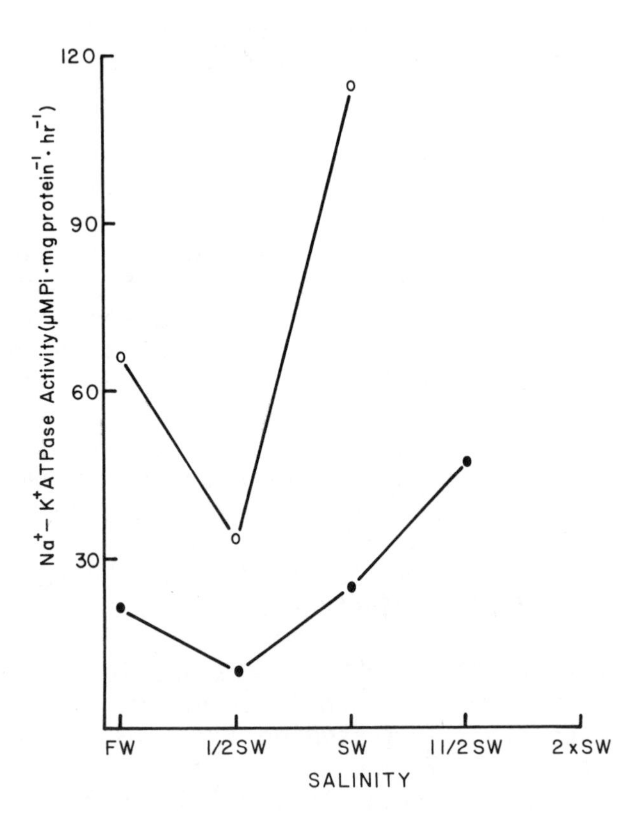

FIGURE 8. The effect of salinity acclimation on the specific activity of Na^+-K^+ ATPase in gill homogenates of Fundulus heteroclitus (open circles, data from Towle et al. 1977), Cyprinodon salinus (closed circles; data from Stuenkel and Hillyard 1980).

pupfish that inhabit constant temperature springs with those of closely related species resident in streams in which T_a, salinity, and flow rate vary. In the light of genetic differences in thermal tolerance of spring versus stream-dwelling pupfish (see Soltz and Hirshfield this volume) physiological differences between populations may become sensitive indexes of evolutionary divergence. In conjunction with VO_2 measurements studies of the properties of hemoglobin are needed, which allow desert fishes to deliver O_2 to their tissues under extremes in T_a and DO, especially in species like *Crenichthys baileyi* found in 37°C springs with DO values less than 1 ppm.

The changes in plasma osmolality, sodium exchange with the environment, and branchial Na^+-K^+ ATPase activity, which are produced in desert pupfish by altering T_a and salinity in their environment, resemble those observed in estuarine species, such as *Fundulus heteroclitus*. This similarity is consistent with evidence presented earlier in this symposium that pupfish species in the deserts of California and Nevada evolved from ancestors resident in estuaries of the Gulf of California which once extended northward into these regions.

ACKNOWLEDGMENTS

I thank Dr. James E. Deacon for his valuable comments and criticisms during the preparation of this manuscript. I also thank Dr. Deacon, Paul Schumann, and Edward Stuenkel for allowing me to use their unpublished data in this review. Finally, the secretarial staff of the Department of Biological Sciences, especially Mrs. Dotty Edelman, have been most helpful in putting the manuscript together.

REFERENCES

Bartholomew, G. A. 1977. Body temperature and energy metabolism. *In*: M. S. Gordon, Animal Physiology: Principles and Adaptations, Third edition. MacMillan, New York, pp. 364-449.

Beamish, F. W. H. 1978. Swimming capacity. *In*: W. S. Hoar and D. J. Randall (Eds.), Fish Physiology, Vol. VII. Academic, New York, pp. 101-189.

Brett, J. R. and T. D. D. Groves. 1979. Physiological energetics. *In*: W. S. Hoar, D. J. Randall and J. R. Brett (Eds.), Fish Physiology, Vol. VIII. Academic, New York, pp. 280-352.

Brown, J. H. 1971. The Desert Pupfish. Scientific American. 225:104-110

Butler, D. G. and F. J. Carmichael. 1972. (Na^+-K^+) - ATPase activity in eel (Anguilla rostrata) gills in relation to changes in environmental salinity: Role of adrenocortical steroids. General and Comparative Endocrinology 19:421-427.

Cross, J. N. 1975. Ecological distribution of the fishes of the Virgin River (Utah, Arizona, Nevada). Master of Science Thesis, University of Nevada, Las Vegas.

Davey, D. L. 1966. A comparative study of the respiratory metabolism of two closely related Nevada desert fish from different thermal environments. Master of Science Thesis, University of Nevada, Reno.

Deacon, J. E. and W. L. Minckley. 1974. Desert Fishes, p. 385-487. In: G. W. Brown, Jr. (Ed.), Desert Biology, Vol. II. Academic, New York.

Dill, D. B., H. T. Edwards and M. Florkin. 1932. Properties of the blood skate (Raja oscillata). Biological Bulletin 62:23-36.

Epstein, F. H., A. I. Katz and G. E. Pickford. 1967. Sodium and potassium activated adenosine triphosphatase of gills: Role in adaptation of teleosts to sea water. Science 156:1245-1247.

Fromm, P. O. 1957. A method for measuring the oxygen consumption of fish. Progressive Fish Culturist 20:137-139.

Greaney, G. S. and D. A. Powers. 1977. Cellular regulation of an allosteric modifier of fish hemoglobin. Nature 270:73-74.

Gunter, G. 1956. A revised list of euryhaline fishes of North and Middle America. American Midland Naturalist 56:345-354.

Houston, A. H. 1973. Environmental temperature and the body fluid system of the teleost. In: W. Chavin (Ed.), Responses of Fish to Environmental Changes. Thomas, Springfield, Ill. pp. 87-162.

Hubbs, C., R. C. Baird and J. W. Gerald. 1967. Effects of dissolved oxygen concentration and light intensity on activity cycles of fishes inhabiting warm springs. American Midland Naturalist 77:104-115.

Karnaky, K. J. Jr., S. A. Ernst and C. W. Philpott. 1967a. Teleost chloride cell. I. Response of pupfish, Cyprinodon variegatus, gill Na, K-ATPase and chloride cell fine structure to various high salinity environments. Journal of Cell Biology 70:144-156.

Karnaky, K. J. Jr., L. B. Kinter, W. B. Kinter and C. E. Stirling. 1976b. Teleost chloride cell. II. Autoradiographic localization of gill Na, K-ATPase in killifish, Fundulus heteroclitus, adapted to low and high salinity environments. Journal of Cell Biology 70:157-177.

Kayser, C. and A. Hensner. 1964. Etude comparative du metabolism energetique dans la serie animale. Journal of Physiology (Paris) 56:489-524.

Kinne, O. 1960. Growth, food intake and food conversion in an euryplastic fish exposed to different temperatures and salinities. Physiological Zoology 33:228-317.

Lomholt, J. P. and k. Johansen. 1979. Hypoxia acclimation in carp: how it affects O_2 uptake ventilation, and O_2 extraction from water. Physiological Zoology 52:38-49.

Maetz, J. 1973. Na^+/NH_4^+ exchanges and NH_3 movement across the gill of Carassius auratus. Journal of Experimental Biology 58:255-275.

Maetz, J. and D. H. Evans. 1972. Effects of temperature on branchial sodium exchange and extrusion mechanisms in the seawater-adapted flounder, Platichthys flesus L. Journal of Experimental Biology 56:565-585.

Marshall, A. G. 1978. Biophysical Chemistry Principles, Techniques, and Applications. Wiley, New York.

Motais, R. and J. Isaia. 1972. Temperature dependence of permeability to water and to sodium of the gill eithelium of the eel, Anguilla anguilla. Journal of Experimental Biology 56:587-600.

Naiman, R. J., S. D. Gerking and R. E. Stuart. 1976. Osmoregulation in the Death Valley pufish Cyprinodon milleri (Pisces: Cyprinodontidae). Copeia 1976:807-810.

Powers, D. A. 1972. Hemoglobin adaptation for fast and slow water habitats in sympatric catostomid fishes. Science 177:360-362.

Powers, D. A. 1980. Molecular ecology of teleost fish hemoglobins: strtegies for adapting to changing environments. American Zoologist 20:139-162.

Powers, D. A., G. S. Greaney and A. R. Place. 1979. Physiological correlation between lactate dehydrogenase genotype and haemoglobin function in fish. Nature 277:240-241.

Prosser, C. L. 1973. Comparative Animal Physiology. 3rd ed. Saunders, Philadelphia.

Rajagopal, P. K. and R. H. Kramer. 1974. Respiratory metabolism of Utah chub, Gila atraria (Girard) and speckled dace, Rhinichthys osculus (Girard). Journal of Fish Biology 6:215-222.

Renfro, J. L. 1970. Osmotic and ionic regulation in the Red River pupfish, Cyprinodon rubrofluviatilis. Ph.D. Dissertation, University of Oklahoma, Norman.

Schmidt-Nielsen, B., Ed. 1980. Biology of The Chloride Cell. Jean Maetz Memorial Symposium. American Journal of Physiology. (238) Regulatory, Integrative and Comparative Physiology 7:139-376.

Schumann, P. B. 1978. Responses to temperature and dissolved oxygen in the roundtail chub, Gila robusta, Baird and Girard. Master of Science Thesis, University of Nevada, Las Vegas.

Silva, P., R. Solomon, K. Spokes and F. H. Epstein. 1977. Ouabain inhibition of gill Na^+-K^+ ATPase: Relationship to active chloride transport. Journal of Experimental Zoology 199:419-426.

Soltz, D. L. and R. J. Naiman. 1978. The Natural History of Native Fishes in the Death Valley System. Natural History Museum of Los Angeles County. Science Sercies 30:1-76.

Stryer, L. 1975. Biochemistry. Freeman, San Francisco.

Stuenkel, E. L. and S. D. Hillyard. 1980. The effects of temperature and salinity gill Na^+-K^+ ATPase activity in the pupfish, Cyprinodon salinus. Comparative Biochemistry and Physiology. 67A:179-182.

Stuenkel, E. L. and S. D. Hillyard. 1981. The effects of temperature and salinity acclimation on metabolic rate and osmoregulation in the pupfish, Cyprinodon salinus. Copeia. 1981:411-417.

Sumner, F. B. and M. C. Sargent. 1940. Some observations in the physiology of warm spring fishes. Ecology 21:45-54.

Sumner, F. B. and V. N. Lanham. 1942. Studies of the respiratory metabolism of warm and cool spring fishes. Biological Bulletin 88:313-327.

Towle, D. W., M. E. Gillman and J. D. Hempel. 1976. Rapid modulation of gill $Na^{+}K^{+}$ - dependent ATPase activity during acclimation of the killifish, Fundulus heteroclitus, to salinity change. Journal of Experimental Zoology 202:179-186.

Utida, S., M. Kamiya and N. Shirai. 1971. Relationship between the activity of Na^{+}-K^{+} activated adenosine triphosphatase and the numer of chloride cells with specieal reference to sea-water adaptation. Comparative Biochemistry and Physiology 38:443-447.

Williams, J. E. 1977. Adaptive responses of woundfin, Plagopterus argentissimus and red shiner, Notropis lutrensis, to a salt spring and their probable effects on competion. Master of Science Thesis, University of Nevada, Las Vegas.

Wood, S. C. 1980. Adaptation of red blood cell function to hypoxia and temperature in ectothermic vertebrates. American Zoologist 20:163-172.

13 The Conservation of Desert Fishes

EDWIN P. PISTER
California Department of Fish and Game, Bishop

ABSTRACT

Recent habitat degradation in deserts of the southwestern United States and adjoining areas of Mexico has seriously depleted endemic fish populations and has resulted in several extinctions and a high percentage of endangerment among those that survive. The urgency and complexity of the problem caused individuals affiliated with several governmental agencies to combine in 1970 with university professors, students, and other concerned parties to form the Desert Fishes Council. This organization improved communications and assisted in the development of an interagency habitat preservation program. Activities to date have been successful in saving at least two species from extinction and in providing a basis for recovery team efforts for southwestern fishes, implemented pursuant to the Endangered Species Act of 1973. Examples of endangerment and recovery are taken from the deserts of Nevada and California, recovery efforts throughout the southwest are summarized, and the current status of California desert fishes is presented to exemplify the problems faced by desert fishes in general. Research potential and the costs and benefits of endangered species preservation are discussed. Because the environmental problems faced by the desert fishes closely parallel those faced by mankind, the methodology developed may assist in our mutual survival.

INTRODUCTION

Recent population expansion and agricultural development throughout the southwestern United States has created an unprecedented demand for already limited water resources. It is not surprising, therefore, that among the many elements that constitute North America's faunal diversity probably none is more seriously threatened than the fishes of the southwestern desert regions.

During the last 40 years man's activities apparently have caused the extinction of four species and six subspecies in six genera in California, Nevada, and Arizona: Ash Meadows killifish (*Empetrichthys merriami*); Raycraft Ranch killifish (*E. latos concavus*); Pahrump Ranch killifish (*E. latos pahrump*); Pahranagat spinedace (*Lepidomeda altivelis*); thicktail chub (*Gila crassicauda*); Monkey Spring pupfish (*Cyprinodon* n. sp.); Shoshone pupfish (*C. nevadensis shoshone*); Tecopa pupfish (*C. n. calidae*); Yaqui shiner (*Notropis formosus mearnsi*); and Grass Valley speckled dace (*Rhinichthys osculus reliquus*). Minckley (1973, p. 190) also listed as extinct a Gila River form of the Arizona pupfish *Cyprinodon macularius*, distinct from other known populations. The Big Spring spinedace (*Lepidomeda mollispinnis pratensis*), feared for several years to be extinct, was recently rediscovered in Big Spring, Lincoln County, Nevada. In addition, at least 50 species and subspecies in 26 genera in eight Great

Basin states and Northern Mexico are considered threatened, 17 of which are currently listed as endangered by the Secretary of the Interior (U.S. Department of the Interior 1980). This situation has been caused primarily by agricultural pumping and diversion of watercourses and has been aggravated by the introduction of predaceous game fishes and other piscine competitors. The wide dispersion of the mosquitofish (*Gambusia affinis*) in many areas of the Southwest has also been a significant factor in the decimation of endemic fishes (Miller 1961; Minckley et al. 1977; Myers 1965). It is significant that among the 43 North American fishes listed as endangered or threatened 28 are found in the southwestern United States and northern Mexico and that among the 31 fishes in the United States listed as endangered 23 are found in desert areas (Soltz 1979; U.S. Department of the Interior 1980; Williams this volume).

This basic problem is well illustrated in the Death Valley area of southeastern California and southwestern Nevada. During the Pleistocene epoch much of this area was covered by large lakes (Hubbs and Miller 1948; Hubbs et al. 1974; Mehringer 1977; Miller 1948). Ancestral fishes were distributed in these waters. As the pluvial period ended the lakes and their tributary streams gradually receded to a fraction of their former size, essentially a few small springs and intermittent watercourses interspersed in an extremely arid desert. None of these watercourses reached the sea. Only the Owens River, fed by the snows of the eastern Sierra Nevada, remained as a substantial permanent stream. Even so, in recent times and before the construction of the Los Angeles aqueduct the Owens River terminated in Owens Lake, a closed basin.

Until recently the concept of a "Death Valley System" was generally accepted, wherein pluvial waters in the area of Death Valley from Lake Mono and the Owens River to the north and west, the Amargosa River to the east, and the Mojave River to the south all ultimately drained during the Pleistocene epoch into Pluvial Lake Manly which covered the floor of Death Valley. It was further surmised that during those periods of the Pleistocene in which greater amounts of precipitation occurred, a fairly general distribution of fishes existed within this Death Valley System.

More recently, however, zoogeographers are theorizing that an east and west drainage system existed in the Death Valley area; the westerly drainages terminated in Lake Searles with only occasional spillover into Lakes Panamint and Manly (Figure 1). There is much evidence, based on mollusc and fish distribution (Dwight W. Taylor personal communication; Miller this volume) and mineral deposition (Smith 1960) to support the latter theory. Furthermore, it is probable that the invasion of fishes into various portions of the

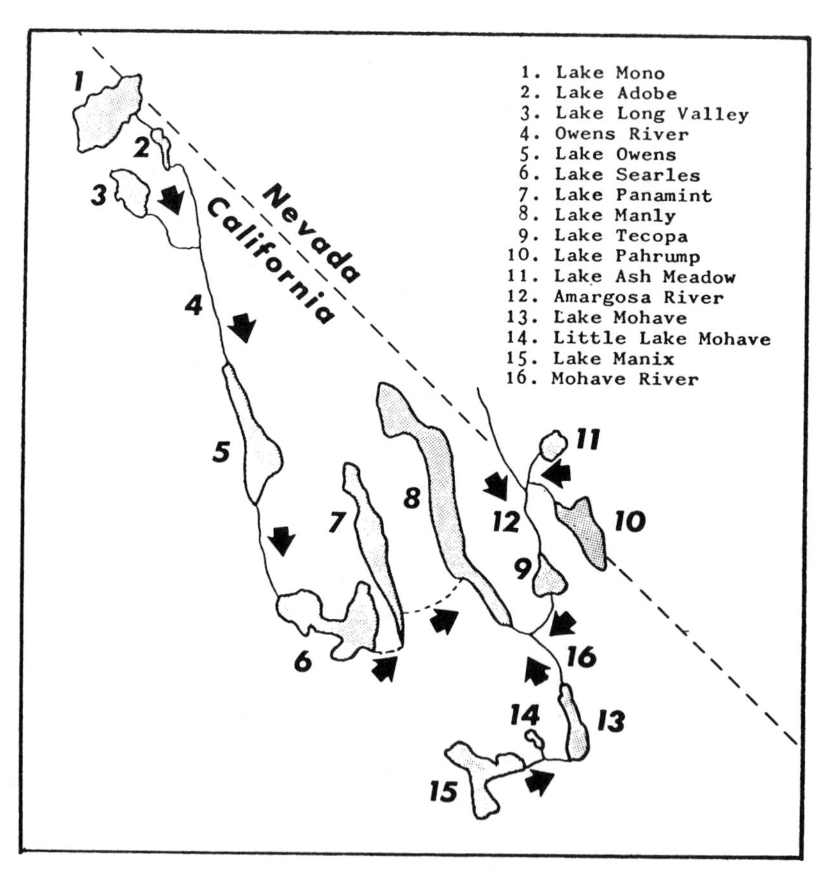

FIGURE 1. Major Pleistocene waters of the Death Valley area. Adapted from Pister 1974. Scale: 1 cm = 37 km.

Death Valley area occurred at widely separated times since late Pliocene (Miller this volume). The more recent concept of a Death Valley System, however, is used as a descriptive term in this chapter.

A relatively limited fish fauna existed in these waters: killifishes (Cyprinodontidae), minnows (Cyprinidae), and suckers (Catostomidae) (La Bounty and Deacon 1972; Miller 1948). These fishes were restricted to the few remaining habitats suitable for their survival. In the case of the killifishes, apparently the most adaptable of the native species, the isolated springs, marshes, and stream courses may be likened to "islands of water in a sea of sand," and the fishes therein, in a manner not unlike that of Darwin's Galapagos finches, evolved into five separate species and several subspecies within the genus _Cyprinodon_ and two distinct species (one with three subspecies) within the genus _Empetrichthys_. The minnows were less successful in adapting to variable water temperatures and to the increased alkalinity and salinity that resulted from evaporation since the pluvial period. Minnows are currently represented by several distinct populations of _Rhinichthys osculus_ (one in Owens Valley and others in the Amargosa River drainage), _Gila bicolor snyderi_ in the Owens River system, and _Gila bicolor mohavensis_ from the Mojave River. In addition, an endemic sucker (_Catostomus fumeiventris_) exists in substantial numbers in the upper Owens River drainage (Table 1).

BACKGROUND

White men had scarcely arrived in the Death Valley area when they began to cast covetous eyes on the limited water resources that followed the pluvial period (Figure 1). In the early 1900s the Los Angeles Aqueduct was completed to supply water to growing populations more than 325 km to the south. This water development project extended into the northernmost reaches of the Death Valley System and effected severe habitat changes throughout the Owens River drainage. Dams minimized flooding that formerly created prime

TABLE 1. A Distributional Checklist of California Desert Fishes[a]

Family and Species		Death Valley Area			Colorado River Area	
	Mono Basin	Owens River	Amargosa River	Mojave River	Salton Sea	Colorado River
Catostomidae						
razorback sucker, *Xyrauchen texanus*						NR
Owens sucker, *Catostomus fumeiventris*	I	N				
flannelmouth sucker, *Catostomus latipinnis*						NE
Centrarchidae						
Sacramento perch, *Archoplites interruptus*	I	I				
black crappie, *Pomoxis nigromaculatus*		I?			I	I
warmouth, *Chaenobryttus gulosus*		I				I
green sunfish, *Lepomis cyanellus*		I		I	I	I
bluegill, *Lepomis macrochirus*		I		I	I	I
redear sunfish, *Lepomis microlophus*		I		I		I
largemouth bass, *Micropterus salmoides*		I	I	I	I	I
smallmouth bass, *Micropterus dolomieui*		I				I
Cichlidae						
Mozambique tilapia, *Sarotherodon mossambicus*					I	I
redbelly tilapia, *Tilapia zillii*					I	I

Clupeidae					
threadfin shad, *Dorosoma petenense*				I	I
Cyprinidae					
carp, *Cyprinus carpio*	I		I	I	I
goldfish, *Carassius auratus*		I	I	I	I
golden shiner, *Notemigonus crysoleucas*					I
Colorado squawfish, *Ptychocheilus lucius*				NE	NE
roundtail chub, *Gila robusta*					NE
bonytail, *Gila elegans*				NE	NR
Owens chub, *Gila bicolor snyderi*	NR				
Mohave chub, *Gila bicolor mohavensis*			NR		
arroyo chub, *Gila orcutti*			I		
speckled dace, *Rhinichthys osculus* subspp.	NR	N			
red shiner, *Notropis lutrensis*				I	I
fathead minnow, *Pimapheles promelas*					I
woundfin, *Plagopterus argentissimus*					NE
Cyprinodontidae					
Trinidad rivulus, *Rivulus marmoratus*				IE	
desert pupfish, *Cyprinodon macularius*				NR	NE
Owens pupfish, *Cyprinodon radiosus*	NR				
Amargosa pupfish, *Cyprinodon nevadensis amargosae*	I	N	I		
Salt Creek pupfish, *Cyprinodon salinus*		N			
Cottonball Marsh pupfish, *Cyprinodon milleri*		N			
Tecopa pupfish, *Cyprinodon nevadensis calidae*		NE			
Shoshone pupfish, *Cyprinodon nevadensis shoshone*		NE			

Table 1. (Continued)

Family and Species	Death Valley Area				Colorado River Area	
	Mono Basin	Owens River	Amargosa River	Mojave River	Salton Sea	Colorado River
Elopidae						
machete, *Elops affinis*					NE	N
Gasterosteidae						
threespine stickleback, *Gasterosteus aculeatus*	I	I		I		
Gobiidae						
longjaw mudsucker, *Gillichthys mirabilis*					I	
Ictaluridae						
channel catfish, *Ictalurus punctatus*		I		I	I	I
yellow bullhead, *Ictalurus natalis*					I	I
brown bullhead, *Ictalurus nebulosus*		I			I	
black bullhead, *Ictalurus melas*						IR
flathead catfish, *Pylodictis olivaris*					I	I
Mugilidae						
striped mullet, *Mugil cephalus*					N	N

Percichthyidae						
striped bass, *Morone saxatilis*						I
Poeciliidae						
mosquitofish, *Gambusia affinis*	I	I	I	I	I	I
sailfin molly, *Poecilia latipinna*					I	I
shortfin molly, *Poecilia mexicana*					I	I
green swordtail, *Xiphophorus helleri*					I	
portal fish, *Poeciliopsis gracilis*					I	
Pomadasyidae						
sargo, *Anisotremus davidsoni*					I	
Salmonidae						
kokanee, *Oncorhynchus nerka*				I		
brook trout, *Salvelinus fontinalis*	I	I		I		
cutthroat trout, *Salmo clarki*	I	I			NE	NE
brown trout, *Salmo trutta*	I	I		I		
golden trout, *Salmo aguabonita*	I	I				
rainbow trout, *Salmo gairdneri*	I	I		I	IE	I
Sciaenidae						
orangemouth corvina, *Cynoscion xanthulus*					I	
gulf croaker, *Bairdiella icistia*						

[a]N = native; I = introduced; E = extinct; R = rare; ? = status uncertain.

cyprinodont habitat. Marsh areas were drained and the permanent watercourses were stocked with game fishes, primarily brown trout (*Salmo trutta*) largemouth bass (*Micropterus salmoides*) and other, less predaceous species, including the mosquitofish.

The once abundant populations of the Owens pupfish *Cyprinodon radiosus* (Kennedy 1916; Wilke and Lawton 1976) were observed by Professor Carl Hubbs in 1934 to have been reduced to only two known locations and, when described (Miller 1948), the species was thought to be extinct until rediscovered by California Department of Fish and Game personnel in 1956. The rediscovery was verified by Hubbs and Miller in 1964. Even then only prompt action taken by the California Department of Fish and Game in 1969 saved the species from extinction when its only remaining habitat dried up (Miller and Pister 1971).

Among the other endemic fishes of the Owens River drainage, in the genera *Catostomus*, *Rhinichthys*, and *Gila*, only the sucker *Catostomus fumeiventris* remained essentially unaffected by man's activities. The *Rhinichthys* populations were greatly diminished by essentially the same factors that affected the Owens pupfish; and *Gila*, although abundant, had hybridized with introduced bait minnows. Only one genetically pure population of the Owens chub is known and it is restricted to a small section of the old channel of the Owens River below Long Valley Dam, which impounds Crowley Lake. Here it is completely isolated from other waters. Intensive life history studies of the Owens chub are now being conducted to provide a basis for additional recovery efforts.

In the Amargosa Valley area near the southeastern end of the Death Valley System agricultural operations were inexorably lowering water tables to a point at which diminishing spring flows began to threaten endemic fish populations. In Ash Meadows, as in the Owens Valley, introduced species were exacting their toll. *Empetrichthys merriami* of Ash Meadows, considered rare more than two decades ago (Miller 1948), had already become extinct. Among the three Pahrump Valley kinds *E. latos concavus* and *E. l. pahrump* followed *E. merriami* into oblivion. Only one member of the entire genus *Empetrichthys* (*E. latos latos*, the Pahrump killifish) survives today, represented by only a few hundred individuals in artificial refugia. In addition, two subspecies of *Cyprinodon nevadensis* (*C. n. shoshone* and *C. n. calidae*) are thought to be extinct, although there is a remote possibility that a remnant population of the latter subspecies may survive in a small spring in the Tecopa (Inyo Co.) area. This cannot be verified, however, until sufficient numbers become available to allow meaningful statistical analysis of meristic data. The Devils Hole pupfish (*C. diabolis*) currently survives in a precarious way as its population slowly responds to an increasing water level in the habitat in which it had existed, unmolested, since the Pleistocene.

Despite the encroachment on their environment, these hardy fishes for the most part were able for many years to maintain populations adequate for their preservation. Miller (1969) made a thorough inventory of the native fishes of the Death Valley System for the U. S. National Park Service in 1967 and reported that only two species were seriously endangered. Within the next two years, however, habitat destruction had progressed to a point at which the number of endangered species and subspecies had grown to seven and by 1972 12 were recognized as endangered. Although recovery efforts have been implemented, they remain in peril. Their status has not improved sufficiently to allow delisting.

The most rapid habitat destruction occurred in the Ash Meadows area of Nye County, Nevada, where during the mid 1960s extensive marsh system draining removed approximately 90 percent of the habitat available to *Cyprinodon* and *Rhinichthys*. Three years of peat mining followed in which even the sand dunes were absorbed into the peat area for marginal agricultural production, following which heavy equipment leveled vast areas of land in preparation for crops. Springs were stripped of riparian vegetation to reduce transpiration loss and to facilitate installation of irrigation structures, and well-drilling equipment was widely used to tap underground aquifers.

In one location (Jackrabbit Spring) a pump was suspended on steel beams directly over the spring and subsequent pumping completely eliminated the fish population (*Cyprinodon nevadensis mionectes* and *Rhinichthys osculus nevadensis*). Mechanical injury (pumping) and siltation were major causes of damage at other spring locations, but by far the greatest impact was manifested by an insidious lowering of the water table and decreased flows throughout the spring system.

Devils Hole, a detached portion of Death Valley National Monument, has no surface outlet. Its elevation is determined by flow from a vast underground aquifer. Here the changes were less dramatic but even more serious. *Cyprinodon diabolis*, which is confined to Devils Hole (the smallest known habitat of any vertebrate species), is completely dependent on water to cover a sufficient portion of a limestone shelf to provide spawning area and substrate for attached algae, which harbors its only source of food. Lowering water levels within Devils Hole gradually exposed the shelf, threatening with eventual doom the most highly differentiated species of fish in the Death Valley System. The lowering water level also reduced the amount of water surface exposed to sunlight and severely reduced the available food supply.

THE DESERT FISHES COUNCIL

It is important to note that during the late 1960s the methodology necessary for handling endangered species preservation was just beginning to develop among the various state and federal resource management agencies. Furthermore, almost all prior preservation efforts had been devoted to birds and mammals. Management of an endangered fish species was an almost entirely new concept (Miller and Pister 1971; Minckley and Deacon 1968).

Lacking an established means of solving the multitude of problems associated with preserving the Death Valley System fishes, representatives of the Fish and Game Departments of California and Nevada, various Bureaus of the U.S. Department of the Interior, and the private conservation sector met in Death Valley in April 1969 to review the situation. During a field trip it became apparent that agricultural development in the Ash Meadows area would soon spell the doom of most of the endemic fishes, not to mention the unique aquatic ecosystems with their associated plants and invertebrates, unless prompt and effective action were taken. At that time basic plans were made for a more formal meeting to draft a sound, integrated plan for species preservation.

Accordingly, on November 18-19, 1969, a group of concerned scientists, representing state and federal resource management agencies and several universities, met at Death Valley National Monument headquarters to discuss the plight of these fishes, several of which were already listed by the Secretary of the Interior as endangered. At this symposium basic preservation programs and guidelines were developed to save these fishes from what, at that time, appeared to be certain extinction (Pister 1970).

A "Pupfish Task Force" was established in May 1970 by the Secretary of the Interior to coordinate Interior effort in the preservation program. Advisory assistance to the Task Force was provided by biologists who represented the Fish and Game Departments of Nevada and California and scientists from the Universities of Nevada, Michigan, and California. This Task Force was instrumental in motivating the preservation effort at this critical time.

The Second Annual Symposium was held on November 17-18, 1970, again at Death Valley National Monument headquarters (Pister 1971). The status of the various endangered species and the progress of specific preservation programs were reviewed and updated. Those who attended established the Desert Fishes Council, whose basic purpose, as stated in its constitution is

> ...to provide for the exchange and transmittal of information on the status, protection and management of the desert fishes and their habitats. For the purpose of the Council the term desert fishes is intended to include any endemic fish, be it species, subspecies, or

race, that inhabits drainages of the North American deserts (Basin and Range Province) and additional drainage areas and endemic fishes as determined by the Council.

I served as the Council's first Chairman.

Formally organized, the Council began to play an even stronger role in coordinating the preservation effort and in assisting government agencies to plan and conduct preservation programs. These programs include hydrologic studies, water-use surveillance, location of transplant sites and refugia, aquarium culture, land reclassification, legal action, and publicity (U.S. Department of the Interior 1971). To expedite the programs the Federal Pupfish Task Force was dissolved in October 1971, when basic direction of Interior's desert fish preservation effort was transferred to the Regional Office of the Bureau of Sport Fisheries and Wildlife (now U.S. Fish and Wildlife Service) in Portland, Oregon.

To date the Council has taken a major part in saving at least one of the endangered fishes from extinction and has made substantial progress in rescuing others. However, our knowledge of the biology and habitat requirements of these fishes must be expanded as only some of the species are responding with uncertain success to preservation efforts.

More recently the Council's activities have been extended beyond the Death Valley System to include all the American deserts of the southwestern United States and adjoining areas of Mexico, including the Colorado River basin and its Pleistocene tributaries. By dividing this region into 12 areas, each under the close scrutiny of a responsible coordinator, the Council's effectiveness has been greatly increased. Assistance is now provided by the Council to determine the fishes endangered or threatened, to implement recommendations for restoration of these desert fishes and their habitats, and to provide philosophical leadership and motivation as resource agencies inevitably evolve into a new management direction.

Council membership currently (1980) exceeds 300 and has been expanded to include, in addition to federal, state, and university scientists, a number of resource specialists and administrators, the private conservation sector, students, and individuals concerned with long-term environmental values. The U. S. Endangered Species Act of 1973 authorized federal funding for recovery team work. Council members take an active role in the activities of recovery teams specifically implemented to improve the status of the Pahrump killifish, Devils Hole pupfish, Warm Springs pupfish (*Cyprinodon nevadensis pectoralis*), Cui-ui (*Chasmistes cujus*), and endangered fishes of the Colorado River, the humpback chub (*Gila cypha*), Colorado River squawfish (*Ptychocheilus lucius*) and bonytail (*Gila elegans*). The Fish and Wildlife Service is currently organizing a Death Valley Area Recovery Team to consolidate and coordinate recovery efforts for the Devils Hole pupfish, Warm Springs pupfish, Pahrump killifish, Owens pupfish, Mohave chub, and Owens chub.

Farther to the southeast recovery teams are working to save the woundfin (*Plagopterus argentissimus*), Arizona trout (*Salmo apache*), Gila trout (*Salmo gilae*), and endangered fishes of the Rio Grande River area, the Clear Creek gambusia (*Gambusia heterochir*), Big Bend gambusia (*Gambusia gaigei*), Pecos gambusia (*Gambusia nobilis*), Comanche Springs pupfish (*Cyprinodon elegans*), and Goodenough gambusia (*Gambusia amistadensis*). In addition, it is planned that the Rio Grande River Recovery Team will eventually assume recovery efforts for the Devils River minnow (*Dionda diaboli*), Leon Springs pupfish (*Cyprinodon bovinus*), and bluntnose shiner (*Notropis simus*).

The Council also encourages and participates in programs designed to preserve other unlisted fishes and the general integrity of desert aquatic ecosystems. It supports a holistic approach to ecosystem management and considers all life forms to be of equal value. The name and constitutional direction of the Council were derived under the concept that if the endemic desert fishes and their habitats are preserved, the various life forms that exist with the fishes and evolved in association with them will also be preserved.

PRESERVATION PROGRAM

A case in point is the Devils Hole pupfish. Because of the extreme urgency of the threat to Devils Hole, transfers of fish were made during the last decade into five seemingly suitable refugia at various locations in the desert region of California and Nevada. An artificial refugium constructed by the Bureau of Reclamation (now Water and Power Resources Service) below Hoover Dam has been stocked with Devils Hole pupfish, and although there has been successful reproduction the pupfish attain larger size and have slightly different morphological characteristics than those in Devils Hole (Williams 1977). Habitats in which species evolved thus act to mold and maintain unique features of these animals; therefore it is urgent that the native habitat be preserved. All attempts by highly qualified aquarists to rear Devils Hole pupfish have failed.

In 1970 the National Park Service and Nevada Department of Fish and Game (now Department of Wildlife) suspended a fiberglass shelf beneath the surface of Devils Hole to provide a spawning area and installed lights to enhance algae production. The long-term success of these ventures appeared doubtful and the shelf has been removed.

Obviously the only known way to guarantee the continued existence of the Devils Hole pupfish and other populations in the Ash Meadows area is to maintain the underground water level. Legal action to prohibit pumping in areas that affect Devils Hole was initiated by the U. S. Department of Justice at the request of the Department of the Interior Solicitor's Office in September 1971. A preinjunction hearing was held in Las Vegas Federal District Court on July 3 and 5, 1972, a temporary injunction was ruled on April 15, 1973, and on June 7, 1976, was made permanent by a unanimous ruling of the U. S. Supreme Court. The population appears to be responding slowly to a more stable habitat. As with most desert fishes, the key to long-term preservation lies in habitat integrity, and this is best assured by land ownership and control.

In an attempt to provide habitat protection for the Ash Meadows area over the last several years, various legislators (primarily Senator Alan Cranston of California) have introduced legislation into past sessions of Congress to create a Desert Pupfish National Monument. In addition, the U. S. Geological Survey has delineated an area large enough to protect the aquifers that supply the springs in Ash Meadows and maintain the water level in Devils Hole. Using this area as a guideline, the U. S. Fish and Wildlife Service and Nevada Department of Wildlife during the last few years have been investigating the possibility of creating a wildlife management area to encompass Ash Meadows. The primary problem remains one of land acquisition. Although the major agricultural effort in Ash Meadows has ceased, much of the land has been sold to a development company that has already submitted a plan for residential subdivision to Nye County officials. Various plans to place Ash Meadows in public ownership are being investigated, although the same political climate that spawned the "Sagebrush Rebellion" makes the likelihood of a purchase remote. The land exchange process appears to offer the greater promise.

Another compelling reason for preserving Ash Meadows is that it is one of only two such desert spring ecosystems remaining on the North American continent. Its counterpart, the Cuatro Cienegas Basin of Coahuila, Mexico, is rapidly and tragically being modified by pumping and drainage for agriculture. The loss of the Ash Meadows biota would therefore be even more tragic. Ash Meadows has also retained a unique flora (Beatley 1971) but much of it has already been destroyed by land development. The fact that various scientists believe that the upper Amargosa River basin will not support sustained agriculture lends a touch of irony (California Department of Water Resources 1964). Rising gasoline prices make the development of Ash Meadows into a Las Vegas "bedroom community" much less likely than a decade ago, although pressure to develop a retirement-type community continues to pose a threat.

Preservation plans for the endangered and threatened Death Valley fishes in other areas are also being actively pursued. Before the extermination of the

Pahrump killifish population in Manse Spring (the type locality for the subspecies) transplants were made cooperatively in 1971 by the U. S. Fish and Wildlife Service and Nevada Department of Wildlife into Corn Creek near Las Vegas and in early 1972 by the Bureau of Land Management and Nevada Department of Wildlife into Shoshone Ponds near Ely. Both ventures appear to be promising, although a resident bullfrog population (*Rana catesbeiana*) is a severe threat to Corn Creek, and land development for the proposed MX Missile system threatens the transplants at Shoshone Ponds.

The recovery program for the Mohave chub involves transplanting it into various permanent water sources in the desert areas. The program has met with varying degrees of success but new populations have been established, thereby greatly improving the general safety of the subspecies. The Bureau of Land Management's habitat management plan for the Mohave chub at Fort Soda (San Bernardino County, California) within the original range gives much cause for optimism. This facility constitutes the major refugium.

In 1969, in cooperation with the Los Angeles Department of Water and Power, the California Department of Fish and Game constructed the Owens Valley Native Fish Sanctuary at the northern end of the Death Valley System to provide a refugium for the four native fishes of the Owens River system, three of which are currently threatened or listed as endangered by the Secretary of the Interior. In 1980 a contiguous lower section, more specifically designed to meet the habitat requirements of *Cyprinodon*, was added to this refugium.

In addition, the Bureau of Land Management has constructed a refugium for the Owens pupfish near the Owens Valley Native Fish Sanctuary, and the California Department of Fish and Game has constructed a refugium for endemic fishes in another location in the northern Owens Valley. Portions of the major refuge area in Fish Slough (Inyo and Mono counties) have been designated as ecological reserves under the provisions of a 1968 act of the California Legislature. The current program to save the native Owens River system fishes appears to be progressing but its success is not yet assured. As in Ash Meadows, the key to long-term preservation lies in habitat integrity, and the program is currently frustrated by threatened development of 202 acres of privately owned land in Fish Slough. Attempts are underway by B.L.M. to acquire this land by the exchange process, but the outcome of this negotiation is questionable. In the meantime the existence or extinction of three endemic fishes hangs in the balance.

CURRENT STATUS OF CALIFORNIA DESERT FISHES

Death Valley

Within the Death Valley area (Figure 1) the historically fishless (post-Pleistocene) Mono Basin now contains nine introduced fishes. Among four native fishes in the Owens River portion of the system two are listed as endangered and one is threatened. Only the Owens sucker remains in substantial numbers, whereas 18 introduced fishes are thriving. Within the Amargosa River drainage in California two of six native fishes are probably now extinct. The remainder is forced to compete with three exotics. The only native fish in the Mojave River drainage, the Mohave chub, is listed as endangered, whereas 15 exotics are firmly established in this arid region (Table 1).

Colorado River System

The California portion of this major drainage system, which includes the Salton Basin and much of the historic Death Valley drainage, constitutes the California Desert Conservation Area of the Bureau of Land Management. In the Colorado River System the native fish fauna has been seriously impacted by the same essential factors that decimated the Death Valley System fishes: habitat alteration and the introduction of exotic species. The Salton Basin fish fauna is comprised of six native and 25 introduced (two introduced species are now extinct) fishes, and the Colorado River fauna contains 10 native and 24 introduced fishes in the California section. Thirteen additional introduced fishes in the lowermost river have become extinct or have not yet reproduced and become established (Minckley 1979). Among the six native Salton Basin fishes four are extinct. The desert pupfish (*Cyprinodon macularius*), which is seriously endangered by habitat changes within its native range, has somehow escaped extinction. Another native, the striped mullet (*Mugil cephalus*) is believed to exist in the Salton Sea. Among the 10 native fishes of the Colorado River in California six are extinct in California and two are considered endangered. Only the striped mullet and Pacific tenpounder (*Elops affinis*) are found there regularly and then only in limited areas.

Interagency preservation measures similar to those described for the Death Valley System fishes have been implemented in an effort to save at least a remnant of the Colorado Basin's native fish fauna. The major responsibility is borne by the Endemic Species Committee (Lower Basin) of the Colorado River Wildlife Council and the Colorado River Fishes Recovery Team established under the Endangered Species Act of 1973. The current distribution and status of fishes throughout the California desert are summarized in Tables 1 and 2.

Among 20 native fishes in the California desert only seven remain in good numbers, whereas six are endangered and seven are extinct. No doubt the presence of 52 introduced species has been a major cause of the *status quo*.

HABITAT PRESERVATION

The goal of the Desert Fishes Council has been to retain the pristine condition of the native habitat and to reestablish the endemic fishes in as much of their original range as possible. Only in situations in which a species or subspecies is on the verge of extinction has the Council recommended a transplant. Even then this is done with the greatest care to minimize the possibility of hybridization or of engendering any of the other problems that inevitably result from the unwise introduction of exotic species.

TABLE 2. Status and Distribution of California Desert Fishes: A Summary

	Death Valley Area				Colorado River Area	
	Mono Basin	Owens River	Amargosa River	Mojave River	Salton Sea	Colorado River
Historic native species or subspecies	0	4	6	1	6	10
Extinct native species or subspecies	0	0	2	0	4	6
Extant native species or subspecies	0	4	4	1	2	4
Introduced species or subspecies	9	18	3	15	25	24
Extinct introduced species or subspecies	0	0	0	0	2	0
Extant introduced species or subspecies	9	18	3	15	23	24
Total extant species or subspecies	9	22	7	16	25	28

By following the principle of habitat preservation three major objectives are accomplished:

1. The continued existence of the species is given further assurance.
2. The evolutionary development of the fish continues uninterrupted, whereas its adaptation to a transplant environment almost certainly creates a shift as the fish begins to adapt to the unusual conditions therein (see above for Devils Hole pupfish).
3. Other forms of life within the habitat are also preserved.

This last item is highly important and is often overlooked as we become involved in the singular purpose of protecting a given species. Generally speaking, the field biologist's knowledge of the earth's fauna is limited to the larger life forms. Consequently, when we talk of the extinction of fish species we are generally unaware of the other forms of life that in all probability will also be destroyed when the habitat is lost. For these reasons all efforts in addition to habitat preservation are viewed simply as a means of preserving the gene pool pending preservation or restoration of the native habitat.

RESEARCH POTENTIAL

Although it is axiomatic that we should endeavor to preserve all endemic fish and wildlife (Ehrenfeld 1976; Nibley 1978; Pister 1976, 1979), it is nevertheless fascinating to discuss a few of the known characteristics of the desert fishes that make them unique and valuable to us. All fishes are of interest and value, but as an example let us consider the cyprinodonts.

The cyprinodonts are of immense importance to scientists specializing in biogeography, physiology, genetics, evolution, and fish behavior. One species has been found to tolerate salinities several times that of sea water. Another occupies an artesian well outflow that ranges from freezing to 47.2°C at the source; the fish enter water as hot as 43.9°C. A population in tiny Mexican Spring (Ash Meadows, Nevada) numbered only 20 to 40 individuals and survived in only 300 liters of water, ranging in temperature from 4.4 to 34.0°C. Mexican Spring recently became overgrown with vegetation and dried up by evapotranspiration. All populations studied have revealed similar abilities to adapt to extreme environmental conditions (Brown 1971; Brown and Feldmeth 1971; Naiman 1976; Naiman et al. 1973, 1976; Soltz and Naiman 1978).

Research on this family of fishes currently includes genetic interaction between an endangered species and an introduced congener, effects of acclimation on reproductive temperature tolerance, effects of temperature and salinity on metabolic rate, basic life history studies, comparative spawning behavior, population dynamics, reproductive performance in relation to salinity, effects of temperature and salinity on electrolyte and energy metabolism, genetic differences in physiological tolerances, species specificity and mating systems, the effect of temperature and salinity on sodium and potassium ATPase activity in gill tissue, thermal responses, morphological variation between isolated populations, breeding systems, and possible use in mosquito control. Many studies have preceded those currently in progress and many will follow if we are successful in our preservation efforts. Research possibilities with the cyprinodont fishes appear to be unlimited, and each researcher reveals new and intriguing fields to the next investigator.

Cyprinodonts possess the capacity to adapt quickly to extremely hostile environmental conditions. Because man is beginning to face a new set of environmental conditions hostile to his own well-being, it would be tragic to lose this resource (or even a part of it) before we fully understand it. The biology of all native Death Valley System fishes, including their evolution, distribution, and conservation, is discussed by Soltz and Naiman (1978). The general subject is further addressed by Soltz (1979).

COSTS AND BENEFITS

The values of endangered species preservation have been touched on in the preceding sections of this chapter, but because this point is of key importance to such a discussion additional treatment may be warranted.

The problems inherent in the preservation of desert fishes seem to typify endangered species in general. Their endangerment is a manifestation of habitat alteration and competition from exotic species, both of which are direct or indirect results of man's self-serving attitudes. Costs of preservation programs are minimal by modern-day standards. In 1978 the entire cost of administering the Endangered Species Act, including aid to the states, was $16.2 million; $19.4 million was the 1979 estimate. This is significantly less than the cost of a single jet fighter plane.

Although it is virtually impossible to place a dollar figure on the preservation of endangered animals (except for game or commercially valuable species), it is possible to list other benefits without delving too deeply into the issues of ethics or morality. For instance, endangered species frequently serve as indicators of larger environmental problems that may adversely affect persons who show no concern for (or are unaware of) the plight of the endangered animal. Examples are found in the Devils Hole pupfish and snail darter (*Percina tanasi*), both of which were considered (and ruled on favorably) by the U. S. Supreme Court.

In the first instance studies conducted during the Devils Hole pupfish campaign revealed a plan contemplated by the Nevada State Engineer to mine 3 million acre feet of water from the Amargosa desert over a period of 30 years (Nevada State Engineer's Office 1971). At that time, the plan admitted, it would no longer be feasible to run the pumps and water quality would probably have deteriorated to a point at which it would no longer be suitable for normal use. The plan implied, however, that during that 30-year period alternate water sources would *probably* be located. What are the costs of such a venture? The loss of an ecosystem (with its largely endemic flora and fauna) unique in the world. The benefits? Questionable, at best, especially when considered in the long term. Sadly enough, current plans to subdivide much of the Ash Meadows area would ultimately have an equally devastating effect.

In the well-publicized matter of the snail darter the media took great delight in pitting a "three-inch minnow" against a $120 million dam project. What the media generally neglected to mention, however, was that at the height of the controversy in 1978 the Chairman of TVA was seriously questioning the wisdom of completing the dam (Los Angeles Times June 24, 1978). No doubt his admirable candor was influenced to some extent, at least, by a report by the General Accounting Office that made five key points in its recommendation to Congress to make a new analysis of costs and benefits before taking further action on the Tellico project (U.S. General Accounting Office 1977).

In summary, these key points were that the project would threaten the survival of an endangered species, destroy a archaeologically rich valley, render useless about 16,500 acres of prime agricultural land, convert the last large flowing river into a reservoir in a region that already contains 20 underutilized reservoirs within 100 miles of the project site, and create a reservoir in lieu of a scenic riverway that would do much to alleviate overcrowding at the nation's fourth most widely used national park, Great Smoky Mountains. Furthermore, it was strongly implied that the benefit/cost ratio of the project, favorable when first calculated during the 1960's, had probably changed to a point at which it would currently be unfavorable, largely because of increased land values. Annual costs would probably exceed annual benefits.

Unfortunately the new studies suggested by the General Accounting Office were not made nor will they ever be. In 1979 Congress exempted the Tellico project from the provisions of the Endangered Species Act. At this point one cannot help but recall a cynical comment attributed to Winston Churchill: "People who sincerely believe in laws and sausages should never watch them being made." The Tellico project, by inference, will be paid for in perpetuity by every taxpayer in the nation.

Additional benefits of endangered species preservation have been the "spinoff" variety, manifested largely in inevitable and highly beneficial philosophical shifts within state and federal resource management agencies as the role of nongame species in tomorrow's society becomes increasingly apparent (Pister 1976, 1979).

Fish and wildlife resources are commodities affected by recognized economic principles, yet they are seldom thought of in that way. Future demand factors for resources currently considered "worthless" (the nonresources described by Ehrenfeld 1976) are certain to increase their value in years to come in the same way that supply and demand determines the price of potatoes and automobiles. It behooves us, then, to keep our species bank as full as possible in order that we may avail ourselves of the accrued interest. We must resist the temptation to draw on the principal that constitutes the world's fish and wildlife resource for exactly the same reason that an individual cannot continue to spend his principal. Sooner or later it runs out, and in this instance we have no way to print new currency.

I hesitate, however, to involve fish and wildlife resources too deeply in the field of economics, other than to apply a few obvious principles. They deserve better than that. I subscribe to the logic of the provost of Oxford's Oriel College who, in the early 1800s, was reluctant to establish a professorship in economics in the fear that to do so would admit into the curriculum a science "so prone to usurp the rest" (Warren 1978). It seems that economics surely plays a disproportionate role in our decision-making process.

It is both interesting and ironic that man's own future is so inextricably involved with his dominion over his fellow creatures. This may prove to be the greatest test ever given him, for if he is successful in exercising a righteous dominion it is reasonable to expect that he will continue to exist. If he is not successful, it becomes only a matter of time until he, too, will perish from the earth. It is a great challenge and a sobering thought.

Nibley (1978; 85-86) states accurately that

> We have taught our children by precept and example that every living thing exists to be converted into cash, and that whatever would not yield a return should be quickly exterminated to make way for creatures that do.

He then points out that this course, in the long run, is no different from going the wrong way on the freeway during rush hour, with the struggle to live becoming a fight <u>against</u> nature. There exists no reasonable or acceptable alternative than for man to change his attitudes so that the animal and vegetable creation will cease to afflict and torment him. Otherwise he will remain in his present course of waging a war of extermination against all that annoys him until he renders the earth completely uninhabitable. He has already made a good start in that direction.

The responsible prediction that the current extinction rate of 1000 species per year could easily rise to 10,000 per year (one species per hour) by the late 1980s, and is expected to accelerate through the 1990s (Wilson 1980), is hardly reassuring. Any thoughtful person should be horrified by the ramifications of such an eventuality, as man's dependence on natural diversity for his long-term existence becomes increasingly clear. Because the extinction rate now exceeds the production rate for new species, any competent biometrician should be able to estimate how much time we have left.

It is becoming clear, however, that the ultimate benefit to be gained by concerning ourselves with the well being of other species lies as much within the realm of sociology as biology and economics. Man is finally beginning to show signs of becoming more honest and introspective in his analysis and understanding of the dominion granted him by his creator. The Endangered Species Act of 1973 is a strong indication.

Fortunately there are those who are blessed with the ability to express themselves with a simplicity and eloquence unfettered by sophistication. In his chapter on conservation in <u>Round River</u>, Leopold (1953; 146-147) handles the question of species value:

> The outstanding scientific discovery of the twentieth century is not television, or radio, but rather the complexity of the land mechanism. Only those who know the most about it can appreciate how little we know about it. The last word in ignorance is the man who says of an animal or plant: "What good is it?" If the land mechanism as a whole is good, then every part is good, whether we understand it or not. If the biota, in the course of aeons, has built something we like but do not understand, then who but a fool would discard seemingly useless parts? To keep every cog and wheel is the first precaution of intelligent tinkering. Have we learned this first principle of conservation: to preserve all parts of the land mechanism? No, because even the scientist does not yet recognize all of them.

At this point in time it seems reasonable to ask if the future holds any hope at all for life on the earth as we know it. Probably the best answer to this is found in a quote by Henry Ford: "You can say it can be done, or you can say it can't be done and be correct either way." Therefore, although the future sometimes looks dark indeed, I persist as an optimist. I like to think there *is* a way. Why? Because the alternative is so completely unacceptable.

CONCLUSION

The desert fishes preservation program indicates clearly that, despite occasionally conflicting philosophies (i.e., preservation versus development), it is possible for governmental agencies to work together effectively in matters of basic resource protection. This is best accomplished by the simple expedient of bringing people together to work toward a common goal. It is an obvious but often overlooked fact that agencies in reality are nothing more than people and when people work well and effectively together so do the agencies.

From the standpoint of the state-federal relationship the preservation effort has been a highly cooperative one, based on the recovery-team concept and competent state-directed species management programs strongly supported and assisted by the federal agencies. Invaluable research programs and technical assistance have been provided by the academic community with support from the private conservation sector.

Although the preservation of the desert fishes is by itself a worthy objective, the efforts expended in their behalf are directed toward higher goals. What we have in the Ash Meadows area, and throughout the southwestern desert, is far more than some unique fishes and a declining water table. We have in this system a tiny microcosm that reflects the same problems faced by the entire Earth, or soon to be faced. We have a unique natural resource that is facing extinction because of encroachment on the habitat by man for economic gain. This is essentially the basis of the world's environmental problems today.

This entire matter is concisely summarized by one of the early pioneers in desert fish preservation (Deacon 1979:41):

> Fishes of the West are affected by the same general kinds of ecological problems that are causing extinctions throughout the world. The interplay of economics with perceived value in society has led us into the numerous ecological problems facing us today. There is some evidence to suggest that society is making some preliminary effort to slow the rate of extermination. Perhaps this is happening because the conclusions of ecologists, philosophers, and theologians regarding the relationship of man and environment are to some extent being translated into legislation as well as into conventional wisdom.

We have come a long way in our work on desert fishes in the 12 years that have elapsed since the burgeoning problem was identified, case histories were presented, and basic preservation philosophies and methodologies were suggested by Minckley and Deacon (1968). We still have a long way to go, however, before knowledge of the resource and methods of conserving it will have evolved to the point at which we can begin to feel comfortable with the status quo. In all probability we will never reach that point and will of necessity have achieved a full realization that eternal vigilance is the price of preservation.

Our conservation efforts have taken us before the highest courts of the land, and both the courts and the law have given us confidence that there is support within "the system" for the work we are doing. Yet I feel we take most of our inspiration and motivation not from Man's law, but from the higher law of Nature, a law stated with the simple eloquence of Aldo Leopold in his essay, "The Land Ethic" (Leopold 1949:224-225): "A thing is right when it tends to preserve the integrity, stability, and beauty of the biotic community. It is wrong when it tends otherwise." It seems that others are beginning to recognize the truth of this law.

In our quest to save the desert fishes it is hoped that we can set up guidelines and procedures that will be helpful to the solution of similar problems elsewhere. Man is only now beginning to realize that the Earth's natural resources are indeed finite and that his consumption of these resources at such an alarming rate simply to maintain (in the United States anyway) a bloated standard of living must soon come to an end. Already elected officials are beginning to delve into the uncertain complex of factors posed by natural law, economic reality, and political popularity. Man must begin, sooner or later, to decide where the line must be drawn between environmental preservation and economic development.

ACKNOWLEDGMENTS

It is impossible to discuss the Great Basin fishes without acknowledging the contributions made by Dr. R. R. Miller and the late Professor Carl L. Hubbs. Charles H. Meacham, former Commissioner, U. S. Fish and Wildlife Service, was quick to assist and was instrumental in establishing the Interior Department's Pupfish Task Force. I am grateful to the California Department of Fish and Game and my immediate supervisor William M. Richardson for support. Dr. W. King provided valuable counsel and encouragement at the beginning of the program, and J. T. McBroom was an inspiration during his chairmanship of the Pupfish Task Force. Dr. W. L. Minckley provided current information on the status of Arizona and California fishes and made helpful suggestions in the preparation of this chapter. G. F. Black provided valuable information on fishes of the Salton Basin. Dr. J. E. Deacon served as a strong inspiration throughout the program. The Los Angeles Department of Water and Power contributed to the recovery of the Owens Valley fishes. At this point is is difficult to make further specific acknowledgments because so many others provided interest, encouragement, and enthusiasm. To all of them I extend my heartfelt thanks.

REFERENCES

Beatley, J. C. 1971. Vascular plants of Ash Meadows, Nevada. University of California Laboratory of Nuclear Medicine and Radiation Biology, Report No. UCLA 12-845.

Brown, J. H. 1971. The desert pupfish. Scientific American 225:104-110.

Brown, J. H. and C. R. Feldmeth. 1971. Evolution in constant and fluctuating environments: thermal tolerances of desert pupfish (*Cyprinodon*). Evolution 25:390-398.

Bunnell, S. 1970. The desert pupfish. Cry California 5:2-13.

California Department of Water Resources. 1964. Groundwater occurrence and quality, Lahontan region. Bulletin 106-1.

Deacon, J. E. 1979. Endangered and threatened fishes of the west. Great Basin Naturalist Memoirs 3:41-64.

Ehrenfeld, D. W. 1976. The conservation of non-resources. American Scientist 64:648-656.

Hubbs, C. L. and R. R. Miller. 1948. Correlation between fish distribution and hydrographic history in the desert basins of western United States. In: The Great Basin, with Emphasis on Glacial and Postglacial Times. Bulletin of the University of Utah 38:17-166.

Hubbs, C. L., R. R. Miller and L. C. Hubbs. 1974. Hydrographic history and relict fishes of the North-Central Great Basin. California Academy of Sciences, Memoirs, Volume 7.

Kennedy, C. H. 1916. A possible enemy of the mosquito. California Fish and Game 2:179-182.

LaBounty, J. F. and J. E. Deacon. 1972. Cyprinodon milleri, a new species of pupfish (family Cyprinodontidae) from Death Valley, California. Copeia 1972:769-780.

Leopold, A. 1949. A Sand County Almanac. Oxford University Press, New York.

Leopold, A. 1953. Round River. Oxford University Press, New York.

Mehringer, P. J. 1977. Great Basin late quaternary environments and chronology. In: Don D. Fowler (Ed.), Models and Great Basin Prehistory: a Symposium. Desert Research Institute Publications in Social Sciences, No. 12, pp. 13-16.

Miller, R. R. 1948. The cyprinodont fishes of the Death Valley system of eastern California and southwestern Nevada. Miscellaneous Publications of the Museum of Zoology, University of Michigan 68: 1-155.

Miller, R. R. 1961. Man and the changing fish fauna of the American southwest. Papers of the Michigan Academy of Science, Arts and Letters 46:365-404.

Miller, R. R. 1969. Conservation of fishes of the Death Valley system in California and Nevada. Transactions of the California-Nevada Section of the Wildlife Society 1969:107-122.

Miller, R. R. 1973. Two new fishes, Gila bicolor snyderi and Catostomus fumeiventris, from the Owens River basin, California. Occasional Papers of the Museum of Zoology, University of Michigan 667:1-19.

Miller, R. R. and E. P. Pister. 1971. Management of the Owens pupfish, Cyprinodon radiosus, in Mono County, California. Transactions of the American Fisheries Society 100:502-509.

Minckley, W. L. 1973. Fishes of Arizona. Arizona Game and Fish Department, Phoenix.

Minckley, W. L. 1979. Aquatic habitats and fishes of the lower Colorado River, southwestern United States. U.S. Department of the Interior, Bureau of Reclamation.

Minckley, W. L. and J. E. Deacon. 1968. Southwestern fishes and the enigma of "endangered species." Science 159:1424-1432.

Minckley, W. L., J. N. Rinne and J. E. Johnson. 1977. Status of the Gila topminnow and its co-occurrence with mosquitofish. U. S. Forest Service Research Paper RM-198.

Myers, G. S. 1965. Gambusia, the fish destroyer. Tropical Fish Hobbyist 13:31-32, 53-54.

Naiman, R. J. 1976. Productivity of a herbivorous pupfish population (Cyprinodon nevadensis) in a warm desert stream. Journal of Fish Biology 9:125-137.

Naiman, R. J., S. D. Gerking and T. D. Ratcliff. 1973. Thermal environment of a Death Valley pupfish. Copeia 1973:366-369.

Naiman, R. J., S. D. Gerking and R. D. Stuart. 1976. Osmoregulation in the Death Valley pupfish Cyprinodon milleri (Pisces: Cyprinodontidae). Copeia 1976:807-810.

Nevada State Engineer's Office. 1971. Water supply for the future in southern Nevada. Prepared by Montgomery Engineers of Nevada. Nevada State Engineer's Office, Carson City.

Nibley, H. W. 1978. On subduing the earth. In: Nibley on the Timely and the Timeless. Brigham Young University Press, Provo, pp. 85-99.

Pister, E. P. 1970. The rare and endangered fishes of the Death Valley system--a summary of the proceedings of a symposium relating to their protection and preservation. California Department of Fish and Game, Sacramento.

Pister, E. P. 1971. The rare and endangered fishes of the Death Valley system--a summary of the proceedings of the second annual symposium relating to their protection and preservation. Desert Fishes Council.

Pister, E. P. 1974. Desert fishes and their habitats. Transactions of the American Fisheries Society 103:531-540.

Pister, E. P. 1976. A rationale for the management of nongame fish and wildlife. Fisheries 1:11-14.

Pister, E. P. 1979. Endangered species:costs and benefits. Environmental Ethics 1:341-352.

Smith, W. C. 1960. Borax and borates. In: Industrial Minerals and Rocks. American Institute of Mining and Metallurgical Engineers, New York, pp. 103-118.

Soltz, D. L. 1979. Our disappearing desert fishes. Nature Conservancy News 29:8-12.

Soltz, D. L. and R. J. Naiman. 1978. The Natural History of Native Fishes in the Death Valley System. Natural History Museum of Los Angeles County, Science Series 30:1-76.

U. S. Department of the Interior. 1971. Status of the desert pupfish. A task force report. U. S. Department of the Interior, Fish and Wildlife Service.

U. S. Department of the Interior. 1980. List of endangered and threatened wildlife and plants. Federal Register 45(99).

U. S. General Accounting Office. 1977. The Tennessee Valley Authority's Tellico Dam project--costs, alternatives, and benefits. Report to the Congress by the Comptroller General of the United States, October 14, 1977.

Warren, C. 1978. Our economics is too small. Unpublished keynote address delivered at joint conference of Western Association of Fish and Wildlife Agencies and Western Division American Fisheries Society, San Diego, California, July 18, 1978.

Wilke, P. J. and H. W. Lawton (Eds.) 1976. The Expedition of Captain J. W. Davidson from Fort Tejon to the Owens Valley in 1859. Ballena Press, Socorro, NM.

Williams, J. E. 1977. Observations on the status of the Devil's Hole pupfish in the Hoover Dam refugium. U.S. Bureau of Reclamation Report No. REC-ERC-77-11.

Wilson, E. O. 1980. Resolutions for the 80s. The Harvard Magazine, January-February, 1980. Harvard University, Cambridge, MA.

14 Threatened Desert Fishes and the Endangered Species Act

JAMES D. WILLIAMS
Fish and Wildlife Service
Washington, D.C.

ABSTRACT

The Endangered Species Act of 1973 is one of the strongest laws ever enacted by the Congress to protect threatened fish, wildlife, and plants. Amendments to the Act, passed by the Congress in 1978, resulted in major changes to several sections. The impact of the 1978 amendments on the listing process, delineation of critical habitat, and consultation process is discussed. The 1966 and 1969 endangered species legislation, important precursors of the 1973 Act, are briefly examined.

The efforts of state and federal agencies and conservation and professional organizations in compiling the first lists of endangered and threatened desert fishes is reviewed. These early lists were important in the development of the present list of threatened desert fishes and the protection of habitat before the passage of the 1973 Act. The 1966, 1973, and 1980 Department of the Interior list of endangered and threatened fishes of the United States and a comprehensive list of threatened desert fishes of the western United States are also presented.

The task of habitat protection and recovery of threatened desert fishes is a cooperative effort between state and federal agencies and conservation organizations. State and federal cooperative agreements and the establishment of recovery teams, as provided by the Act, are instrumental in the conservation of desert fishes.

INTRODUCTION

The Endangered Species Act of 1973 was signed into law by the President on December 28, 1973. This act was the strongest legislation ever enacted to protect and preserve endangered and threatened animals and plants. It expanded coverage of two preceding laws: the Endangered Species Preservation Act of 1966 and the Endangered Species Conservation Act of 1969. These three endangered species statutes passed the Congress over the last 15 years are the culmination of conservationist efforts that began around 1900. Early wildlife protection laws were aimed primarily at the problem of excessive killing and commercialization of terrestrial wildlife. Although the problem of habitat deterioration was realized at that time, it was overshadowed by the excessive killing and selling of wildlife to the point that it was rarely discussed. Publications such as Our Vanishing Wildlife (Hornaday 1913), and Conserving Endangered Wildlife Species (Jackson 1945) and references therein served to alert the public to the scope of the problem and recommended courses of action to prevent extinction of wildlife.

ENDANGERED SPECIES LEGISLATION

The Acts

The 1966 and 1969 Acts. The Endangered Species Preservation Act of 1966 called for the establishment of a list of endangered species by the Secretary of the Interior. It permitted the use of land and water conservation funds to acquire habitat for their protection. The Fish and Wildlife Service was also authorized to expend funds for the management of listed endangered species. The most serious shortcomings of the 1966 Act was the absence of federal prohibitions against trading in and taking endangered species and the lack of provisions for habitat protection.

Three years later the Congress passed the Endangered Species Conservation Act of 1969, which called for the establishment of two lists, one for foreign and the other for native species. The 1969 Act prohibited importation of these species except under special permits. It also banned the purchase or sale of any animal illegally taken. Like the earlier legislation, it provided for the acquisition of habitat and the use of funds for the protection and management of these species. It also lacked provisions for habitat protection.

The 1973 Endangered Species Act and Amendments. In drafting the 1973 Endangered Species Act (ESA) the Congress expanded the ideas incorporated in the 1966 and 1969 Acts and added several new sections. They also split the jurisdiction between the departments of Commerce and the Interior but made Interior's Fish and Wildlife Service generally responsible for terrestrial species (including inland and freshwater species). It also gave Commerce's National Marine Fishery Service general jurisdiction over marine species. The following is a brief summary of some of the provisions pertinent to the conservation of desert fishes. Bean (1977) presents a comprehensive analysis of the 1973 Act and preceding endangered species legislation.

In passing the ESA of 1973 the Congress noted that because of man's activities some species of wildlife had become extinct and others were threatened by it. They also recognized that endangered species have educational, scientific, recreational, historical, and esthetic values and should be preserved as part of the nation's natural heritage. The Congress found that endangered and threatened wildlife needed protection on a worldwide basis and provided a tool, within the ESA, to implement international treaties and conventions. It was realized that states and other interested parties would have to be part of any successful endangered species program therefore financial assistance and other incentives were provided to those participating.

The Congress stated that the purposes of the Act were to conserve the ecosystems on which endangered and threatened species depend and to provide a program for their conservation. The Act also ensures that the United States will live up to the international treaties and conventions on conservation to which it is a party. Finally, Congress declared it their policy that all federal departments and agencies should seek to conserve endangered and threatened species and should use their authority to further the provisions of the Act.

Unlike earlier legislation, the 1973 Act established two conservation status categories: endangered and threatened. In addition to fish and wildlife it provided protection to plants and all invertebrates, not just mollusks. An *endangered* species was defined as any species in danger of extinction throughout all or a significant portion of its range, whereas a *threatened* species is one that is likely to become an endangered species throughout all or a significant portion of its range within the foreseeable future. The term "species" was redefined in the 1978 amendments to exclude populations of invertebrates and now reads "any subspecies of fish or wildlife or plants, and any distinct population segment of any species of vertebrate fish or wildlife which interbreeds when mature."

A total of 39 species and subspecies of native and foreign fishes were designated as endangered under the 1966 and 1969 legislation. After passage of the 1973 Act these fishes were carried forward and given protection as endangered species. Under Section 4 of the

1973 Act the determination of a species as endangered or threatened is based on one or more of the following factors:

1. The present or threatened destruction, modification, or curtailment of its habitat or range.
2. Utilization for commercial, sporting, scientific, or educational purposes at levels that detrimentally affect it.
3. Disease or predation.
4. Absence of regulatory mechanisms adequate to prevent the decline of a species or degradation of its habitat.
5. Other natural or man-made factors affecting its continued existence.

When species are classified as threatened, regulations necessary for their protection and management may be issued by the Secretary. It may not be necessary, however, to regulate some species. All listing and related regulations proposed by the Secretary are published in the *Federal Register* for review and comments by appropriate state and federal agencies, interested persons, and organizations.

Critical Habitat

The 1978 and 1979 ESA amendments increased the number of steps in the process of listing endangered and threatened species significantly. Briefly, the new requirements included delineation of critical habitat at the time of listing, publication of a proposal summary in the local newspaper, scheduling of public meetings and (when requested) public hearings, an examination of the economic and other impacts of any critical habitat determination, notification of local government officials, and offering the proposal for publication in scientific journals. Overall, these new requirements have significantly increased the cost and more than doubled the time and effort required to list a species as endangered or threatened and determine its critical habitat. Final Section 4 regulations for listing were published February 27, 1980, in the *Federal Register*, Vol. 45, No. 40, pp. 13010-13026.

The concept of critical habitat was first put forth in the 1973 Act but was not defined. In an effort to clarify and standardize the use of the term, the FWS published its definition in the *Federal Register* in 1975. The Congress further clarified the term in the 1978 amendments, defining it as

> the specific areas within the geographical area occupied by the species, at the time it is listed in accordance with the provisions of Section 4 of this Act, on which are found those physical or biological features (I) essential to the conservation of the species and (II) which may require special management considerations or protection; and specific areas outside the geographical area occupied by the species at the time it is listed in accordance with the provisions of Section 4 of this Act, upon a determination by the Secretary that such areas are essential for the conservation of the species.

The amendments further clarified the limits of critical habitat, stating that "except in those circumstances determined by the Secretary, critical habitat shall not include the entire geographical area which can be occupied by the threatened or endangered species."

State Cooperation

A totally new provision in the 1973 ESA strongly supports cooperation with states in the establishment of a state endangered species program that provides for management and cooperative agreements. Cooperative agreements also provide for Federal assistance to the states for implementation of state endangered species and threatened species programs. For a state to be eligible for a cooperative agreement with the Fish and Wildlife Service, the state agency must have the following:

1. Authority to conserve species that have been determined by the state or the FWS to be endangered or

threatened. This authority should be broad enough to cover additional species that may be listed in the future.

2. Acceptable conservation programs for all resident fish or wildlife species in the state that are deemed endangered or threatened by the FWS.

3. Authority to conduct investigations.

4. Authority to acquire land or aquatic habitats for conservation of resident endangered and threatened species.

5. Provisions for public participation in designating resident endangered or threatened species of fish and wildlife.

Under this matching fund program the federal share shall not exceed two-thirds of the estimated program costs; however, this share can be increased to 75 percent for conservation efforts shared by two or more states.

The 1978 amendments relaxed the cooperative agreement requirements to the extent that to qualify for an agreement, states can now develop conservation programs for selected or priority species which the Secretary of Interior and the state agree are in need of urgent conservation programs. Under the original Act, states were required to have authority to protect and conserve all endangered species added to the federal list which occurred under their jurisdiction.

Section 7

Much of the protection provided endangered species by the 1973 Act was centered under Section 7, which charges the Secretary to review all Department of the Interior programs and to use them to further the Act. All other federal agencies in consultation with the Secretary and with his assistance were to use their authority to implement the Act by carrying out programs for conservation of endangered and threatened species. These agencies were also to ensure that actions authorized, funded, or carried out by them did not jeopardize the continued existence of these species or result in the destruction or adverse modification of critical habitat.

While maintaining the thrust of Section 7, the 1978 and 1979 amendments did make substantial changes. These changes, briefly summarized here, were explained in detail by Finnley (1978, 1980). The 1978 amendments strengthened the consultation process with other federal agencies by expediting (90-day limit) the consultation process, requiring jeopardy biological opinion to suggest reasonable and prudent alternatives that would avoid injury to the species and adverse modification of its critical habitat, and requiring federal agencies to prepare a biological assessment for listed species that occur in the area in which a Federal project is planned. The most drastic change in Section 7 is the addition of a two-tiered exemption process. This complicated process provides for the establishment of a review board and cabinet-level committee and authorizes them under specified conditions to exempt federal agencies from compliance with some of the Act's protective provisions. To date the exemption process has been used only twice; one review involved the snail darter (*Percina tanasi*).

LIST OF ENDANGERED FISHES

Early Efforts

Early scientific and endangered species conservation literature (pre-1950) contains little to no information on threatened fishes and other aquatic species. The reports in the fishery literature are numerous, dating back to the late 1800s, on the impact of pollution and dams on the sport and commercial fisheries of the United States. They generally describe the destruction that resulted from pollution and disruption of spawning runs by dams blocking stream channels. Later, Lachner (1956) discussed the impact of man's activities on the upper Ohio River and the changes in the native fish fauna in the eastern United States. Among the earliest and most comprehensive reports on the decline of western fishes were the publications, *Man and the Changing Fish Fauna of the American Southwest* (Miller 1961) and *Extinct, Rare and Endangered*

American Freshwater Fishes (Miller 1964). Miller (1961) reported habitat destruction by the physical alteration of streams and their associated riparian vegetation, depletion of ground water, mining operations, and the introduction of nonnative fishes. The second article by Miller (1964) reports 38 fishes as endangered or in need of protection, 21 of which were "urgently threatened." All of these were native to the arid lands of the western United States and Mexico except one, the Maryland darter. Miller (1963) reported on the impact of chemicals used for the removal or reduction of "undesirable" fishes as part of a fishery management operation. His example was the Green River poison operation which took place in September of 1962 in Wyoming, Utah, and Colorado before the flooding of Flaming George Reservoir.

The concern shown by the American Society of Ichthyologists and Herpetologists during the 1950s culminated in the formation of a Committee on Fish Conservation in 1961. This committee immediately began to list the most endangered fishes of the United States. During the first year they identified more than 65 species in need of protection and by 1963 had expanded the list to approximately 100 species and subspecies of fishes from the United States, Mexico, and Canada (Miller 1964). The Endangered Species Committee of the American Fishery Society (AFS) joined with the committee on fish conservation to compile a single list of "endangered" fishes. The list (Threatened Freshwater Fishes of the U.S.) prepared by these committees (Miller 1972) included species threatened at the state and national levels. The Endangered Species Committee of the AFS recently published a comprehensive list of threatened fishes of North America (Deacon et al. 1979).

In 1969 the International Union for the Conservation of Nature (IUCN) released its first set of looseleaf Red Data Book status reports for fishes (Volume IV), compiled by Miller (1969). Additional status report sheets for fish were released by IUCN in 1971. Among the 91 fishes reported in that Red Data Book, 58 species are native to the United States. The remaining 33 species are foreign. Thirty-eight species of the 58 native fishes inhabited desert waters. The completely revised IUCN Red Data Book also compiled by Miller (1977) contains conservation status reports on 194 fishes throughout the world. Included in the 194 species are 120 native fishes, 66 species of which inhabit aquatic systems of the arid western United States.

Federal and State Efforts

The data base for the first list of endangered fishes was compiled for the U.S. Bureau of Sport Fisheries and Wildlife (presently the U.S. Fish and Wildlife Service) by the Committee on Rare and Endangered Wildlife Species, appointed by the Director in 1963. This data base was released by the Department of the Interior in 1966 in a publication entitled "Rare and Endangered Fish and Wildlife of the United States," often referred to as the "U.S. Redbook." The first edition, limited to vertebrate animals, contained data sheets on 38 species and subspecies of fishes considered rare or endangered (Table 1). In a separate list without data sheets 45 fishes were entered as status undetermined, four as peripheral to the United States and six as extinct. The 93 species and subspecies of fishes listed in the various categories were fairly evenly distributed between southwestern and southeastern states, with a few in the northeast and Great Lakes areas. Rare and Endangered Fish and Wildlife of the United States (1966) was revised, updated, and reissued in 1968, although no additions or revisions were reflected in the fish section.

The next U.S. Redbook, issued by the Department of the Interior, Fish and Wildlife Service (1973), was entitled "Threatened Wildlife of the United States" and contained a list of 100 species and subspecies of fishes, with data sheets for 55 native fishes (Table 1). An additional 35 species and subspecies were listed as "status undetermined," four as peripheral, and six as extinct. Some species listed as rare and endangered in the 1966 edition were dropped from consideration in the 1973 edition. Other (approximately 20 species and subspecies) fishes listed as status undetermined or not listed at all in the 1966 U.S.

Redbook were listed with a brief account of their status in the 1973 U.S. Redbook.

During the early 1970s the development of a state endangered species list was a major contribution toward the identification of endangered fishes. California was the first state to compile and publish a list. Their 1972 publication _At the Crossroads_ was an excellent example that was followed by other states during the mid- to late 1970s. Today more than half the states have publications that describe their endangered species. A few outstanding state publications on fishes are those by Ramsey (1976) for Alabama, Bailey (1977) for North Carolina, Gilbert (1978) for Florida, Schmitt (1978) for New Mexico, Nordstrom et al. (1977) for Missouri, Jenkins (1979) for Virginia, Starnes and Etnier (1980) for Tennessee and _At the Crossroads_ 1978 by California Department of Fish and Game.

There is at present a total of 57 species and subspecies of fishes on the official Federal List of Endangered and Threatened Species. Among the 57 listed, 11 species are foreign and the remaining 46 are native fishes (Table 1). Thirty-four of the 46 are classified as endangered and 12 are threatened. A breakdown by distribution reveals that 31 (67 percent; those preceded by an asterisk in Table 1) of the 46 species are desert fishes, 26 of which are endangered and five are threatened.

Current Lists

The most recent effort at compiling a list of endangered fishes on a national basis was carried out by the Endangered Species Committee of the American Fisheries Society (See Table 2). Fishes included in this list were placed in one of three conservation status categories on a national basis: endangered, threatened, or special concern. Definition of these terms were as follows: (1) _endangered_, any species or subspecies in danger of extinction throughout all or a significant portion of its range; (2) _threatened_, any species or subspecies likely to become an endangered species within the foreseeable future throughout all or a significant portion of its range; (3) _special_

TABLE 1. Department of the Interior 1966, 1973, and 1980, List of Endangered and Threatened Fishes of the United States. Asterisk indicates that species occur in the desert regions of the United States.

Common Name	Scientific Name	1966 List	1973 List	1980 Official Federal List
	Sturgeons: Acipenseridae			
Shortnose sturgeon	*Acipenser brevirostrum*	X	X	X
Lake sturgeon	*Acipenser fulvescens*	X	X	
Atlantic sturgeon	*Acipenser oxyrhynchus*	X		
	Trouts: Salmonidae			
Longjaw cisco	*Coregonus alpenae*	X	X	X
Deepwater cisco	*Coregonus johannae*	X	X	
Blackfin cisco	*Coregonus nigripinnis*	X	X	
*Arizona trout	*Salmo apache*	X	X	X
*Little Kern golden trout	*Salmo aguabonita whitei*		X	X
*Lahontan cutthroat trout	*Salmo clarki henshawi*	X	X	X
*Paiute cutthroat trout	*Salmo clarki seleniris*	X	X	X
*Greenback cutthroat trout	*Salmo clarki stomias*	X	X	X
*Rio Grande cutthroat trout	*Salmo clarki virginalis*		X	
*Humboldt cutthroat trout	*Salmo clarki subsp.*		X	

*Montana Westslope cutthroat trout	_Salmo clarki subsp._	x		
*Gila trout	_Salmo gilae_	x	x	x
Atlantic salmon	_Salmo salar_	x		
Sunapee trout	_Salvelinus aureolus_	x	x	
Blueback trout	_Salvelinus oquassa_	x	x	
Arctic grayling	_Thymallus arcticus_	x	x	
	Mudminnow: Umbridae			
Olympic mudminnow	_Novumbra hubbsi_	x	x	
	Minnows: Cyprinidae			
*Desert dace	_Eremichthys acros_	x	x	
*Borax Lake Chub	_Gila boraxobius_			x
*Humpback chub	_Gila cypha_	x	x	x
*Bonytail chub	_Gila elegans_			x
*Mohave chub	_Gila mohavensis_		x	x
*Pahranagat bonytail	_Gila robusta jordani_		x	x
Slender chub	_Hybopsis cahni_			x
Spotfin chub	_Hybopsis monacha_			x
*Little Colorado spinedace	_Lepidomeda vittata_	x	x	
*Moapa dace	_Moapa coriacea_	x	x	x
*Woundfin	_Plagopterus argentissimus_		x	x

(Continued)

Table 1. (Continued)

Common Name	Scientific Name	1966 List	1973 List	1980 Official Federal List
*Colorado River squawfish	_Ptychocheilus lucius_	x	x	x
*Kendall Warm Springs dace	_Rhinichthys osculus thermalis_		x	x
*White River (Mtn.) sucker	_Catostomus clarki intermedius_		x	
	Suckers: Catostomidae			
*Modoc sucker	_Catostomus microps_		x	
*Cui-ui	_Chasmistes cujus_	x	x	x
	Catfishes: Ictaluridae			
Yellowfin madtom	_Noturus flavipinnis_			x
Scioto Madtom	_Noturus trautmani_			x
	Cavefishes: Amblyopsidae			
Ozark cavefish	_Amblyopsis rosae_	x	x	
Alabama cavefish	_Speoplatyrhinus poulsoni_			x
	Killifishes: Cyprinodontidae			
*Leon Springs pupfish	_Cyprinodon bovinus_			x
*Devil's Hole pupfish	_Cyprinodon diabolis_	x	x	x

*Comanche Spring pupfish	*Cyprinodon elegans*	x	x	x
*Tecopa pupfish	*Cyprinodon nevadensis calidae*		x	
*Nevada pupfish	*Cyprinodon nevadensis mionectes*		x	
*Warm Springs pupfish	*Cyprinodon nevadensis pectoralis*		x	x
*Owens River pupfish	*Cyprinodon radiosus*	x	x	x
*Pahrump killifish	*Empetrichthys latos*	x	x	x
	Livebearers: Poecilidae			
*Goodenough gambusia	*Gambusia amistadensis*			x
*Big Bend gambusia	*Gambusia gaigei*	x	x	x
*San Marcos gambusia	*Gambusia georgei*		x	x
*Clear Creek gambusia	*Gambusia heterochir*	x	x	x
*Pecos gambusia	*Gambusia nobilis*		x	x
*Gila topminnow	*Poeciliopsis occidentalis*	x	x	x
	Sticklebacks: Gasterosteidae			
*Unarmored threespine stickleback	*Gasterosterus aculeatus williamsoni*		x	x
	Sunfishes: Centrarchidae			
Roanoke bass	*Ambloplites cavifrons*		x	

(Continued)

Table 1. (Continued)

Common Name	Scientific Name	1966 List	1973 List	1980 Official Federal List
Suwannee bass	*Micropterus notius*	x	x	
Sharphead darter	*Etheostoma acuticeps*	x	x	
Slackwater darter	*Etheostoma boschungi*			x
	Perches: Percidae			
*Fountain darter	*Etheostoma fonticola*		x	x
Niangua darter	*Etheostoma nianguae*	x	x	
Watercress darter	*Etheostoma nuchale*		x	x
Okaloosa darter	*Etheostoma okaloosae*		x	x
Bayou darter	*Etheostoma rubrum*			x
Maryland darter	*Etheostoma sellare*	x	x	x
Trispot darter	*Etheostoma trisella*	x	x	
Tuscumbia darter	*Etheostoma tuscumbia*	x	x	
Leopard darter	*Percina pantherina*			x
Snail darter	*Percina tanasi*			x
Blue pike	*Stizostedion vitreum glaucum*	x	x	x
	Sculpins: Cottidae			
Pygmy sculpin	*Cottus pygmaeus*		x	
Total number of species		38	55	46

concern, species or subspecies that could become endangered by relatively minor disturbances to their habitat or that require additional information to determine their status. Species or subspecies threatened only at the state or local level were not included in this list; existing state lists, however, were instrumental in all stages of development of the AFS list.

The AFS report (Deacon et al. 1979) lists 251 species and subspecies of fishes from the United States, Canada, and Mexico. Two hundred species and subspecies (80 percent) are found in the United States and the remaining 51 species (20 percent) are distributed in Canada and Mexico. Among the 200 fishes of the United States 50 species (25 percent) were listed as endangered, 89 species and subspecies (44 percent) as threatened, and 61 species and subspecies (31 percent) as special concern. The species and subspecies in the western United States are listed in Table 2.

PROTECTION AND RECOVERY OF ENDANGERED FISHES

Fish and Wildlife Service Activities

Having identified and prepared lists of endangered and threatened fishes and having developed species priorities, the big task is to effect the recovery of these fishes to the point of which they will again become viable components of their ecosystems. To accomplish this monumental task the U.S. Fish and Wildlife Service has appointed recovery teams to develop individual species recovery plans. These plans are prepared by the team (or under contract) on a cooperative effort with State and Federal agencies and individuals who have a responsibility for an interest in the species. As of January 1981 there was a total of 40 approved recovery plans, 13 of which are for fishes. Eleven are for desert fishes: Arizona trout (*Salmo apache*); Gila trout (*Salmo gilae*); greenback cutthroat trout (*Salmo clarki stomias*); humpback chub (*Gila cypha*); woundfin (*Plagopterus argentissimus*); Colorado River squawfish (*Ptychocheilus lucius*); Cui-ui (*Chasmistes cujus*); Devils Hole pupfish (*Cyprinodon*

TABLE 2. Threatened Desert Fishes of the Western United States: 1980[a]

Historical Common Name	Scientific Name	Status[b]	Threat[c]	Present Distribution
	Trouts: Salmonidae			
Little Kern golden trout	*Salmo aquabonita whitei* Jordan	T	1,4	CA
Arizona trout	*Salmo apache* Miller	T	1,4	AZ
Lahontan cutthroat trout	*Salmo clarki henshawi* Gill and Jordan	T	1,4	CA, NV, UT, WA
Colorado River cutthroat	*Salmo clarki pleuriticus* Cope	SC	1,4	CO, UT, WY
Paiute cutthroat trout	*Salmo clarki seleniris* Snyder	T	1	CA
Greenback cutthroat trout	*Salmo clarki stomias* Cope	T	1,4	CO
Utah cutthroat trout	*Salmo clarki utah* Suckley	T	1,4	UT, WY, NV
Rio Grande cutthroat trout	*Salmo clarki virginalis* (Girard)	SC	1,4	CO, NM
Alvord cutthroat trout	*Salmo clarki* ssp.	SC	1	OR
Humboldt cutthroat trout	*Salmo clarki* ssp.	SC	1	NV
Gila trout	*Salmo gilae* Miller	T	1	NM, AZ
Redband trout	*Salmo* sp.	SC	1,4	CA, OR, ID, NV
Montana Arctic grayling (stream form)	*Thymallus arcticus montanus* (Pallas)	T	1	MT
	Minnow: Cyprinidae			
Mexican stoneroller	*Campostoma ornatum* Girard	SC	1,3	AZ, TX, Mexico
Devils River minnow	*Dionda diaboli* Hubbs and Brown	T	1	TX
Desert dace	*Eremichthys acros* Hubbs and Miller	T	1,5	NV
Alvord chub	*Gila alvordensis* Hubbs and Miller	SC	1	NV, OR

Fish Creek Springs tui chub	*Gila bicolor euchila* Hubbs and Miller	E	1,4,5	NV
Independence Valley tui chub	*Gila bicolor isolata* Hubbs and Miller	T	1,4,5	NV
Mohave tui chub	*Gila bicolor mohavensis* (Snyder)	E	1,4	CA
Newark Valley tui chub	*Gila bicolor newarkensis* Hubbs and Miller	SC	1,5	NV
Oregon Lakes tui chub	*Gila bicolor oregonensis* (Snyder)	SC	1	OR
Lahontan tui chub	*Gila bicolor obesa* (Girard)	SC	1	NV
Owens tui chub	*Gila bicolor snyderi* Miller	E	1,4,5	CA
Catlow tui chub	*Gila bicolor* ssp.	SC	1	OR
Sheldon tui chub	*Gila bicolor* ssp.	SC	5	NV
Cowhead Lake tui chub	*Gila bicolor* ssp.	SC	1	CA
Hutton Spring tui chub	*Gila bicolor* ssp.	T	1	OR
Borax Lake chub	*Gila boraxobius*	T	1,5	OR
Thicktail chub	*Gila crassicauda* (Baird and Girard)	E	1	CA
Humpback chub	*Gila cypha* Miller	E	1	AZ, CO, UT, WY
Bonytail	*Gila elegans* Baird and Girard	E	1,4	AZ, CA, CO, NV, UT, WY
Gila chub	*Gila intermedia* (Girard)	SC	1,4	AZ, NM
Chihuahua chub	*Gila nigrescens* (Girard)	E	1,4	NM, Mexico
Yaqui chub	*Gila purpurea* (Girard)	E	1,4	AZ, Mexico
Gila roundtail chub	*Gila robusta grahami* Baird and Girard	T	1,4	AZ, NM
Pahranagat roundtail chub	*Gila robusta jordani* Tanner	E	1,4	NV
Virgin River roundtail	*Gila robusta seminuda* Cope	E	1	AZ, NV, UT
Oregon chub	*Hybopsis crameri* Snyder	SC	1,4	OR
Least chub	*Iotichthys phlegethontis* (Cope)	T	1,4	UT

(Continued)

Table 2. (Continued)

Historical Common Name	Scientific Name	Status[b]	Threat[c]	Present Distribution
White River spinedace	Lepidomeda albivallis Miller and Hubbs	T	1,4	NV
Virgin spinedace	Lepidomeda mollispinis mollispinis Miller and Hubbs	T	1,4	AZ, UT
Big Spring spinedace	Lepidomeda mollispinis pratensis Miller and Hubbs	E	1,4,5	NV
Little Colorado spinedace	Lepidomeda vittata Cope	SC	1	AZ
Spikedace	Meda fulgida Girard	T	1,4	AZ, NM
Moapa dace	Moapa coriacea Hubbs and Miller	E	1,3,4,5	NV
Yaqui beautiful shiner	Notropis formosus mearnsi Snyder	SC	1	AZ, Mexico
Rio Grande shiner	Notropis jemezanus (Cope)	SC	1,4	NM
Proserpine shiner	Notropis proserpinus (Girard)	T	1	TX
Bluntnose shiner	Notropis simus (Cope)	E	1,4	NM, TX
Woundfin	Plagopterus argentissimus Cope	E	1	AZ, NV, UT
Splittail	Pogonichthys macrolepidotus (Ayres)	SC	1	CA
Colorado squawfish	Ptychocheilus lucius Girard	E	1,3,4	AZ, CO, UT, CA NM, NV, WY
Relict dace	Relictus solitarius Hubbs and Miller	SC	1	NV
Independence Valley speckled dace	Rhinichthys osculus lethoporus Hubbs and Miller	E	1,4,5	NV
Ash Meadows speckled dace	Rhinichthys osculus nevadensis Gilbert	E	1,4	NV
Clover Valley speckled dace	Rhinichthys osculus oligoporus Hubbs and Miller	E	1,4,5	NV

Kendall Warm Springs dace	Rhinichthys osculus thermalis Hubbs and Kuehne	SC	5	WY
Moapa speckled dace	Rhinichthys osculus moapae Williams	T	1,3,4	NV
Foskett Spring speckled dace	Rhinichthys osculus ssp.	T	1,5	OR
Loach minnow	Tiaroga cobitis Girard	SC	1,4	AZ, NM
	Suckers: Catostomidae			
Yaqui sucker	Catostomus bernardini Girard	SC	1	AZ, Mexico
White River desert sucker	Catostomus clarki intermedius (Tanner)	T	1	NV
Webug sucker	Catostomus fecundus Cope and Yarrow	SC	1,4	UT
Zuni bluehead sucker	Catostomus dicobolus yarrowi Cope	T	1	NM
Lost River sucker	Catostomus luxatus (Cope)	SC	1,4	CA, OR
Modoc sucker	Castostomus microps Rutter	E	1,4	CA
Warner sucker	Castostomus warnerensis Snyder	E	1,4	OR
Shortnose sucker	Chasmistes brevirostris Cope	T	1,4	CA, OR
Cui-ui	Chasmistes cujus Cope	E	1	NV
June sucker	Chasmistes liorus Jordan	SC	1,4	UT
Razorback sucker	Xyrauchen texanus (Abbott)	T	1,4	AZ, CA, CO, NV UT, WY
	Freshwater Catfishes: Ictaluridae			
Yaqui catfish	Ictalurus pricei (Rutter)	SC	1	AZ, Mexico
Widemouth blindcat	Satan eurystomus Hubbs and Bailey	T	1	TX

(Continued)

Table 2. (Continued)

Historical Common Name	Scientific Name	Status[b]	Threat[c]	Present Distribution
Toothless blindcat	*Trogloglanis pattersoni* Eigenmann	T	1	TX
	Killifishes: Cyprinodontidae			
Preston White River springfish	*Crenichthys baileyi* ssp.	T	4,5	NV
Southern White River springfish	*Crenichthys baileyi* ssp.	T	1,3,4	NV
Warm Springs White River springfish	*Crenichthys baileyi* ssp.	SC	1,4,5	NV
Railroad Valley springfish	*Crenichthys nevadae* Hubbs	SC	1	NV
Leon Springs pupfish	*Cyprinodon bovinus* Baird and Girard	E	1,4,5	TX
Devils Hole pupfish	*Cyprinodon diabolis* Wales	E	1,5	NV
Comanche Springs pupfish	*Cyprinodon elegans* Baird and Girard	E	1	TX
Devils River Conchos pupfish	*Cyprinodon eximius* spp.	T	1	TX
Gila desert pupfish	*Cyprinodon macularius macularius* Baird and Girard	T	1,4	AZ, Mexico
LeConte desert pupfish	*Cyprinodon macularius* ssp.	E	1,4	CA
Quitobaquite desert pupfish	*Cyprinodon macularius* ssp.	SC	1,5	AZ
Valley Amargosa pupfish	*Cyprinodon nevadensis amargosae* Miller	SC	1,4	CA

Ash Meadows Amargosa pupfish	*Cyprinodon nevadensis mionectes* Miller	E	1,4	NV
Warm Springs Amargosa pupfish	*Cyprinodon nevadensis pectoralis* Miller	E	1,4,5	NV
Owens pupfish	*Cyprinodon radiosus* Miller	E	1,4	CA
White Sands pupfish	*Cyprinodon tularosa* Miller and Echelle	SC	1,4,5	NM
Pahrump killifish	*Empetrichthys latos latos* Miller	E	1,4,5	NV
	Livebearers: Poecilidae			
Amistad gambusia	*Gambusia amistadensis* Peden	E	1,5	TX
Big Bend gambusia	*Gambusia gaigei* Hubbs	E	1,5	TX
San Marcos gambusia	*Gambusia georgei* Hubbs and Peden	E	1,4,5	TX
Clear Creek gambusia	*Gambusia heterochir* Hubbs	E	4	TX
Pecos gambusia	*Gambusia nobilis* (Baird and Girard)	SC	1,4	TX, NM
Gila topminnow	*Poeciliopsis occidentalis* (Baird and Girard)	E	1	AZ, NM
	Sticklebacks: Gasterosteidae			
Unarmored threespine stickleback	*Gasterosteus aculeatus williamsoni* Girard	E	1,4	CA
	Sunfishes: Centrarchidae			
Guadalupe bass	*Micropterus treculi* (Vaillant and Bocourt)	SC	1	TX

(Continued)

Table 2. (Continued)

Historical Common Name	Scientific Name	Status[b]	Threat[c]	Present Distribution
	Perches: Percidae			
Fountain darter	*Etheostoma fonticola* (Jordan and Gilbert)	E	1,5	TX
	Gobies: Gobiidae			
Tidewater goby	*Eucyclogobius newberryi* (Girard)	SC	1	CA
	Sculpins: Cottidae			
Rough sculpin	*Cottus asperrimus* Rutter	SC	1	CA
Malheur mottled sculpin	*Cottus bairdi* ssp.	SC	1	OR
Utah Lake sculpin	*Cottus echinatus* Bailey and Bond	E	1,4	UT
Shoshone sculpin	*Cottus greenei* (Gilbert and Culver)	T	1	ID

[a]Updated from Deacon et al. 1979.

[b]E = endangered, T = threatened, SC = special concern.

[c]Threats,
1 = Present or threatened destruction of habitat.
2 = Overutilization.
3 = Disease.
4 = Hybridization or competition with exotic or translocated species.
5 = Restricted natural range.

diabolis); Warm Springs pupfish (*Cyprinodon nevadensis pectoralis*); Pahrump killifish (*Empetrichythys latos*); and unarmored threespine stickleback (*Gasterosteus aculeatus williamsoni*). In addition to the approved plans, there are 24 draft recovery plans, five of which are for fishes and three [Moapa dace (*Moapa coriacea*); Clear Creek gambusia (*Gambusia heterochir*); and Yaqui topminnow (*Poeciliopsis occidentalis sonoriensis*)] are desert fishes. The implementation of these plans will be accomplished, in most cases, by a cooperative effort of State and Federal agencies.

Much of the protection and recovery of endangered species is carried out by states in cooperative agreements with the Fish and Wildlife Service. Since passage of the 1973 Act 35 states and one territory have entered into cooperative agreements. Eleven states (Arkansas, California, Colorado, Florida, Illinois, Maryland, Missouri, North Carolina, Tennessee, Utah, and Wisconsin) now have projects on 48 species of fishes which are on the federal list or are candidates for the endangered species list. Examples of the activities presently being carried out by states include studies on status, distribution, life history, ecology, habitat needs of these species, and evaluation of streams for possible reintroduction. Other projects include monitoring known populations, developing educational materials on endangered fishes and their habitats, implementing recovery plan tasks, and identifying essential habitat.

Like the 1966 and 1969 endangered species legislation, the 1973 Act authorizes the Secretary of the Interior to acquire habitat for endangered and threatened species. Through FY 1979 the FWS had acquired more than 70 thousand acres (for fish and wildlife) at a cost of almost 10 million dollars. Only one purchase of approximately 13 acres involved a desert fish (Moapa dace). Habitat acquisition for endangered species is one of the best means of protecton. This is rarely feasible for desert stream fishes because of the large watershed areas usually required for the protection of the stream habitat. Habitat acquisition, (land area and water rights) has proved to be effective in the protection of desert spring fishes.

Since its inception in 1970 the Desert Fishes Council has played a major role in encouraging the conservation of desert aquatic ecosystems and in supporting research effort and pertinent land use policy and legislation. The history and activities of the Desert Fishes Council are described by Pister (this volume).

Outlook For The Future

In view of the persistent efforts of federal agencies, states, conservation organizations, and individual biologists the future of endangered species looks promising. Development and increased consumption of natural resources will undoubtedly continue to pressure endangered fishes. Fortunately the mechanisms to react to and, in some cases, mitigate action that threatens these fishes are readily available. These mechanisms depend on basic biological data that are available for most species. By cooperative effort our knowledge of the life history and habitat requirements of endangered fishes is rapidly increasing. With better information our management and recovery efforts will be greatly enhanced.

Although our efforts to conserve endangered and threatened fishes have increased significantly in recent years, our strategy has changed very little. In fact, it is interesting to note that the National Wildlife Federation in a 1956 publication on *Our Endangered Wildlife* suggested the following courses of action:

1. Promote coordinated research on endangered forms of wildlife to determine methods of restoring them to safe population levels.
2. Encourage programs of federal and state agencies designed to protect endangered wildlife and restore natural environments. Habitat improvement and enforcement of laws can be greatly furthered by strong public support.
3. Stop the wanton and accidental slaughter of remnant populations of endangered animals and the destruction of habitat by unwise drainage and pollution of wetlands, burning and other abuses of forests and grasslands, and misuse of agricultural lands.

4. Work for the establishment of comprehensive use policies at the national, state, and local levels that will ensure protection and improvement of wildlife habitat in connection with the development of agricultural, mineralogical, and industrial resources.

5. Initiate educational campaigns to create public awareness of the plight of endangered wildlife. These are especially important at the local, state, and regional levels within the ranges of those animals that are on the verge of extinction.

6. Protect the integrity and defend from encroachment those state, federal, and private sanctuaries, refuges, parks, forests, wilderness, and management areas that are maintained for the benefit of endangered species. Forestall the invasion of primitive habitats essential to the welfare of those endangered species that are dependent on undeveloped areas for living space.

7. Work for effective pollution control for our streams, lakes, marshes, and coastal waters.

8. Support the efforts of the Survival Service Commission of the International Union for the Protection of Nature in Brussels, Belgium, in protecting endangered forms in North America and throughout the world.

Although progress has been made during the last 24 years since these courses of action were suggested, much remains to be done to save our endangered fishes in their natural habitats. If we continue in this direction with the determination and legal and political backing of recent years, I am sure we can recover our endangered and threatened fishes.

REFERENCES

Bailey, J. R. 1977. Freshwater Fishes. In: Endangered and Threatened Plants and Animals of North Carolina. North Carolina State Museum of Natural History, Raleigh, pp. 265-298.

Bean, M. 1977. The evolution of national wildlife law. Report to Council on Environmental Quality, Washington, D.C.

California Fish and Game. 1972. At the crossroads 1972: A report on California's endangered and rare fish and wildlife.

California Fish and Game. 1978. At the crossroads 1978: A report on California's endangered and rare fish and wildlife.

Deacon, J. E., G. Kobetich, J. D. Williams and S. Contreras. 1979. Fishes of North America, endangered, threatened or of special concern: 1979. Fisheries 4:29-44.

Finnley, D. 1978. President signs endangered species amendments. Endangered Species Technical Bulletin 3:1-11.

Finnley, D. 1980. Endangered species act extended and amended. Endangered Species Technical Bulletin 5:1-4.

Gilbert, C. R. 1978. Fishes. In: Rare and Endangered Biota of Florida. Vol. 4. University Presses of Florida.

Hornaday, W. T. 1913. Our Vanishing Wildlife: Its Extermination and Preservation. New York Zoological Society.

Jackson, H. T. 1945. Conserving endangered wildlife species. Annual Report Smithsonian Institution 1945, Washington, D.C., pp. 247-272.

Jenkins, R. E. and J. A. Musick. 1979. Freshwater and Marine Fishes. In: Endangered and Threatened Plants and Animals of Virginia. Center for Environmental Studies Virginia Polytechnic Institute and State University, pp. 319-367.

Lachner, E. A. 1956. The changing fish fauna of the Upper Ohio Basin. In: Man and the Waters of the Upper Ohio Basin. Pymatuning Laboratory. Field Biology, Special Publications 1:64-78.

Miller, R. R. 1961. Man and the changing fish fauna of the American Southwest. Papers of the Michigan Academy Science, Arts and Letters 46:365-404.

Miller, R. R. 1963. Is our underwater life worth saving? National Parks and Conservation Magazine 37:4-9.

Miller, R. R. 1964. Extinct, rare and endangered American freshwater fishes. In: The Protection of Vanishing Species. Proceedings International Congress of Zoology, Vol. 8, XVI, Washington, D.C., pp. 4-16.

Miller, R. R. 1969. Freshwater fishes. Red Data Book, Vol. 4: Pisces. International Union for the Conservation of Nature, Morges, Switzerland: 9 pp., 79 status sheets.

Miller, R. R. 1972. Threatened freshwater fishes of the United States. Transactions of the American Fisheries Society. 101:239-252.

Miller, R. R. 1977. Red Data Book, Vol. 4: Pisces. International Union for the Conservation of Nature, Morges, Switzerland.

National Wildlife Federation. 1956. Our Endangered Wildlife. National Wildlife Federation, Washington, D.C.

Nordstrom, G. R., W. L. Pflieger, K. C. Sadler and W. H. Louis. 1977. Rare and Endangered Species of Missouri. Missouri Department of Conservation and U.S. Department of Agriculture, Soil Conservation Service.

Ramsey, J. S. 1976. Freshwater fishes. In: Endangered and Threatened Plants and Animals of Alabama. Bulletin Alabama Museum of Natural History 2:53-65.

Schmitt, G. 1978. Fishes. In: Handbook of Species Endangered in New Mexico. New Mexico Department of Game and Fish, pp. E1-E59.

Starnes, W. C. and D. A. Etnier. 1980. Fishes. In: Tennessee's Rare Wildlife, Vol. 1, The Vertebrates. Tennessee Wildlife Resources Agency, pp. B1-B134.

U.S. Department of the Interior. 1966. Rare and endangered fish and wildlife of the United States. U.S. Bureau of Sport Fish and Wildlife Resource Publication 34:F1-F41.

U.S. Department of the Interior. 1973. Threatened wildlife of the United States. U.S. Bureau of Sport Fish and Wildlife Resource Publication 114: 1-289.

15 Habitats of North American Desert Fishes

GERALD A. COLE
Arizona State University, Tempe

ABSTRACT

Most natural habitats of desert fishes are derived from subterranean sources or surface waters that rise in more humid regions. They exist precariously, having been modified in the past and are subject to gross alteration in the near future. Thus some permanent large rivers that pass through desert basins from greater elevations have been changed by impoundment from warm and turbid to cold and clear flows. Other streams that were permanent and relatively transparent have become deeply entrenched, turbid, and intermittent. It is predicted that several springs and their outflows, including adjacent marshes, will disappear in the future as others have done already. Many support or once supported endemic faunas. Impoundment of the larger streams has created the open lake, an aquatic habitat alien to closed-basin desert regions. Canalization, similarly, has created a new type of lotic habitat and has served to destroy the insular qualities and isolated nature of desert aquatic habitats and their biotas.

Environmental factors range from stability in limnocrenes and the uppermost reaches of rheocrenes to extreme astaticism in other habitats where water level, temperature, pH, and salinity vary markedly, both temporally and spatially. Intermittent streams are inhabited by some North American desert fish species because of their vagility and ability to colonize pooled refugia and to survive flash flooding. Such streams lack malacostracans but this could be a function, in part, of the lack of interstitial water in the dry stream bed, a feature that may also preclude piscine aestivation.

In arid regions in which annual evaporation rates surpass the precipitation, highly concentrated waters of diverse ionic composition result. Many are poikilosaline, defined by variable salinity. Osmotic problems are inherent in such waters and few, if any, fishes are present in concentrations with twice the salinity of seawater. Moreover, somewhere near or above 5°/oo salinity there are direct ecologic effects of the relative ionic composition, something not always evident in dilute waters. High levels of bicarbonate and monocarbonate are believed to preclude fish survival in some desert environments, but other factors must be considered. In some instances total salinity and/or high pH may be limiting. Probably no fish survival is possible in waters in which the pH is consistently in the neighborhood of 10.3. In some desert limnocrenes and rheocrenes high levels of bicarbonate may be associated with unusually high tensions of gaseous CO_2. The stresses placed on fishes may be complex and include the effects of bicarbonate on acidbase balance, the effect of CO_2 gradients on respiratory exchanges and, in some cases, the possibility of gas-bubble disease from CO_2 tensions 1000 times the saturation value. Despite harsh vicissitudes of some desert aquatic environments, there are ameliorating phenomena:

evaporative cooling, two types of diel mixing, the freshening effects of seepage in or out of a closed basin and removal of marginal evaporites by deflation during times of low water level.

INTRODUCTION

The habitats of North American desert fishes have been described from various viewpoints by many authors. Most of the pertinent papers have been cited I am sure by other contributors to this symposium; only a few are mentioned here. Hubbs and Miller (1948) emphasized geologic and hydrologic history that pertains to modern desert fish habitats. Summaries by Edmondson (1963) and Cole (1963) brought together earlier work on waters of the Great Basin and the American Southwest. Cole (1968) discussed limnologic aspects of world-wide desert waters, some of which apply to North American fish habitats. Minckley and Cole (1968) detailed the chemical limnology of Cuatro Ciénegas, Mexico, and Minckley (1969) approached the same inland basin from a broad ecological viewpoint. Deacon and Minckley (1974) included excellent environmental data for fishes found in the deserts of North America, and Soltz and Naiman (1978) discussed the Death Valley setting in their treatment of the natural history of fishes that occur there. In addition, individual habitats have been treated in detail by many ichthyologists as they described new species or published accounts of populations and communities.

The generalities made about desert waters have stressed meteorologic-hydrologic conditions that create great areas of internal drainage (e.g., closed basins). Some desert streams rise in more humid regions and pass through the desert, terminate in closed lakes, or are lost in the sand. Other streams owe their origins to subterranean sources, emerging from springs that commonly are thermal.

The open lake is a man-made feature of desert tracts. Impoundment of large rivers that flow through arid regions on their way to the sea creates open lakes that would not, by definition, occur in deserts. Their fish faunas are usually nonnative species, and in a sense these desert lakes are not part of desert ichthyology.

The extreme instability of aquatic ecosystems in the desert is well documented, fluctuation being the key word. Moreover, the fragility of these habitats is characteristic and many have been altered grossly or have disappeared completely in the present century (Miller 1961).

Ephemeral desert waters are common but on our continent most are devoid of fishes. An exception might be the temporary stream with a few persistent pools that serve as refugia for forms like *Agosia chrysogaster*, a vagile species capable of reaching them when the stream has essentially ceased to exist. There seem to be no adaptations, however, for aerial respiration, aestivation, or survival as partly desiccated zygotes or embryos among the fishes of the young North American deserts.

Even when desiccation is not part of the annual regimen desert waters can be harsh, especially with respect to salinity, temperature, and low oxygen tensions. The literature on North American desert fishes contains many accounts of stress and the physiologic problems of osmoregulation and temperature tolerance.

Other generalizations made about aquatic habitats of the desert emphasize the high annual primary productivity, the importance of benthic algae and some higher plants as primary producers, and the unusually long growing season in thermal waters.

Also, aquatic settings in our deserts have an insular quality that makes them particularly appealing to students of speciation. Taxonomy becomes especially dynamic when the biota of desert lakes and streams are studied; the deme becomes an important category and endemism is common. Unfortunately this island nature is subject to anthropogenic modification and as a result hybridization and/or extinction are relatively common events.

North American desert waters are extremely varied. The lentic environments range from tiny shallow springs to large terminal lakes that can be exemplified, respectively, by some of the pitlike pozos at Cuatro

Ciénegas, and deep Pyramid Lake (104 m) which occupies a Nevada graben.

The lotic settings include first-order rivulets that emerge from springs, larger, but often transitory streams, and at the other extreme the tremendous Colorado River which has drained almost 280,000 km^2 by the time it reaches Lee's Ferry in northern Arizona, still more than 1000 km from its mouth in the Gulf of California. In addition, irrigation canals and ditches abound.

Impoundment is always a threat to a desert waterway. In some instances it ends the stream's existence; this, for example, has happened to the lower reaches of the Salt and Gila rivers in Arizona. On the Colorado River the closing of Glen Canyon Dam about 25 km above Lee's Ferry eliminated the striking seasonal variation in the river's discharge, ionic composition, temperature, and sediment load in the gorge of the Grand Canyon. Impoundment did not destroy the river but grossly altered it as a fish habitat.

The morphology of tiny bodies of water is of little interest to most limnologists, but desert ichthyologists find this facet of the fish habitat to be important, perhaps crucial; for example, a major threat to the endemic Devils Hole pupfish (*Cyprinodon diabolis*), as the water level fell in its unique habitat, was the exposure of a subsurface shelf on which reproductive activities normally take place. There must be many potential aquatic refugia for endangered fishes that are adequate with respect to all physicochemical characteristics of the water as well as nutritional requirements but lack shallow areas necessary for perpetuating the population. There might be opportunity for research on some desert fishes based on altering the dimensions of experimental ponds. The complicated social and breeding systems seen in the species of *Cyprinodon* may depend on morphologic features of their habitats.

The writings of desert limnologists often do not concern the desert ichthyologist. The former are fascinated by hypersaline waters from which fish are excluded; they write of *Artemia* and brine-fly habitats, in which total salinities approach 250 $^{o}/oo$. There are relatively few animals in salinities of 100, and 70‰ (about twice that of seawater) is probably near the limit for fish populations to maintain themselves.

GENERAL WATER CHEMISTRY OF FISH HABITATS

Subject to high evaporation rates, and the resultant concentration, desert waters have more diversity in ionic composition than those from humid areas. There are various ways of classifying them, but a simple scheme is followed here.

Typical dilute freshwater is a calcium-carbonate type, mostly in the form of $Ca(HCO_3)_2$. The predominance of calcium and bicarbonate among the cations and anions, respectively, is common in humid regions and represents the mean of the world's rivers (Livingstone 1963). In desert climates the aquatic picture changes, for the solubility of $CaCO_3$ is extremely low; under some conditions perhaps 20,000 times less soluble than the common sodium chloride. As total salinity mounts via concentration, the solubility product of $CaCO_3$ is reached and it precipitates. Thus other types of water are formed. The evaporation-precipitation series usually leads from calcium bicarbonate water to sodium chloride water. If the original dilute calcium water contains abundant sulfate, termed loosely gypsum water ($CaSO_4$), the series leads to sodium sulfate water as it approaches the sodium chloride category. In some instances, depending on the composition of the initial water, the concentration pathway leads to a sodium carbonate-bicarbonate type. Occasionally the three principal anions (carbonate, sulfate, and chloride) are present in almost equal amounts; this was labeled "triple" water by Clarke (1924).

In arid regions, however, $CaCO_3$ waters paradoxically are fairly high in total salinity. They are usually found in limestone springs and their effluent rheocrenes. Calcium carbonate is held temporarily in solution as $Ca(HCO_3)_2$ by unusual amounts of CO_2-H_2CO_3, also derived from subterranean sources. The equilibrium is upset downstream as CO_2 is lost to the atmosphere, and therefore $CaCO_3$ is precipitated. This fish habitat is found in the lower reaches of the short surviving segment of Fossil Creek, Arizona. As discussed later, some of these hard-carbonate desert waters lack fish. They are different from the lake and stream waters of nondesert regions, despite the similarities in ionic proportions.

The unusual population of *Cyprinodon macularius* found in the pool at Quitobaquito Spring, Arizona, occurs in a sodium water that is slightly lower in chloride than the water of Pyramid Lake (Cole and Whiteside 1965). Quitobaquito seems to represent an earlier stage than Pyramid Lake in the ideal trend from calcium-carbonate to sodium waters. The great difference in the diversity of the fish faunas of the two habitats, however, is not related to different water chemistry but is primarily a function of habitat size.

Salton Sea (California) fishes live in briny water with an ionic composition much like seawater, although it has varied in the past and still does (Walker 1961). This depends on the balance struck between freshwater input and evaporation. The *Cyprinodon* of Tecopa Bore, California, lives in a "triple" water (Naiman 1976). The fishes of Cuatro Ciénegas occur in a spectrum, from dilute gypsum water to more concentrated sodium sulfate pools. The latter resemble some of the Bottomless Lakes, New Mexico, where *Cyprinodon pecosensis* is found.

Many depressions in arid areas of North America, especially in the Great Basin, contain alkaline soda waters (Na_2CO_3, $NaHCO_3$). Probably Moses Lake, Washington, is one of the best examples of an American soda lake that contains a fish fauna (Edmondson 1963).

Under the relatively dilute conditions necessary for fish survival it is not apparent that the water type plays a paramount role in the lives of desert fishes. The Colorado River and the tributaries entering it in the Grand Canyon present an array of chemical types (Kubly and Cole 1979). The river is much like a "triple" water, although $SO_4^=$ and $CO_3^=$ are usually more abundant that chloride. The tributaries that enter the river include: dilute dolomitic waters, $CaMg(CO_3)_2$; impure dolomitic waters with significant amounts of chloride and sulfate; sulfate waters, some of the gypsum variety; sodium bicarbonate streams; and some more concentrated NaCl influents. In the less concentrated streams little evidence is shown that the fauna varies in response to the different ionic compositions. Other factors such as turbidity, temperature, rate of flow, and stream size could be more important to the fishes.

SPECIAL FEATURES OF DESERT WATERS

Although waters from arid North America have been described in some detail and contrasted with the waters of mesic regions (Naiman, this volume) certain aspects have not been accentuated. They are now pointed out.

Gradients, both spatial and temporal, are striking in some desert waters. A reconnaissance at one time of day will not always reveal the true nature of the habitat. Similarly, a sample taken from one depth or from one point at the surface of a desert pool or stream does not assess the habitat accurately. Moreover, factors such as dissolved CO_2, abundant carbonate and bicarbonate ions, and high pH are rarely significant in humid regions but should not be neglected in desert ichthyology.

Gradients

Because vertical and horizontal gradients are sharp in the waters of arid regions, great care must be taken to determine just where fishes occur; for example, some fishes (probably doomed) were present in the upper level of a small shallow pit (45 to 50 cm) in Cuatro Ciénegas (Cole and Minckley 1968). The lower layer of the pit was 47°C; only 46 cm higher the water was a pleasant 24°C. Similarly, the bottom stratum contained salinity amounting to 111.3^{o}/oo (about 3.3 times seawater), whereas the upper layer was dilute. Carelessness in sampling this pool might have led to the erroneous conclusion that its fishes survived extremely high temperature and salinity.

Barlow's (1958) article underscored both horizontal and temporal differences in the Salton Sea. The diel horizontal movements of Cyprinodon showed that the complexities of the Salton Sea as a fish habitat could not be described from a single sample taken at a specific time. Reports on the California thermal stream called Tecopa Bore (Naiman 1976) show distinct downstream changes. Cyprinodon uses only 83 percent of the stream, finding temperatures in the upper reaches too high for survival.

Free Carbon Dioxide

Occasionally, a large amount of free CO_2 is a factor that precludes fish existence. The best example is probably Montezuma Well, Arizona, and its effluent stream, but the phenomenon may be common. The Well is a good example of a desert calcium bicarbonate water; its filtrable residue is about 616 mg $liter^{-1}$.

Because desert waters are often derived from subterranean sources, abundant CO_2 in the headspring region may force the fishes to remain in the lower segments of a rheocrene. Two thermal waters in Arizona, known as Monkey Springs, have free CO_2 amounting to about 49 mg $liter^{-1}$ and Cyprinodon survived there at least for a while (Minckley 1973). This may be close to the tolerance limit.

Diel contrasts also occur. At dawn in Montezuma Well the pH may be only 6.0 or 6.2, the attendant dissolved CO_2 amounting to 865 and 680 mg $liter^{-1}$, respectively. On sunny summer afternoons pH values as high as 6.9 have been observed, and in the Well that is equivalent to 109 mg $liter^{-1}$ of free CO_2. Such high levels of CO_2 are rarely met in natural aquatic habitats and would be most typical of springs.

The effects of gaseous CO_2 on a fish are not clear-cut and may be linked to other environmental factors. Many years ago Doudoroff (1957) pointed out that most fishes show distress in the presence of 50 mg $liter^{-1}$ CO_2 and do not survive 100 mg $liter^{-1}$. In these instances unfavorable gradients probably interfere with normal respiration, despite adequate ambient oxygen.

The possibility of gas-bubble disease from high CO_2 tensions is usually dismissed. Natural concentrations rarely attain 30 mg $liter^{-1}$ and, compared with O_2 and N_2, carbon dioxide is not important when entrainment of air occurs as water cascades over dams and escarpments. Only in unusual habitats like Montezuma Well could there be a possibility that hyperbaric pressures of CO_2 would play a lethal role. The adverse effects on respiration, however, would prevail in such instances before evidence of gas-bubble stress.

Although CO_2 might deny spring sources to fishes, loss of the gas from the outflow could permit their existence downstream. Gradients of CO_2 are common in desert rheocrenes (Cole 1975) and biotic zonation accompanies such clines. There can be adverse effects, however, at least on the invertebrates as a spring outflow loses CO_2. In the Montezuma Well effluent degasification is accompanied by precipitation of calcite and this is lethal (in experimental situations) to hyalellid amphipods that survive upstream and in the Well proper. The effect of rising pH may also be involved; the calcareous coating of the crustaceans' respiratory surfaces may not be the sole factor in their deaths. In some anaerobic outflows the precipitation of ferric hydroxide following oxygenation downstream is known to have adverse effects on the fauna. Anaerobic springs, however, are surprisingly rare in our deserts.

Alkalinity, pH, and Bicarbonate

The interrelations of monocarbonate-bicarbonate alkalinity, pH, and total salinity may be complex in some desert habitats. McCarraher (1962) discussed these complexities in relation to *Esox* survival in some alkaline Nebraska ponds. No successful transplants of the pike occurred in six ponds in which the salinity was 2.48^{o}/oo and higher. These waters, however, also had high pH values (9.5 to 10.8). Thus the effect of bicarbonate is difficult to isolate: as its concentration increases in soda waters (dominated by Na_2CO_3 and $NaHCO_3$), the pH rises and further increase is accompanied, of course, by intolerable salinities. In calcium bicarbonate waters, such as Montezuma Well, the presence of bicarbonate at unusual levels (689 mg $liter^{-1}$) is accompanied by unusual and perhaps lethal amounts of CO_2-H_2CO_3.

Fishes occur in some of the North American soda waters, but species diversity is low when compared with tropical Africa (Beadle 1974). Many species are found in alkaline African soda lakes, probably the result of a long evolutionary history not matched in North America. It appears from Beadle's data that no fish are present when the pH is consistently near 10.3. Fishless Lake Nakuru has a high pH (up to 12.0 at times), but its effect may be masked by a salinity of $45^o/oo$. Beadle drew a line between the fresh and saline waters of Africa at $5^o/oo$.

Haswell et al. (1980) suggest further ramifications in the relationships of CO_2, bicarbonate, and salinity. Studies on branchial carbonic anhydrase suggest a link between acid-base regulation and salt transport. Thus unfavorable ambient bicarbonate concentrations and gradients might inhibit normal osmoregulation and acid-base regulation.

AMELIORATION OF STRESSFUL ENVIRONMENTAL FACTORS

Diel fluctuations reestablish favorable conditions in many desert waters, and conditions for fish life are less harsh than might appear at first glance. The presence of fortuitous negative feedback mechanisms serve to shorten periods of stress.

Carpelan (1967) described regular nocturnal lowering of pH in canyon pools in the California desert but they were not fish habitats. Diurnal photosynthesis raised the pH and lowered the specific conductance as $Ca(HCO_3)_2$ was converted to $CaCO_3$ and precipitated. At night respiration prevailed and the CO_2-H_3CO_3 increase lowered the pH and put $CaCO_3$ back in solution as bicarbonate.

As a result of photosynthesis waters of unusually high pH occur in some Arizona and New Mexican fishing lakes during the summer and fish die-offs ensue. Whatever nocturnal recovery (lowering of pH) occurs is not adequate for survival; the recovery is seasonal and too late to protect the fishes. These lakes, although dilute, contain appreciable amounts of $NaHCO_3$ and support flourishing populations of bluegreen algae.

In other instances diel fluctuations serve to protect the fishes. Foster (1973) studied two shallow, nutrient-rich ponds in Phoenix, Arizona, and found that they supported flourishing populations of introduced *Tilapia*. Diurnal temperature gradients were sharp, stratifying the ponds effectively. In these unusually rich ponds such stratification, if prolonged, could have been disastrous to the bottom fauna and fishes. The stability was transient, however, because nocturnal cooling and evaporation at the surface and the convection currents that resulted destroyed the diurnal stratification. The polymictic character of the ponds was important in their general metabolism. Oxygen produced in the upper strata during daylight hours was distributed throughout the water column at night.

A shallow temporary pond in Arizona was monitored one August day by Dr. Denton Belk (personal communication) and striking diel changes were exhibited. Another type of polymixis was revealed, an additional example of fortuitous negative feedback. No fish were present in the pond. Stability was slightly negative at dawn because heat, accumulated in the sediments the day before, was warming the bottom water. Stability rose to a maximum at 1400 hr as temperature stratification was brought about; it then began to decline, reaching zero at 2000 hr. Presumably, the pond circulated completely for the next 10 or 12 hr. In this shallow pond daytime heat penetrating from above warmed the sediments, which subsequently gave off heat. Heat fluxes from the surface and the bottom met to bring about isothermy in the early evening.

Evaporation was also an important factor in ameliorating harsh conditions for the biota in this pond. The ambient air temperature rose more than 10°C above the mean temperature of the pond. Belk estimated that of the 5610 kcal m^{-2} that arrived at the pond surface that day, 2312 were lost by evaporation. Air temperature increased about 29°C; the pond temperature rose no more than 7.5°C.

A subsequent phenomenon observed during the annual regimen of Belk's temporary pond may apply to only a few desert fish habitats, but it is worth noting. The desiccated basin was gradually stripped of saline evaporites by deflation. When the rain filled the depression once again there was no evidence of increased salinity. Perhaps in some desert fish ponds also, where falling water levels expose marginal areas, deflation lessens the cumulative buildup of salinity. Other mechanisms prevent this buildup. Climatic fluctuations that change a closed basin to an open one bring about salt loss and dilution. This may have happened to Pyramid Lake in the past; its relatively low salinity could be the result of overflow (Hutchinson 1937). The closed lakes that occupy depressions in volcanic soils near Flagstaff, Arizona, are surprisingly dilute (Cole 1963). This is probably a function of seepage through the cinder substratum; thus the lakes are open in a sense, despite the lack of effluent streams.

CONCLUSIONS

Although a great deal is known about the habitats of desert fishes, there is still much to be learned and correlated with behavior and survival. Assessment of the water should not be limited to a single spot at a single time. Temporal changes are important and occur vertically, horizontally and, of course, downstream. The fishes escape the harshest factors of their environments by taking advantage of diel (or seasonal) events that ameliorate the stresses. The aquatic habitat in an arid area should be investigated for what it is: a dynamic entity. Furthermore, environmental factors other than temperature, oxygen, salinity, and ionic composition of the water await study. This is especially true for free CO_2 and in some regions bicarbonate-carbonate effects and high pH. These are factors that are far less important in humid areas. The behavior of a small desert water can become as engaging to the researcher as the behavior of its fish inhabitants. The two are closely related.

REFERENCES

Barlow, G. W. 1958. Daily movements of desert pupfish in the shore pools of the Salton Sea, California. Ecology 39:580-587.

Beadle, L. C. 1974. The Inland Waters of Tropical Africa. An Introduction to Tropical Limnology. Longman, London.

Carpelan, L. H. 1967. Daily alkalinity changes in brackish desert pools. International Journal Oceanology and Limnology 1:165-193.

Clarke, F. W. 1924. The Data of Geochemistry, 5th ed. United States Geological Survey, Bulletin 710.

Cole, G. A. 1963. The American Southwest and Middle America. In: D. G. Frey (Ed.)., Limnology in North America. University of Wisconsin Press, Madison, pp. 393-434.

Cole, G. A. 1968. Desert limnology. In: G. W. Brown, Jr. (Ed.), Desert Biology, Vol. 1. Academic, New York, pp. 423-486.

Cole, G. A. 1975. Calcite saturation in Arizona waters. Verhandlungen Internationale Vereinigung Limnologie 19:1675-1685.

Cole, G. A. and W. L. Minkley. 1968. "Anomalous" thermal conditions in a hypersaline inland pond. Journal of the Arizona Academy of Sciences 5:105-107.

Cole, G. A. and M. C. Whiteside. 1965. An ecological reconnaissance of Quitobaquito Spring, Arizona. Journal of the Arizona Academy of Sciences 3:159-163.

Deacon, J. E. and W. L. Minkley. 1974. Desert fishes. In: G. W. Brown, Jr. (Ed.), Desert Biology, Vol. 2. Academic, New York, pp. 385-487.

Doudoroff, P. 1957. Water quality requirements of fishes and effects of toxic substances. In: M. E. Brown (Ed.), The Physiology of Fishes, Vol. 2. Academic, New York, pp. 403-430.

Edmondson, W. T. 1963. Pacific Coast and Great Basin. In: D. G. Frey (Ed.), Limnology in North America. University of Wisconsin Press, Madison, pp. 371-389.

Foster, J. M. 1973. Limnology of two desert recharged-groundwater ponds. Ph.D. Dissertation, Arizona State University, Tempe.

Haswell, M. S., D. G. Randall and S. F. Perry. 1980. Fish gill carbonic anhydrase: acid-base regulation of salt transport? American Journal of Physiology 238:R240-R245.

Hubbs, C. L. and R. R. Miller. 1948. The zoologic evidence: correlation between fish distribution and hydrographic history in the desert basins of western United States. In: The Great Basin, with Emphasis on Glacial and Postglacial Times. Bulletin of the University of Utah, Biological Series 10(7), pp. 17-166.

Hutchinson, G. E. 1937. A contribution to the limnology of arid regions primarily founded on observations made in the Lahontan Basin. Transactions of the Connecticut Academy of Arts and Sciences 33: 47-132.

Kubly, D. M. and G. A. Cole. 1979. The chemistry of the Colorado River and its tributaries in Marble and Grand Canyons. In: R. M. Linn (Ed.), Proceedings of the First Conference on Scientific Research in the National Parks, Vol. 1. National Park Service Transactions and Proceedings Series. No. 5, pp. 565-572.

Livingstone, D. A. 1963. Chemical composition of rivers and lakes. United States Geological Survey Professional Paper 440=G: 1G-64G.

McCarraher, D. B. 1962. Northern pike, *Esox lucius*, in alkaline lakes of Nebraska. Transactions of the American Fisheries Society 91:326-329.

Miller, R. R. 1961. Man and the changing fish fauna of the American Southwest. Papers of the Michigan Academy of Science, Arts, and Letters 46:365-404.

Minckley, W. L. 1969. Environments of the bolson of Cuatro Ciénegas, Coahuila, Mexico. Texas Western Press, University of Texas, El Paso. Science Series 2:1-63.

Minckley, W. L. 1973. Fishes of Arizona. Arizona Game and Fish Department, Phoenix.

Minckley, W. L. and G. A. Cole. 1968. Preliminary limnologic information on waters of the Cuatro Ciénegas Basin, Coahuila, Mexico. Southwestern Naturalist 13:421-431.

Naiman, R. J. 1976. Primary production, standing stock, and export of organic matter in a Mohave desert thermal stream. Limnology and Oceanography 21:60-73.

Soltz, D. L. and R. J. Naiman. 1978. The Natural History of Native Fishes in the Death Valley System. Natural History Museum of Los Angeles County, Science Series 30:1-76.

Walker, B. W. 1961. The Ecology of the Salton Sea, California, in Relation to the Sport Fishery. California Department of Fish and Game, Fish Bulletin No. 113.

16 An Ecosystem Overview: Desert Fishes and Their Habitats

ROBERT J. NAIMAN
Woods Hole Oceanographic Institution
Woods Hole, Massachusetts

ABSTRACT

The predominant trophic base, level of productivity, detrital characteristics, and food web diversity of streams in mesic and desert regions are summarized and compared at the ecosystem level. It is shown that desert streams differ from their mesic counterparts by their predominantly autotrophic base, lack of permanent headwater channels, absence of significant amounts of woody debris, and intermittent and variable flow patterns. The main departure of desert streams from the river continuum concept, which is developed from lotic ecosystems in mesic regions, is speculated to be that most desert streams, by virtue of their position in the watershed, would be ranked as possibly fourth-order streams in mesic regions. The structural and functional properties of spring fed desert streams differ from those subject to flash floods, and the role of catastrophic flooding is reflected in community development and life history strategies. The basic nature of aquatic ecosystems in the American southwest, however, has been drastically altered since 1850 by grazing and concomitant erosion, and these changes are reviewed briefly to add perspective to the current situation. Finally, general morphological and lifehistory adaptations of fishes to the characteristics of desert waters are illustrated.

INTRODUCTION

Despite the extreme aridity of North American deserts, they encompass many widely spaced springs, streams, rivers, and occasionally lakes. Considerable research has been directed toward organisms that use these bodies of water, especially the fishes, whereas little is known about the characteristics of the ecosystems themselves. Nevertheless it is not surprising that most of our present understanding of aquatic ecosystem dynamics has come from studies in mesic regions.

Aquatic ecosystems in deserts appear to differ from their mesic counterparts in several respects because the sparse terrestrial vegetation in deserts reduces the importance of allochthonous inputs; there is an extended growing season because of the climate; diversity of fishes, and possibly invertebrates, is low because of the difficulties of colonization and vagaries of the habitat; the infrequent but occasionally severe rain storms exert considerable pressure on species and community structure; and primary producers appear to provide a major portion of the trophic base.

In this brief overview current concepts of stream ecosystem dynamics, as they have been developed for mesic regions, are summarized; how desert streams fit these concepts are discussed, and some basic adaptations of fishes to the nature of desert springs, streams, and rivers are considered.

THE DESERT ENVIRONMENT

The three main characteristics of arid ecosystems are the following: precipitation is so low that water is the dominant controlling factor for biological processes, it is highly variable throughout the year in infrequent and discrete events, and its variation has a large unpredictable component (Noy-Meir 1973). The importance of water as a limiting factor in desert ecosystems is clear. Like energy, but unlike most nutrients, water is not recycled in the system but moves through it. Water is a periodically exhaustible resource, replenished only by new input. In arid regions only 10 to 50 rainy days per year occur in 3 to 15 rain events or clusters of rainy days, of which probably no more than 5-6 (sometimes only one) are large enough to affect biotic parts of the system (Noy-Meir 1973).

Thus the input that drives the system comes in pulses of short duration. The response of a desert system, or any of its parts, to a single input may also be a pulse of activity, especially in streams. After a long dry period some parts of the ecosystem (e.g., stream fishes) are at a low ebb of activity or may have been destroyed by complete desiccation. An effective rain triggers biological processes, in particular production and reproduction, and the biomass of plants and animals increases. These processes are slowed as available water supplied by the rain is exhausted. This is especially true in streams and temporary pools in which, after a short growth period, water becomes limiting and processes and biomass decrease to a steady state that may or may not be equal to the preceding one. Noy-Meir noted that a flexible transition between a resistant and active state is highly adaptive in an intermittently favorable environment and the prevalence of this pattern among many desert organsms explains the long-term stability of the system, despite its extreme short-term variability.

CHARACTERISTICS OF DESERT WATERS

Streams and Rivers

Lotic habitats are characterized by a unidirectional movement of water and by considerable spatial and temporal variability (Table 1). Desert streams and rivers are exemplified at one extreme by large rivers, such as the Colorado and Salt Rivers in the southwestern United States, and at the other extreme by thousands of small, almost dry creeks (Figure 1).

Major streams that flow through deserts normally originate at higher elevations that have comparatively high levels of precipitation. Discharge depends on rains or melting snow. Major tributaries of large

FIGURE 1. Illustrated above are some aquatic habitats typical of hot southwestern deserts. (A) Tecopa Bore, California; (B) Salt Creek in Death Valley, California;

FIGURE 1. (continued) (C) an unnamed stream in Canyon de Chelly, Arizona; (D) Cottonball Marsh in Death Valley.

TABLE 1. A Summary of the Generalized Characteristics of Springs, Streams, Marshes, Rivers, and Lakes in Desert Regions that Contain Fish Populations

Parameter	Spring	Streams	Marshes	Rivers	Lakes
Predominant water source	Aquifer	Aquifer/runoff	Aquifer/runoff	Runoff	Runoff
Discharge	Nearly constant	Highly variable	---	Seasonally variable	---
Water level	Nearly constant	Highly variable	Seasonally variable	Seasonally variable	Seasonally variable
Temperature	Often constant	Seasonally variable	Maximum seasonal and daily variability	Seasonally variable	Seasonally variable
Salinity	Low, constant	Low, variable in drying pools	Relatively high, varies seasonally	Low, nearly constant	Low, nearly constant
Oxygen	Depends upon site, usually constant	Adequate, except during low discharge at night	Highly variable	Adequate	Adequate

Light	Seasonal changes in photoperiod	Some shading from terrain, vegetation	Seasonal changes in photoperiod	Seasonal changes in photoperiod	Seasonal changes in photoperiod
Turbidity	Low	Low, except during freshets	Low	High at most times	Seasonally variable

streams also rise in uplands but they subside into sand or evaporate. The mainstream on the surface is reached only in times of flood. Small creeks contribute water only during local storms unless they are directly springfed. Large rivers are more predictable than the smaller ones because the smaller are greatly influenced by the cyclonic storms so characteristic of most desert regions (Deacon and Minckley 1974).

Gross variability of desert rivers is common; most are subject to long periods of reduced flow and are then scoured by major floods; for instance, the Colorado River at Yuma, Arizona, before upstream dams were built, varied from a low discharge of 0.3 $m^3 \cdot s^{-1}$ in 1934 to a peak discharge of 7083 $m^3 \cdot s^{-1}$ in 1916 (Deacon and Minckley 1974; Dill 1944). This former value was not only a function of variable precipitation but also of high evaporation rates that can exceed precipitation by 30 to 40 times (Logan 1968; McGinnes et al. 1968; Naiman 1976a).

Annual temperature variations in desert rivers are highly modified by climate to the extent that winter water temperatures are rarely <10°C in the Colorado River and in summer rarely exceed 30°C, despite intense solar radiation which produces air temperatures that average ~40°C and range to >50°C. This maximum water temperature is scarcely greater than that achieved in large temperate rivers elsewhere and is presumably maintained by rapid evaporation (Weatherby 1963).

Highest temperatures in small creeks occur generally in July and August when maxima range from 24 to 40°C. Slightly higher temperatures may occur in backwaters (33 to 43°C), where flow is reduced. Inflow from springs may ameliorate temperatures, as may shading by stream-side vegetation or canyon walls. Great differentials in temperature, however, may occur in flowing creeks. In Sycamore Creek, Arizona, summer temperature variations commonly exceed 12°C in different habitats at a given moment at depths ranging from 0.5 to 5.0 cm. Deep pools stratify strongly in Fish Creek, Arizona, where surface temperatures of 29°C contrast with 18°C at 1 m in August; riffle temperatures approach 32°C (Deacon and Minckley 1974).

Turbidity and sedimentation have considerable impact on aquatic habitats and fishes situated in areas of fine soils, defoliated basins, or degrading streams. High turbidity shades aquatic habitats supressing production by algae and other plants. Sedimentation of suspended solids, however, may have a detrimental effect on bottom-dwelling organisms or may physically smother fish eggs or larvae. High turbidity poses severe problems for lotic desert fishes, which avoid it by moving to quiet side pools (Trautman 1957) or by having adaptations (e.g., reduced eyes, modified scales) for the situation (Deacon and Minckley 1974).

Springs and Marshes

Desert springs range from quiet pools or trickles to active aquifers. Many larger springs, often positioned along recent fault zones, may emit warm water heated by passage through the geothermal gradient which causes a temperature increase of 1°C for each 15 to 30 m of depth (Waring 1965).

Thermal springs, those that maintain water temperatures somewhat above the mean annual air temperatures, are often remarkably constant (Table 1). Springs issuing from shallower aquifers may be constant at a temperature that approximates the prevailing annual means for air or may fluctuate with season, sometimes in a delayed manner that reflects winter precipitation by the issuance of cooler water in summer and vice versa. Certain areas (e.g., Tecopa, California) have warm springs interspersed among cool springs, which indicate different sources of water for the two types (Deacon and Minckley 1974).

Spring waters range from fresh to highly mineralized and some that carry large amounts of dissolved materials, especially bicarbonates, often deposit extensive formations of travertine downstream (Cole and Batchelder 1969). Chemical features of some springs also include extremely low dissolved oxygen or considerable free carbon dioxide which may exclude some fishes.

As spring waters flow away from the source, they often develop extensive marshes. Marshes are an integral part of the spring ecosystem and at one time also occurred in association with most other aquatic habitats in deserts (Bradley 1970, Ohmart et al. 1975). Many differ radically from their water sources which tend to have considerable temperature, chemical and water level fluctuations (Table 1). Desert marshes can be highly saline at times of low water but fresh at others. As water flows downstream from the spring source, the physicochemical variability may become so extreme that only salt grass (*Distichlis*) may exist, whereas other plants normally associated with these marshes (*Scirpus*) are entirely excluded. In many respects desert marshes resemble those of marine environments in which the biota is often limited to those organisms with considerable resistance or resilience to physicochemical stress.

Lakes

In some arid regions water from rivers or springs may accumulate in basins to form major lakes. Such lakes depend on a fine balance of inflow, evaporation, and seepage to maintain their existance (Cole 1968). In the not too distant past tremendous lakes that occupied large expanses of western North America and northern Mexico fluctuated with changes in climate to form the broad saline flats common today (Hubbs et al. 1974). A dramatic example can be seen in Death Valley which, during the last glacial period, contained historic Lake Manly (see Soltz and Naiman 1978). At its height the lake was about 160 km long, 10 to 17 km wide, and 185 m deep. Few natural lakes presently exist in any desert and most of those that do are too saline or unstable to support fishes. An interesting exception is Pyramid Lake, Nevada, which historically had a tremendous population of cutthrout trout (*Salmo clarki*) (LaRivers 1962).

A HISTORICAL PERSPECTIVE

Many contemporary streams of the American desert are highly modified from the original habitats (Miller

1961). There has been a shift from clear, dependable streams to those of intermittent flow, subject to flash floods that carry heavy loads of silt. As a result of decreasing water volume and destruction of vegetation, there has been a trend toward rising temperatures in remaining waters. Smaller creeks and marshes associated with rivers have largely disappeared, due in a large part to a severe disruption of the hydrologic cycle (Ohmart et al. 1975).

The conspicuous arroyos of semiarid regions, so prominent in Arizona and New Mexico, were cut largely between about 1880 and 1910. These streams are now mostly ephemeral with flat channel bottoms and vertical walls; they range in size to 30 m deep, many hundreds of meters wide, and nearly 200 km long (Antevs 1952). It has been demonstrated that these erosive gashes were brought about by excessive grazing of livestock (Miller 1961; Thornber 1910). Miller described this process as follows:

Over much of the Southwest large expanses were originally covered by a luxuriant growth of grasses and other plants, which during the rainy season, reached a height of nearly 1 m (Leopold 1951). During the dry season dead vegetation formed a protective mulch and its roots effectively bound the soil. This cover was a potent defense against erosion. With the establishment of livestock, vegetation was eaten progressively shorter and severely trampled to expose soil within the daily movement to and from water (Gregory 1950). Such denudation by herds became particularly noticeable after 1875 (Antevs 1952) and led to increased floods and severe erosion. It is well known that destructive floods may originate in a disturbed region, whereas adjacent vegetated areas may be highly resistant to the climatic event. Destruction of climax grass lands has been described and illustrated by Gregory (1950) for Kanab Creek, Arizona, and the San Pedro River, Arizona, which was incised from the mouth for 210 km upstream in just 10 years (Bryan 1925). Cutting of timber in forested headwater areas also promoted rapid runoff and a lowering of the water table and resulted in damaging erosion from floods. The filling in of deep pools by sand and salt has eliminated species dependent on these habitats.

With these alterations the survival of aquatic organisms has been profoundly affected by the concomitant reduction or elimination of aquatic plants and increasing water temperatures. The water table in southwestern United States is now so low that even when exceptionally heavy runoff occurs additional recharge of the water table by downward percolation fails to be effective, probably because of shallow penetration and subsequent evaporation (Miller 1961; Smith 1940). It is within this context that we must view the present situation in order to appreciate that many of the habitats we are describing, studying and managing today have been drastically altered since 1850.

THE WATER SUPPLY

In mesic areas, in which an ample supply of water is usually available, discharge volume per unit area is relatively great and the timing of seasonal discharge regimes is largely predictable. This may not necessarily be the case in deserts in which the availability of water is often sporadic.

To illustrate this point consider two watersheds with nearly identical annual discharges: the Moisie River in Quebec (mean annual discharge 466 $m^3 \cdot s^{-1}$) and the Colorado River (mean annual discharge 571 $m^3 \cdot s^{-1}$) in southwestern United States (Figure 2). The Moisie River has a watershed area of 19,811 km^2 which drains a virgin forest of black spruce (*Picea mariana*) and balsam fir (*Abies balsamea*). Within the watershed are 32,341 km of streams which, if placed end to end, would stretch 79 percent of the way around the equator (Naiman and Sedell 1981). The annual discharge pattern for this river is remarkably regular; the spring flood occurs within 10 to 15 days of the same date each year, there are minimum flows during the cold winter months, and the annual discharge is nearly constant from year to year (Figure 3).

This is in sharp contrast to the Colorado River which has a watershed about 30 times larger than the Moisie River. The Colorado River watershed encompasses

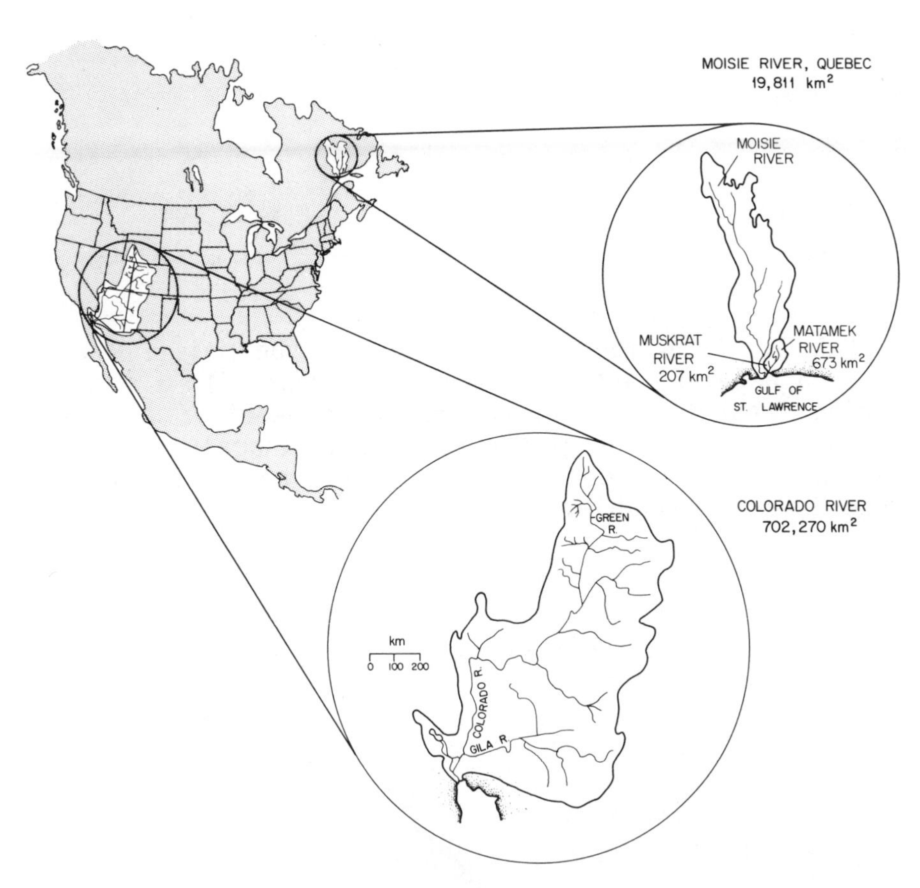

FIGURE 2. The watershed area of the Moisie River, a major river system mesic region, is compared with the Colorado River. Mean annual discharge in these two river systems is nearly identical, despite the much larger size of the Colorado River watershed (Moisie River: 466 $m^3 \cdot s^{-1}$; Colorado River: 571 $m^3 \cdot s^{-1}$).

at least 200 to 300 times the number of stream kilometers of the Moisie River watershed, many of which are dry. The terrestrial vegetation is sparse and large annual variations in discharge result from the unpredictable climate (Figure 4). These variations are more pronounced in individual tributaries than in the main river channel. Discharge per unit area is low and more erratic than in mesic regions.

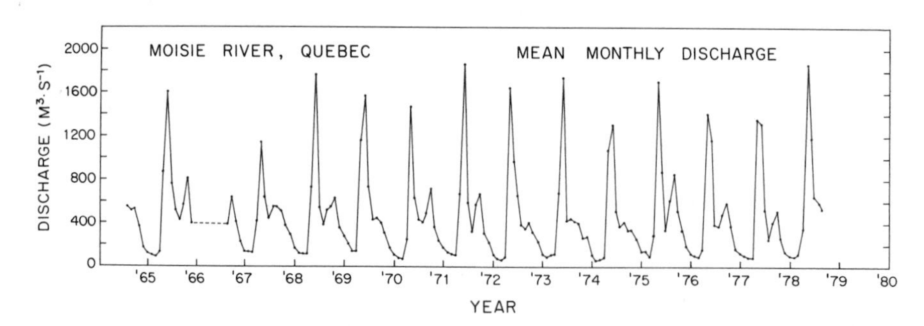

FIGURE 3. Mean monthly discharge of the Moisie River shows a regular pattern from year to year with a strong spring freshet in April and May when about 75 percent of the annual discharge occurs. Total annual discharge has been nearly constant for the years for which records are available.

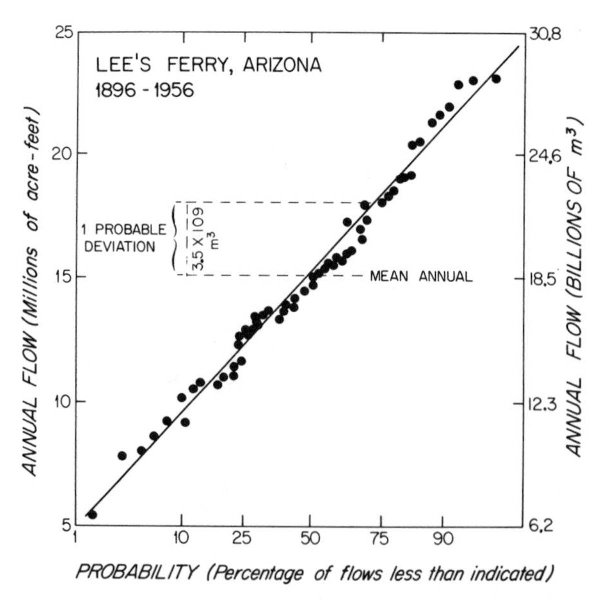

FIGURE 4. Historically, the annual discharge of the Colorado River at Lee's Ferry, Arizona, has been highly variable compared with the relative consistancies of the Moisie River. Discharge events during a single year were also relatively unpredictable (from Leopold et al. 1964, with permission).

Another water supply, which may be of only minor importance in mesic regions but is of considerable importance in deserts, is fossil water remaining from wetter climatic periods that is held in long-term storage beneath the desert floor. Water that supplies desert springs often originates from hydrologic basins located hundreds of kilometers away and is delivered to the spring by extensive underground interbasin movements of water through highly fractured carbonate rocks. Discharge is therefore relatively independent of local precipitation patterns.

A clear example has been summarized by Soltz and Naiman (1978) for a major portion of the Death Valley, California, hydrologic basin (Figure 5). Today the Death Valley watershed covers about 22,500 km^2 with about 6 percent of the basin below sea level (Hunt et al. 1966). The principle surface runoff into Death Valley is by the Amargosa River, which drains about 13,700 km^2 of arid desert land. The Amargosa River is dry most of the year where it enters the southern part of Death Valley. Despite the enormous size of the Death Valley watershed, little surface water is available for aquatic organisms.

Direct precipitation in the Death Valley watershed is not sufficient to account for water discharge in several groups of major springs (Winograd and Thordarson 1975). Most springs and marshes are now dependent on underground interbasin flow of water. Major springs in Ash Meadows, Nevada, have a mean annual discharge of 0.7 $m^3 \cdot s^{-1}$ (Dudley and Larson 1976), collected from an area that encompasses no less than 11,600 km^2 and includes parts of 10 intermontane valleys hydraulically connected by interbasin movement of water through the carbonate rocks. Seven of these valleys are topographically closed. The major uplands in the Ash Meadows Groundwater Basin are Spring Mountains and Sheep Range. Both are near Las Vegas, Nevada and both reach an altitude of more than 3000 m and receive as much as 64 cm of precipitation annually. Precipitation however, may not be the source of the water. It has been demonstrated that water issuing from the springs is between 8000 and 12,000 years old (I. J. Winograd, personal communication). Apparently this vast

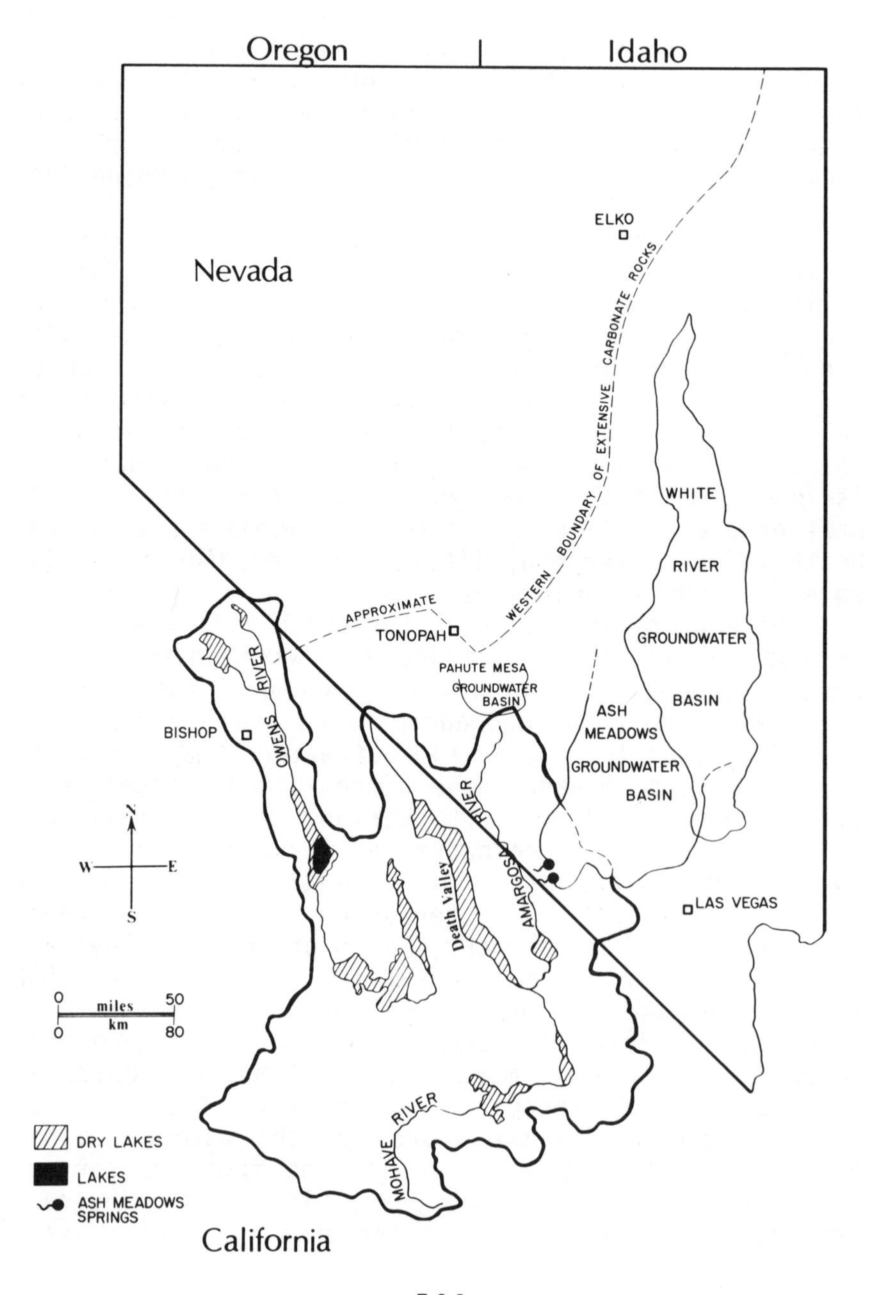
Oregon
Idaho
Nevada
ELKO
WESTERN BOUNDARY OF EXTENSIVE CARBONATE ROCKS
APPROXIMATE
TONOPAH
PAHUTE MESA GROUNDWATER BASIN
WHITE RIVER GROUNDWATER BASIN
ASH MEADOWS GROUNDWATER BASIN
BISHOP
OWENS RIVER
AMARGOSA RIVER
Death Valley
LAS VEGAS
N
W
E
S
0 miles 50
0 km 80
MOHAVE RIVER
DRY LAKES
LAKES
ASH MEADOWS SPRINGS
California

underground reservoir is not being recharged by surface precipitation, as was once suspected, but is a huge body of water that has existed since pluvial times. It is now slowly draining along faults to emerge as springs and marshes in the Ash Meadows-Death Valley area. The underground basin may be much larger than that shown in Figure 5 because available data are not yet adequate to delineate the northern boundaries of the water-bearing carbonate rocks.

Discharge from various Ash Meadows springs range from <0.001 $m^3 \cdot s^{-1}$ to about 0.177 $m^3 \cdot s^{-1}$. It has been estimated that as much as 35 percent of the groundwater that reaches Ash Meadows travels about 145 km from the adjacent White River Groundwater Basin north of Las Vegas (Dudley and Larson 1976; Winograd and Thordarson 1975). So far no evidence of a groundwater division between these basins has been found. In addition, spring discharge in and around Furnace Creek in Death Valley appears to be derived from the Ash Meadows Groundwater Basin, although other sources are possible.

Despite considerable evidence of the antiquity of such spring systems, geological disturbances may greatly influence the temporal and spatial relations of the water supply. An indication of northward tilting of the floor of Death Valley, California, since the time of pluvial Lake Manly strongly suggests that many springs have been shut off and others created (Hunt and Mabey 1966). These recent changes offer one explanation for the absence of fishes in some apparently suitable

---

FIGURE 5. Large amounts of water enter the Death Valley region from the northeast via underground movement through permeable carbonate rocks. The Pahute Mesa Groundwater Basin supplies the headwaters of the Amargosa River, and the Ash Meadows Groundwater Basin supplies the numerous springs in Ash Meadows and probably the eastern side of Death Valley. Limits of both groundwater basins are not yet completely known but their areas are extensive. Water issuing from springs in Ash Meadows is 8000 to 12,000 years old. (From Soltz and Naiman 1978, with permission.)

habitats while fish persist in others (LaBounty and Deacon 1972). Numerous fossil springs, marked by travertine deposits, sinter, or peat mounds attest to a general reduction in surface flow throughout most desert areas over the past few thousand years (Brues 1928; Hubbs et al. 1974; Hubbs and Miller 1948; Mehringer 1967; Waring 1965).

DESERT WATERS AND THE RIVER CONTINUUM CONCEPT

Vannote et al. (1980) presented a concept describing stream ecosystems as a physiochemical and biological continuum from headwaters to the sea. Major portions of this concept have been tested in Oregon and elsewhere with stream order (Strahler 1957) as a basis for comparison (Naiman and Sedell 1979a,b, 1980; Richey et al. 1980; Sedell et al. 1978). A first-order stream has no tributaries, a second-order stream has at least two first-order tributaries, a third-order stream has at least two second-order tributaries, and so on. In general, the scenario that has developed, and currently forms a large basis of our understanding of stream ecosystems is as follows:

1. Primary production is low in headwater streams because of shading by the forest canopy; it increases in mid-order (4th to 6th order) streams where the canopy above the stream is relatively open so extensive periphyton and macrophyte communities to develop; and it decreases in streams of larger order which become deep and turbid, thus restricting primary production to phytoplankton and a narrow band of macrophytes near the shore.
2. The amount of benthic organic matter is large in headwater streams (order 1 to 3) and decreases downstream. Small streams have considerable coarse particulate organic matter (CPOM: >1 mm) in relation to downstream reaches. Woody debris however, >10 cm diameter, constitutes the major portion of organic matter in streams to at least sixth order.
3. Drifting organic matter (seston) has a mean particle size of 3 to 15 μm, regardless of stream order

or discharge regime, but its nutritional quality may increase downstream. Usually <10 percent of the organic carbon exported from a undisturbed watershed is CPOM, despite the fact that most organic inputs to the stream may be in the form of woody debris, leaves, and other large particles. The contribution of allochthonous CPOM in the size range of 1 mm to about 1 cm has not been adequately measured in any program; thus the export figure for CPOM actually may be higher.

4. Invertebrate communities show predictable food acquisition patterns relative to the nature of organic inputs (Cummins 1974). In small streams in which most organic inputs are CPOM (e.g., leaves) a majority of the invertebrate community can be classified as shredders and collectors (e.g., filter feeders) and very few grazers or scrapers. In midorder streams, in which primary production is high and there is considerable fine material in suspension, the dominant functional feeding groups are herbivores and collectors; shredders are relatively are. In larger rivers, in which seston is the principle organic component, nearly the entire community is comprised of collectors.

Although desert streams have not been studied with the same intensity or in a similar manner to mesic waterways, from visual inspection it is clear that they differ from mesic streams in three basic respects. Desert streams have considerable primary production that occurs in headwaters due to a lack of a developed forest canopy, they also lack extensive accumulation of woody debris, and many streams have no continuously flowing water between upstream and downstream reaches.

In mesic regions periphyton and macrophyte communities usually begin extensive development in third to fourth-order streams. In deserts these communities extend into the headwaters in which they often form the dominant trophic base for consumers because of their considerable productivity. In Sycamore Creek, Arizona, gross photosynthesis by periphyton approaches 8.5 g $O_2 \cdot m^{-2} \cdot d^{-1}$ in summer, which is high for a small stream (Busch and Fisher 1981). In some desert streams (e.g., Deep Creek, Idaho) aquatic macrophytes have high production rates but are not directly grazed by consumers. Macrophytes enter the food web mostly after death as detritus. The high rate of gross photosynthesis, low litter inputs, high biomass of algae, and the intermittent but severe floods that characterize Sycamore Creek and other desert streams indicate that these streams are not exporters of organic matter and are, thereby, truly autotrophic stream ecosystems.

Woody material forms an important and persistant component of aquatic ecosystems in mesic regions (Anderson and Sedell 1979; Bilby and Likens 1981). In Oregon streams more than 90 percent of the benthic organic matter is woody debris, regardless of stream order, and it can account for as much as 24 kg $\cdot m^{-2}$ in smaller streams (Naiman and Sedell 1979b). Wood provides physical integrity to small streams by acting as a retention device to hold potentially eroding materials in the system (Keller and Swanson 1979). These debris jams also create a stairstep effect by which 30 to 80 percent of the elevational drop of the stream will be over small waterfalls that dissipate the waters' potential erosional energy. By this same mechanism woody debris creates fish habitat by adding spatial heterogenity to the system. Woody material appears to break down slowly, perhaps over 100 to 200 years, to provide a dependable source of fine particulate organic matter (FPOM: 0.5 μm to 1 mm) to downstream communities. In Quebec streams fine wood particles account for as much as 50 percent of the seston (Naiman, unpublished data).

In desert streams the probable consequences of not having extensive accumulations of woody debris are speculated to be increased rates of erosion, a shorter retention time for organic matter, possibly less community stability, and a loss of a major allochthonous food source for filter feeders. This latter consideration may be offset by higher rates of primary production in headwaters.

The occasional discontinuity between upstream and downstream reaches causes abrupt changes in the seston characteristics as this material is filtered out by the sand. The consequences of these aquatic discontinuities to downstream communities have not been investigated.

The effect of these three major departures from streams in mesic regions requires a slight adjustment of viewpoint. This is easily accomplished if it is considered that orders 1 to 3 of mesic regions are normally dry the year round in most deserts. These channels certainly carry water during wetter or colder climatic periods, developing a streamside vegetation that shades the substrate and provides abundant allochthonous inputs to the system. With the current climate desert streams have dried and the vegetation has died back. Downstream, vegetation cannot normally encroach on the channel because of infrequent but powerful flash floods. In effect, with regard to the river continuum concept, most desert streams can be envisioned as beginning at about fourth order, and larger streams with discontinuous flow may have only a well-developed drainage network during storms.

THE AQUATIC ECOSYSTEM

From the preceding discussion desert regions can be characterized as having low and erratic rainfall, a large watershed area per unit of discharge, considerably more evaporation than precipation, and poor to excellent environmental predictability depending on the habitat (Table 2). It is in this setting that the aquatic community has developed. A relatively simple method of examining the nature of the aquatic ecosystem is by input-output analysis of major organic components.

To illustrate basic differences between the biological components of stream ecosystems in mesic and desert regions data from several streams are presented in Tables 3 and 4. All are small order streams with nearly identical base flow discharges. Watersheds of desert streams are much larger, however, direct precipitation is considerably less, annual degree days are two to four times that of mesic streams, and annual organic inputs are two to five times that of mesic streams (Table 3). This suggests that desert streams may generally be more productive and have a longer growing season for the biotic community than their

TABLE 2. A Comparison of Aquatic Habitats that Contain Fishes in Desert Regions with Those in Mesic Regions

Parameter	Desert Regions	Mesic Regions
Precipitation	Low, erratic	High predictible
Watershed area/unit discharge	Large	Relatively small
Evaporation/precipitation (desiccation)	Large	Relatively small
Terrestrial vegetation	Generally sparce	Abundant
Dominant energy base	Autochthonous	Allochthonous
Environmental predictability	Poor to excellent, depending on habitat	Generally good
Dispersal ability	Poor, discontinuous habitat	Good, continuous habitat

ABLE 3. Some Characteristics of the Study Sites Used for the nput-Output Analysis in Table 4[a]

arameters	Mesic Regions	Desert Regions
tream order	1	1-2
atershed area (km^2)	0.1-3.3	350-447[b]
ase flow discharge ($m^3 \cdot s^{-1}$)	0.002-0.013	0.001-0.090
nnual precipitation (cm)	123-245	<10-41
nnual degree days ($^\circ C \cdot yr^{-1}$)	3165-3285	5547-13870
nnual organic inputs ($kcal \cdot m^{-2}$)	3510-6033	10,910-16,823

[a]Technically, the desert streams are designated a low order, ut if placed in mesic regions they would be about fourth order ecause of the size of their watersheds. Data are from Fisher nd Likens (1973), J. R. Sedell (personal communication), K. W. ummins (personal communication), Naiman (1976a), G. W. Minshall personal communication), and C. E. Cushing (personal communica-ion).

[b]Excludes Tecopa Bore, California, a spring-fed stream ithout a well-developed watershed (Naiman 1976a).

TABLE 4. Organic Energy Budgets of Small Streams in Mesic Regions are Compared With Those in Desert Regions

PARAMETERS	Mesic Regions	Desert Regions
Inputs		
CPOM[b]	8.2-64.0	0.0- 12.1
FPOM	7.0-45.3	0.0- 15.0
DOM	26.4-46.3	0.0- 36.7
GPP	0.2- 2.5	36.2-100.0
Outputs		
CPOM	4.6-19.8	0.0- 13.8
FPOM	2.3- 4.5	1.0- 13.8
DOM	19.6-46.3	0.9- 29.4
RESP	38.8-56.6	43.0- 98.1
STORAGE	0?-31.8	0?- 0.0

[a]Values are in percent. Data are from Fisher and Likens (1973); J. R. Sedell (personal communication), Watershed 10, Oregon; K. W. Cummins (personal communication), Augusta Creek Michigan; Naiman (1976a); G. W. Minshall (personal communication), Deep Creek, Idaho; and C. E. Cushing (personal communication), Rattlesnake Creek, Washington.

[b]CPOM is coarse particulate organic matter >1 mm diameter; FPOM is fine particulate organic matter >0.45 μm but <1 mm diameter; DOM is dissolved organic matter <0-45 μm; GPP is gross primary production; RESP is respiration; and STORAGE is organic matter stored or accrued in sediments for periods >1 year.

counterparts elsewhere, and thus may be able to support larger or more productive fish populations, even with an almost complete lack of coarse allochthonous inputs. In mesic streams CPOM accounts for 45 to 64 percent of the annual organic input but in desert streams it is only 0 to 12 percent (Table 4). Correspondingly, primary producers account for 36 to 100 percent of the organic input in desert streams but only 0 to 2.5 percent in the heavily shaded mesic streams.

Outputs from all communities are primarily by respiration, although dissolved organic matter and short-term aerobic and long-term anaerobic storage in the benthos can be considerable. In the only mesic stream in which storage was examined 32 percent of the annual inputs were buried. In the only desert stream in which this examination was made no organic matter was stored on an annual basis, and despite the fact that most inputs to mesic streams are CPOM, relatively little (7 to 20 percent) CPOM or FPOM is exported. As a result both mesic and desert streams appear to be highly efficient in processing and retaining available organic inputs in the absence of catastrophic events.

The input-output method of ecosystem analysis for desert streams may not adequately emphasize the importance of flash flooding in moving large amounts of CPOM and FPOM through the stream channel. Flash flooding certainly carries abundant allochthonous detritus, but because of the extreme difficulties of reliably sampling these events the extent to which terrestrial detritus is added to the aquatic ecosystem remains unknown. Most of this material appears to move through the system rather than being retained and it is often deposited in downstream impoundments (McConnell 1968, Rinne 1975). Particulate materials increase markedly during flash flooding to exceedingly high levels; peak concentrations occur at the leading edge of the initial flood wave (Fisher and Minckley 1978). These events also decimate the biota and reset successional sequences. Although they are important in shaping the stream communities, considerable emphasis should still be placed on understanding the structural and functional processes of the community during more normal flow regimes.

Only a single detailed bioenergetic description of a desert stream community has been published to date (Naiman 1976 a,b). This system (Tecopa Bore, California) may be slightly atypical because of its warmwater source but it does serve to illustrate basic energy flow pathways typical of many desert spring systems (Figure 6). There are no allochthonous inputs and; the entire system is driven by primary production from blue-green algae (Oscillatoria). Microbial production was not measured but it may be considerable in the warm sediments with abundant hydrogen sulfide. (I suspect sulfur bacteria may be a major metabolic pathway in many desert waters in which organic matter can accrue in anoxic sediments.) Compared with the annual production, the standing stock of detritus and primary producers is relatively small, about 7 percent of the annual input, and in a steady state. Over the 18 months of the study no appreciable net accumulation of organic matter could be detected. Invertebrates are

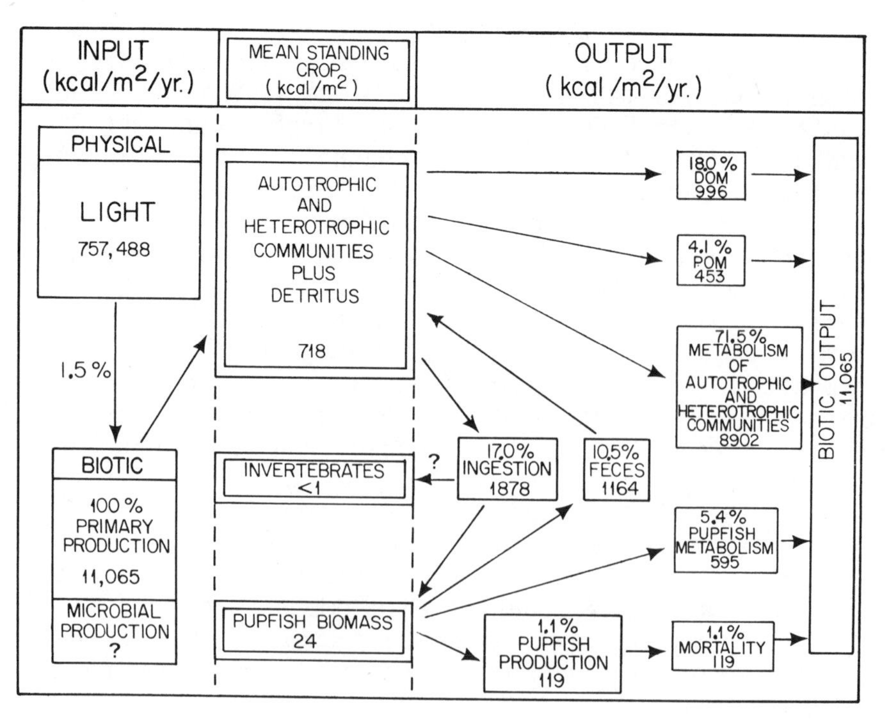

FIGURE 6. The bioenergetics of Tecopa Bore, California, a desert stream near Death Valley, is presented in the schematic diagram. (Adapted from Naiman 1976a, with permission.)

rare, probably as a result of intense predation by pupfish (Cyprinodon). Because few invertebrates are available, pupfish feed almost exclusively on the algal-detritus substrate; 17 percent of the annual primary production is ingested by them (Naiman 1975). Pupfish assimilate 6.5 percent of the annual primary production and return the remainder (10.5 percent) as feces to the benthos. Most assimilated energy is used for metabolism; only 1.1 percent of the annual primary production is invested in fish growth. This may seem relatively minor, but because of the high algal production, extended growing season due to the warm water, and dense pupfish population, annual fish production is 5 to 10 times that of carnivores in mesic areas and comparable to herbivorous fishes in tropical regions (Chapman 1978; Naiman 1976b).

Export from the system as seston is significant; 18.0 percent of the annual output is dissolved organic matter (DOM: <0.5 μm) and 4.1 percent is particulate organic matter (POM: >0.5 μm). These values are comparable to stream ecosystems in mesic regions. Metabolism by the autotrophic and heterotrophic benthic community remains high because of the warm water, accounting for 71.5 percent of the energetic output from the stream.

This desert stream community is highly efficient in use of available organic matter, as are undisturbed streams in mesic regions. The stream community may actually be more efficient than that presented here because of internal spiralling of nutrients and carbon (Wallace et al. 1977). The number of times a molecule is assimilated, released, and reassimilated is not known, but it may be considerable in a community with such a high metabolic rate.

The Tecopa Bore ecosystem, by virtue of being a spring-fed stream without an extensive watershed, has developed an ecosystem structure somewhat different than other desert streams in which flash floods sporatically scour and reset the system. In the latter streams organic matter generally does not accumulate to any appreciable degree because of the variable flow regime. As a consequence the role of sulfur bacteria in these systems may thereby be reduced. Most of the organic transport occurs during rare events of short duration and the life histories of fauna and flora often reflect this unpredictability (Constantz, this volume).

In humid regions terrestrial vegetation exerts a considerable influence on the aquatic ecosystem (Table 5). As a result it is almost a paradigm that stream ecosystems have a heterotrophically based food web (Cummins 1974; Fisher and Likens 1973; Hynes 1975), despite evidence that many streams can have an autotrophic base, especially those in desert regions (Minshall 1978; Naiman 1976a). In desert springs and streams or in larger mesic streams in which the forest canopy is open, autotrophic processes are important (Minshall 1978, Naiman 1976a, Naiman and Sedell 1980).

Desert systems, especially springs, often develop relatively simple food webs with a few dominant primary producers and consumers (Table 5). The number of species at any one location over an annual period may be large but the number at any one time may be very low, with only one or two dominate types of producers and consumers. This is carried to its extreme at Tecopa Bore, where nearly all the primary production is accomplished by a single species of *Oscillatoria* and nearly all consumption is by a single species of pupfish. Aquatic desert systems are highly productive because of the availability of nutrients, lack of shading, and an abundance of sunlight (Foster 1973; Naiman 1976a). At Tecopa Bore mean daily productivity by periphyton is about 3.0 g C $\cdot m^{-2} \cdot d^{-1}$ (Naiman 1976a). As many as 196 pupfish per square meter have been recorded with a mean of 89 fish per square meter throughout the year. Correspondingly, biomass estimates ranged as high as 4.2 g C $\cdot m^{-2}$; and the mean standing crop (2.4 g C $\cdot m^{-2}$) had a turnover rate of five times annually (Naiman 1976b). It was concluded in this study that primary factors responsible for high annual fish production were the extended growing season and the high *Oscillatoria* production.

These findings are consistent with Deacon and Deacon's (1979) study of the Devils Hole pupfish (*Cyprinodon diabolis*) in which they found, with a one month lag, that the population size followed the level

TABLE 5. Some General Characteristics of Aquatic Ecosystems in Desert and Mesic Regions are Summarized and Contrasted

Desert Regions	Mesic Regions
1. Primarily autotrophic	1. Primarily heterotrophic
2. High productivity	2. Low to high productivity
3. Efficient processing	3. Efficient processing
4. Low faunal diversity (?)	4. High faunal diversity
5. Predominately fine detritus	5. Relatively more coarse detritus
6. Simple food webs	6. Complex food webs

of primary production ($r = 0.86$, $p < 0.01$). Increasing sunlight during the summer months stimulated primary production and resulted in an increased number of fish.

At Tecopa Bore primary production was positively correlated with the amount of food eaten by the population ($r = 0.68$, $p < 0.01$) but was negatively correlated with fish biomass two and three months later ($r = -0.64$ and -0.55, respectively). Although the growth rates of pupfish were high, they were not significantly correlated with primary production. This suggests that primary production may be more than adequate to support the large population. Fish density may not be food-limited but controlled instead by the amount of space available, and the high fish production may be largely due to the extended growing season.

ADAPTATIONS TO THE ECOSYSTEM

Discharge and Current

Deacon and Minckley (1974) have correctly pointed out that emphasis on flash floods and their effects detract from the far greater importance of the normal discharge regime for fishes. Flash flooding seldom completely removes fishes from their habitats, despite its impressive physical force.

In deserts of the American Southwest longfin dace (*Agosia*) and some species of *Cyprinodon* are the only native fishes that presently penetrate far into temporary waters at times of high precipitation and runoff. These genera are considered highly resistant to environmental extremes within their native ranges (Deacon and Minckley 1974).

In periods of almost complete desiccation of water courses evaporation may demand essentially all the surface waters of a small stream during the day (Cambell and Green 1968). Mass mortalities of most fishes occur at these times. Longfin dace, however, persist beneath water saturated mats of algae and debris, and at night when flow is resumed under reduced evaporation the fish swim about and feed in a few millimeters of water. For a two-week period in 1965 a declining population of longfin dace persisted in this manner in Sycamore Creek, Arizona, under alternating wet and dry diel cycles until a small rain increased the water supply (Minckley and Barber 1971).

Some impressive morphological adaptations to strong currents have occurred in desert fishes, especially the humpback chub (*Gila cypha*) and the razorback sucker (*Xyrauchen texanus*) of the Colorado River (Figure 7). The unique morphological features of the humpback chub center on the well-developed hump, flattened head and belly, and narrow caudal peduncle which are ideal for life in fast waters. Almost nothing is known about the biology of this species because of the difficulty in collecting it, but it is thought that the hump acts as a barrier to passing water that forces the fish against the bottom (Minckley 1973). The hump is also well developed in the razorback sucker.

Life History Strategies

One of the best examples of adjustment of fish communities to ecosystem dynamics is presented by Constantz in this volume. He relates generalized life-history patterns to the set of environmental variables encountered, and shows that different patterns characterize spring fishes, stream fishes, and species that live in both types of habitat.

Spring fishes usually live in a small, simple ecosystem with a water supply that is temporally and spatially homogeneous. The absence of catastrophic events in this stable environment allows individuals to invest energy in reproduction by closely tracking a predictable food supply. Constantz concludes that spring fishes have generally been selected for a chubby body shape, posterior rounded dorsal and anal fins for maintaining position in slowly moving water, omnivorous feeding on the few dominant food species, an extended breeding season, small body size, and a longevity of about one year. These fishes produce few offspring but the energy investment per offspring is fairly high. They are characterized by brightly colored males, which defend territories, and fewer males than females. Population densities undergo relatively minor fluctuations.

FIGURE 7. The humpback chub (*Gila cypha*) and razorback sucker (*Xyrauchen texanus*) show strong morphological adaptations to the swift and turbulent Colorado River with pronounced humps, presumably for hydrodynamic stability, and the narrow caudal peduncle of *Gila*.

This scenario is in sharp contrast to stream species that must contend with an often highly variable environment. Compared with springs, streams have a complex structure that supports more species and therefore more biotic and physical interactions.

Constantz (this volume) characterizes fishes of desert streams as having a fusiform shape which promotes persistance in fast flowing water and wide dispersal. Most are omnivorous, but the complex ecosystem has allowed a few species to specialize on plant or animal foods. The breeding season is restricted and there are some large-bodied species, although most are small. Some species produce a large number of small eggs and invest little energy per offspring. Males show little territoriality and are less brightly colored than spring species. Their morphological and behavorial adaptations certainly contribute to wide geographical ranges. There is generally an equal number of males and females with wide density fluctuations. The complexity of streams appears to promote a rich diversity of life history traits.

Generalist species which inhabit both springs and streams, such as *Cyprinodon nevadensis* and *Poeciliopsis occidentalis*, show life-history patterns characteristic of the habitat within which they occur (Constantz, this volume).

Physiological Adaptations

Other authors in this volume have reviewed extensively the physiological adaptations to temperature, salinity, and dissolved gases and many have placed their conclusions in an ecological context. I refer interested readers to these chapters and to the excellent reviews of Deacon and Minckley (1974) and Cole (1968; this volume).

CONCLUSIONS

Aquatic ecosytems in desert regions are not unique or unusual, but they are different from aquatic ecosystems in mesic regions by virtue of their physical and climatic settings. It is only because they are little studied that they do not seem to fit current theories of system dynamics developed for aquatic habitats elsewhere. Lack of woody debris and significant allochthonous inputs, high primary production in headwaters, large watersheds with sparse vegetation, and infrequent but intense rain events offer the aquatic ecologist an abundant array of interestng questions to pursue through comparative studies with lotic ecosystems elsewhere.

The behavior, genetics, physiology, and life-history adaptations of desert fishes are adjustment to, or have been advantageously selected for within the context of the aquatic ecosystem. Once the nature of the aquatic ecosystem is appreciated and understood, many of the causes underlying adaptations by these fishes will be clearer.

ACKNOWLEDGEMENTS

I thank W. L. Montgomery, G. D. Constantz, S. G. Fisher, and K.W. Cummins for reading and commenting on the manuscript and K. W. Cummins, J. R. Sedell, G. W. Minshall, and C. E. Cushing for permission to use data in Tables 3 and 4 from unpublished reports. Special thanks to Dianne Steele and Elaine Ellis for typing the many drafts.

REFERENCES

Anderson, N. H. and J. R. Sedell. 1979. Detritus processing by macroinvertebrates in stream ecosystems. Annual Review of Entomology 24:351-377.

Antevs, E. 1952. Arroyo-cutting and filling. Journal of Geology 60:375-385.

Bilby, R. E. and G. E. Likens. 1981. Importance of organic debris dams in the structure and function of stream ecosystems. Ecology 61:1107-1113.

Bradley, W. G. 1970. The vegetation of Saratoga Springs, Death Valley National Monument, California. Southwestern Naturalist 15:111-129.

Brues, C. T. 1928. Studies on the fauna of hot springs in the western United States and the biology of thermophilous animals. Proceedings of the American Academy of Arts and Sciences 63:139-228.

Bryan, K. 1925. Date of channel trenching (arroyo cutting) in the arid southwest. Science 62:338-344.

Busch, D. E. and S. G. Fisher. 1981. Metabolism of a desert stream. Freshwater Biology. In Press.

Campbell, C. J. and W. Green. 1968. Perpetual succession of streamchannel vegetation in a semiarid region. Journal of the Arizona Academy of Sciences 5:86-98.

Chapman, D. W. 1978. Production in fish populations. In: S. D. Gerking (Ed.), Ecology of Freshwater Fish Production. Wiley, New York, pp. 5-25.

Cole, G. A. 1968. Desert limnology. In: G. W. Brown (Ed.), Desert Biology, Vol. I. Academic, New York, pp. 423-486.

Cole, G. A. and G. L. Batchelder. 1969. Dynamics of an Arizona travertine-forming stream. Journal of the Arizona Academy of Sciences 5:271-283.

Cummins, K. W. 1974. Structure and function of stream ecosystems. BioScience 24:631-641.

Cummins, K. W., J. R. Sedell, F. J. Swanson, G. W. Minshall, S. G. Fisher, C. E. Cushing and R. L. Vannote. 1981. Problems in evaluating organic matter budgets for stream ecosystems. Unpublished manuscript.

Deacon, J. E. and M. J. Deacon. 1979. Research on endangered fishes in the national parks with special emphasis on the Devils Hole pupfish. In: Proceedings of the First Conference on Scientific Research in the National Parks. United States Department of the Interior, National Park Service Transactions and Proceedings Series No. 5, pp. 9-19.

Deacon, J. E. and W. L. Minckley. 1974. Desert Fishes. In: G.W. Brown (Ed.), Desert Biology, Vol. II. Academic, New York, pp. 385-488.

Dill, W. A. 1944. The fishery of the lower Colorado River. California Fish and Game 30:109-211.

Dudley, W. W. and J. D. Larson. 1976. Effect of irrigation pumping on desert pupfish habitats in Ash Meadows, Nye County, Nevada. United States Geological Survey, Professional Paper 927:1-52.

Fisher, S. G. and G. E. Likens. 1973. Energy flow in Bear Brook, New Hampshire: an integrative approach to stream ecosystem metabolism. Ecological Monographs 43:421-439.

Fisher, S. G. and W. L. Minckley. 1978. Chemical characteristics of a desert stream in flash flood. Journal of Arid Environments. 1:25-33.

Foster, J. M. 1973. Limnology of two desert recharged-groundwater ponds. Ph.D. Dissertation. Arizona State University, Tempe.

Gregory, H. E. 1950. Geology and geography of the Zion Park region, Utah and Arizona. United States Geological Survey, Professional Paper 330:1-200.

Hubbs, C. L. and R. R. Miller. 1948. Correlation between fish distribution and hydrographic history in the desert basins of the western United States. Bulletin of the University of Utah, Biological Series 38:17-166.

Hubbs, C. L., R. R. Miller and L. C. Hubbs. 1974. Hydrographic history and relict fishes of the north-central Great Basin. Memoirs of the California Academy of Sciences 7:1-259.

Hunt, C. B. and D. R. Mabey. 1966. Stratigraphy and structure, Death Valley, California. United States Geological Survey, Professional Paper 494A:A1-A138.

Hunt, C. B., T. W. Robinson, W. Bowles and A. L. Washburn. 1966. Hydrologic basin, Death Valley, California. United States Geological Survey, Professional Paper 494B:B1-B138.

Hynes, H. B. N. 1975. The stream and its valley. Verhandlungen Internationale Vereinigung Limnologie 19:1-15.

Keller, E. A. and F. J. Swanson. 1979. Effects of large organic material on channel form and fluvial processes. Earth Surface Processes 4:361-380.

LaBounty, J. F. and J. E. Deacon. 1972. Cyprinodon milleri, a new species of pupfish (family Cyprinodontidae) from Death Valley, California. Copeia 1972:769-780.

LaRivers, I. 1962. Fishes and Fisheries of Nevada. Nevada State Printing Office, Carson City.

Leopold, L. B. 1951. Vegetation of southwestern watersheds in the nineteenth century. Geographical Reviews 41:295-316.

Leopold, L. B., M. G. Wolman and J. P. Miller. 1964. Fluvial Processes in Geomorphology. Freeman, San Francisco.

Logan, R. F. 1968. Causes, climates, and distribution of deserts. In: G.W. Brown (Ed.), Desert Biology, Vol. I. Academic, New York, pp. 21-51.

McConnell, W. J. 1968. Limnological effects of organic extracts of litter in a southwestern impoundment. Limnology and Oceanography 13:343- 349.

McGinnes, W. G., B. J. Goldman and P. Paylore (eds.). 1968. Deserts of the World, an Appraisal of Research into their Physical and Biological Environments. University of Arizona Press, Tucson.

Mehringer, P. J. 1967. Pollen analysis of the Tule Springs site, Nevada. In: H.M. Wormington and D. Ellis (Eds.), Pleistocene Studies in Southern Nevada. Anthropology Papers No. 13, Nevada State Museum, Carson City, pp. 130-200.

Miller, R. R. 1961. Man and the changing fish fauna of the American Southwest. Papers of the Michigan Academy of Science, Arts, and Letters. 46:365-404.

Minckley, W. L. 1973. Fishes of Arizona. Arizona Game and Fish Department, Phoenix.

Minckley, W. L. and W. E. Barber. 1971. Some aspects of the biology of the longfin dace, a cyprinid fish characterisitc of streams in the Sonoran Desert. Southwestern Naturalist 15:459-464.

Minshall, G. W. 1978. Autotrophy in stream ecosystems. BioScience 28:767-771.

Naiman, R. J. 1975. Food habits of the Amargosa pupfish in a thermal stream. Transactions of the American Fisheries Society 104:536-538.

Naiman, R. J. 1976a. Primary production, standing stock, and export of organic matter in a Mohave Desert thermal stream. Limnology and Oceanography 21:60-73.

Naiman, R. J. 1976b. Productivity of a herbivorous pupfish population (Cyprinodon nevadensis) in a warm desert stream. Journal of Fish Biology 9: 125-137.

Naiman, R. J. and J. R. Sedell. 1979a. Characterization of particulate organic matter transported by some Cascade Mountain streams. Journal of the Fisheries Research Board of Canada 36:17-31.

Naiman, R. J. and J. R. Sedell. 1979b. Benthic organic matter as a function of stream order in Oregon. Archiv fur Hydrobiologie 87:404-422.

Naiman, R. J. and J. R. Sedell. 1980. Relationships between metabolic parameters and stream order in Oregon. Canadian Journal of Fisheries and Aquatic Sciences 37:834-847.

Naiman, R. J. and J. R. Sedell. 1981. Stream ecosystem research in a watershed perspective. Verhandlungen Internationale Vereinigung Limnologie 21. In press.

Noy-Meir, I. 1973. Desert ecosystems: Environment and producers. Annual Review of Ecology and Systematics 4:25-51.

Ohmart, R. D., W. O. Deason and S. J. Freeland. 1975. Dynamics of marsh land formation and succession along the lower Colorado River and their importance and management problems as related to wildlife in the arid southwest. In: Transactions of the 40th North American Wildlife and Natural Resources Conference. Wildlife Management Institute, Washington, D.C., pp. 240-251.

Richey, J. E., J. T. Brock, R. J. Naiman, R. C. Wissmar and R. F. Stallard. 1980. Organic carbon: Oxidation and transport in the Amazon River. Science. 207:1348-1351.

Rinne, J. N. 1975. Hydrology of the Salt River and its reservoirs, central Arizona. Journal of the Arizona Academy of Sciences 10:75-86.

Sedell, J. R., R. J. Naiman, K. W. Cummins, G. W. Minshall and R. L. Vannote. 1978. Transport of particulate organic material in streams as a function of physical processes. Verhandlungen Internationale Vereinigung Limnologie. 20:1366-1375.

Smith, G. E. P. 1940. The groundwater supply of the Eloy District in Pinal County, Arizona. Unversity of Arizona Aquacultural Experiment Station, Technical Bulletin 87:1-42.

Soltz, D. L. and R. J. Naiman. 1978. The Natural History of Native Fishes in the Death Valley System. Natural History Museum of Los Angeles County, Science Series 30:1-76.

Strahler, A. N. 1957. Quantitative analysis of watershed geomorphology. Transactions of the American Geophysical Union 38:913-910.

Thornber, J. J. 1910. The grazing ranges of Arizona. Arizona Aquacultural Experiment Station, Bulletin 65:335-338.

Trautman, M. B. 1957. The Fishes of Ohio. Ohio State University Press.

Vannote, R. L., G. W. Minshall, K. W. Cummins, J. R. Sedell and C. E. Cushing. 1980. The river continuum concept. Canadian Journal of Fisheries and Aquatic Sciences 37:130-137.

Wallace, J. B., J. R. Webster and W. R. Woodall. 1977. The role of filter feeders in flowing waters. Archiv fur Hydrobiologie 79:506-532.

Waring, G. A. 1965. Thermal springs of the United States and other countries of the world: A summary. United States Geological Survey, Professional Paper 492:1-383.

Weatherley, A. H. 1963. Zoogeography of *Perca fluviatilis* (Linnaeus) and *Perca flavescens* (Mitchell), with special reference to the effects of high temperature. Proceedings of the Zoological Society of London 141:557-576.

Winograd, I. J. and W. Thordarson. 1975. Hydrogeologic and hydrochemical framework, south-central Great Basin, Nevada-California, with special reference to the Nevada Test Site. United States Geological Survey, Professional Paper 712C:C1-C126.

Index

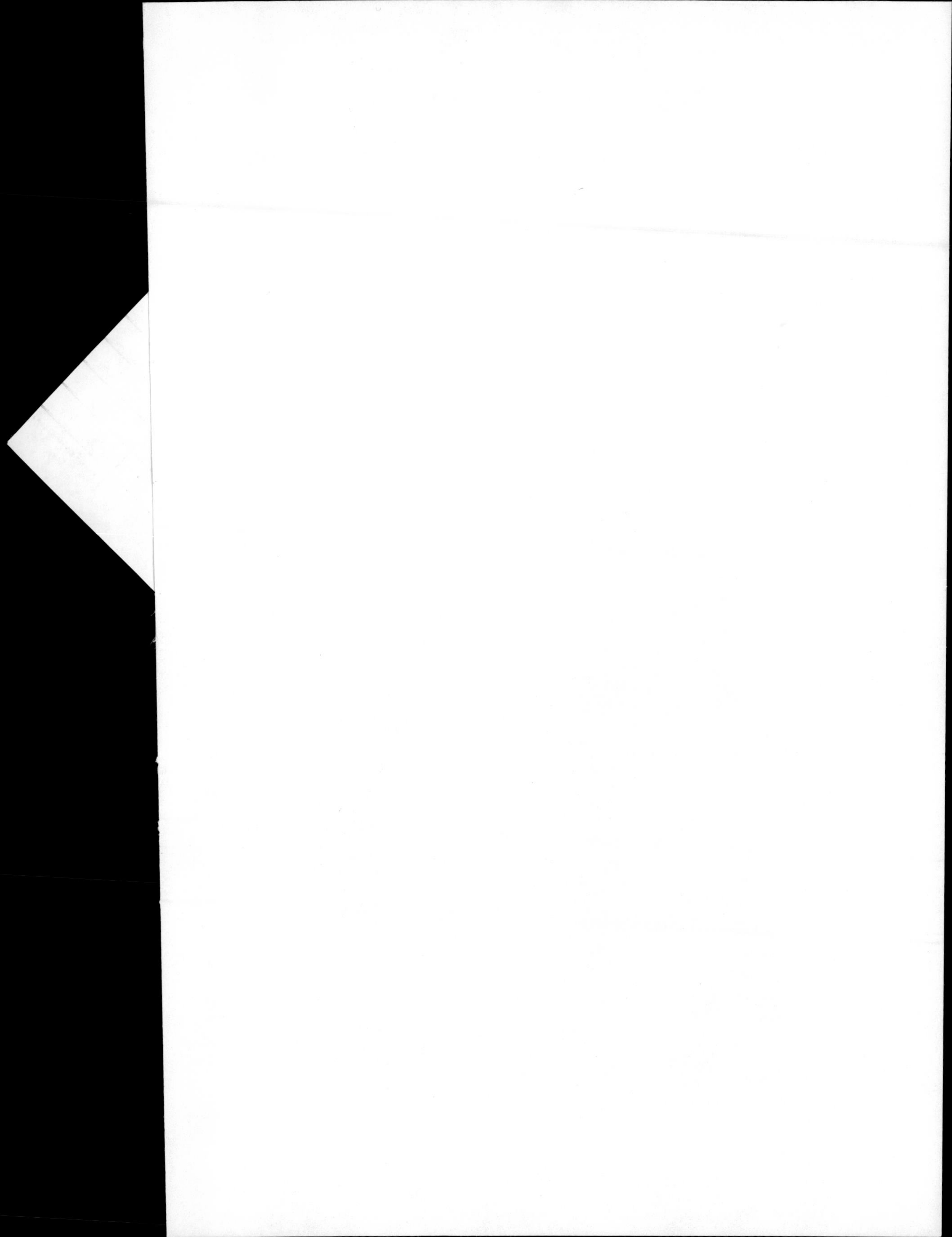

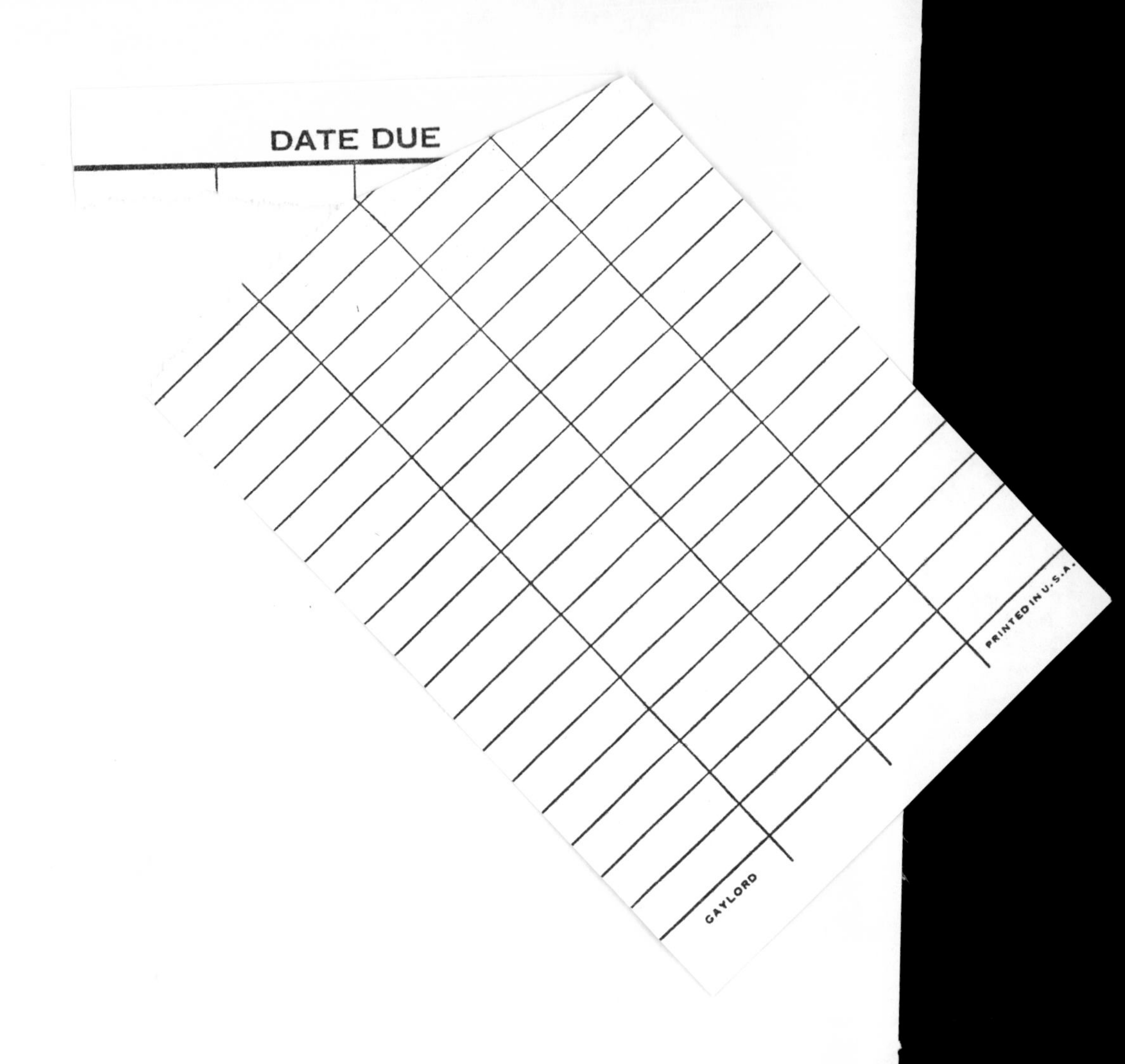
GAYLORD
PRINTED IN U.S.A.